INTERMEDIATE STATISTICS

This book is dedicated to Bill (both of you), Brooklyn, Dan (both of you), LJ, and Lincoln (both of you).

INTERMEDIATE STATISTICS

A CONCEPTUAL COURSE

BRETT W. PELHAM

Los Angeles | London | New Delhi
Singapore | Washington DC

Los Angeles | London | New Delhi
Singapore | Washington DC

FOR INFORMATION:

SAGE Publications, Inc.
2455 Teller Road
Thousand Oaks, California 91320
E-mail: order@sagepub.com

SAGE Publications Ltd.
1 Oliver's Yard
55 City Road
London EC1Y 1SP
United Kingdom

SAGE Publications India Pvt. Ltd.
B 1/I 1 Mohan Cooperative Industrial Area
Mathura Road, New Delhi 110 044
India

SAGE Publications Asia-Pacific Pte. Ltd.
3 Church Street
#10-04 Samsung Hub
Singapore 049483

Acquisitions Editor: Vicki Knight
Associate Editor: Lauren Habib
Editorial Assistant: Kalie Koscielak
Production Editor: Eric Garner
Copy Editor: Gillian Dickens
Typesetter: C&M Digitals (P) Ltd.
Proofreader: Wendy Jo Dymond
Indexer: Jeanne Busemeyer
Cover Designer: Anupama Krishnan
Marketing Manager: Nicole Elliot
Permissions Editor: Jason Kelley

Copyright © 2013 by SAGE Publications, Inc.

All rights reserved. No part of this book may be reproduced or utilized in any form or by any means, electronic or mechanical, including photocopying, recording, or by any information storage and retrieval system, without permission in writing from the publisher.

Printed in the United States of America

A catalog record of this book is available from the Library of Congress.

978-1-4129-9498-9

This book is printed on acid-free paper.

12 13 14 15 16 10 9 8 7 6 5 4 3 2 1

Brief Contents

Preface	xiii
Acknowledgments	xvii
About the Author	xix
Chapter 1: A Review of Basic Statistical Concepts	1
Chapter 2: Descriptive Statistics	45
Chapter 3: Linear and Curvilinear Correlation	81
Chapter 4: Nonparametric Statistics (Tests Involving Nominal Variables)	101
Chapter 5: Reliability (and a Little Bit of Factor Analysis)	125
Chapter 6: Single-Sample and Two-Sample t Tests	161
Chapter 7: One-Way and Factorial Analysis of Variance (ANOVA)	191
Chapter 8: Within-Subjects and Mixed Model Analyses	213
Chapter 9: Multiple Regression	231
Chapter 10: Examining Interactions in Multiple Regression	263
Chapter 11: ANCOVA, Covariate-Adjusted Means, and Predicted Scores	303
Chapter 12: Suppressor Variables	315
Chapter 13: Mediation and Path Analysis	325
Chapter 14: Data Cleaning	355
Chapter 15: Data Merging and Data Management	387
Chapter 16: Avoiding Bias: Characterizing Without Capitalizing	395
References	407
Author Index	413
Subject Index	417

Detailed Contents

Preface	xiii
Acknowledgments	xvii
About the Author	xix

Chapter 1: A Review of Basic Statistical Concepts — 1
- Introduction — 1
- How Numbers and Language Revolutionized Human History — 2
- Descriptive Statistics — 5
  - Central Tendency and Dispersion — 6
  - The Shape of Distributions — 9
- Inferential Statistics — 13
  - Probability Theory — 16
  - A Study of Cheating — 20
- Things That Go Bump in the Light: Factors That Influence the Results of Significance Tests — 24
  - Alpha Levels and Type I and II Errors — 24
  - Effect Size and Significance Testing — 25
  - Measurement Error and Significance Testing — 25
  - Sample Size and Significance Testing — 26
  - Restriction of Range and Significance Testing — 26
- The Changing State of the Art: Alternate Perspectives on Statistical Hypothesis Testing — 28
  - Estimates of Effect Size — 28
  - Meta-Analysis — 31
- Summary — 33
- Appendix 1.1: Some Common Statistical Tests and Their Uses — 34
- Notes — 42

Chapter 2: Descriptive Statistics — 45
- The Very Small Survey of Moderately Large Shoe Sizes — 46
- Estimating Spending in the U.S. Population — 51
- Missing Data and Variable Values — 52
- Describing the Ethnic Diversity of U.S. States — 55
- Descriptive Statistics in Public Opinion Polls — 64

Shape Matters: The Normal Distribution, Skewness, and Kurtosis	67
Sometimes Shape *Really* Matters	74
How Much Skewness or Kurtosis Is Too Much (or Too Little)?	77
Correcting for Skewness and Kurtosis	78
For Further Thought	79

Chapter 3: Linear and Curvilinear Correlation — **81**

Introduction: A Brief Tribute to Karl Pearson	81
A Hypothetical Study of How Unfair Life Is	82
A Hypothetical Correlational Study of Afrocentrism	90
A Study of Freedom of the Press and Perceived Corruption in Europe	91
The Power of Impossible Outliers	93
A Look at Brandeis's Hypothesis Through a Curved Lens	96
Appendix 3.1: A Primer for Predicting Scores on Y From Scores on X	100

Chapter 4: Nonparametric Statistics (Tests Involving Nominal Variables) — **101**

Introduction: The Correlation Coefficient's Nominal Cousins	101
A Pilot Study of Name-Letter Preferences	102
A Second Pilot Study of Name-Letter Preferences	107
The Chi-Square Statistic, Phi Coefficients, and Odds Ratios	113
A Correlational Study of Interpersonal Attraction	116
A Small Change of Pace: From Marriage to Mental Illness	118
How to Report the Results of a Chi-Square Analysis of Nominal Variables	120
Appendix 4.1: How to Report the Results of a Chi-Square Analysis	121
Notes	123

Chapter 5: Reliability (and a Little Bit of Factor Analysis) — **125**

Chapter Overview	125
Introduction: The Concept of Reliability	126
"Just the Factors, Ma'am"	127
Caveats Regarding Real Data	131
Principal Components Analysis With Real Data	133
Checking Out the Eigenvalues	135
Reliability Analysis	138
Adding Items Together to Make a Scale	140
A Comparison of Cronbach's Alpha and Split-Half Reliability	142
Applying What You Learned to a Hypothetical Study of Self-Esteem	143
A Return to Extraversion: Reliability Analysis as a Tool for Item Development	145
Limitations of Cronbach's Alpha	149
Appendix 5.1: Why Psychological Scales Are More Reliable Than the Average of Their Imperfectly Reliable Components	152

Appendix 5.2: Reporting the Results of a Factor Analysis and a Reliability Analysis	157
Notes	160

Chapter 6: Single-Sample and Two-Sample *t* Tests — 161

Introduction	161
Bending the Rules About Happiness	163
Simplifying the Outcome	166
The Independent Samples (Two-Samples) *t* Test	171
Results of the Teacher Expectancy Study	173
More Simplification	173
Yet Another Name-Letter Preference Study	174
An Archival Study of Heat and Aggression	176
A Blind Cola Taste Test	177
Appendix 6.1: Reporting the Results of One-Sample and Two-Sample *t* Tests	181
Appendix 6.2 Some Useful SPSS Syntax Statements and Logical Operands	183
Appendix 6.3: Running a One-Sample Chi-Square Test in Older Versions of SPSS (SPSS 19 or Earlier)	189

Chapter 7: One-Way and Factorial Analysis of Variance (ANOVA) — 191

Introduction: The Trouble With Levels	191
Understanding One-Way ANOVAs by Experimenting With Alcohol	193
Finding Meaning in Means: Using Contrasts	195
Looking at More Than One Independent Variable: Factorial ANOVAs	198
A Hypothetical Example of When and How "It Depends"	201
More Practice Understanding Main Effects and Interactions	205
Practice Study 1: A Lab Study of Aggression Among Kids	205
Practice Study 2: A Lab Study of Self-Pay	206
Three-Way ANOVAs and Beyond	207
Putting It All Together	209
Appendix 7.1: Results of a Unique Memory Study That Used Planned Contrasts	210

Chapter 8: Within-Subjects and Mixed Model Analyses — 213

Introduction: Controlling for Individual Differences	213
Some Bogus Within-Subjects Studies of Bogus Traits	214
Examining Three Within-Subjects Versions of the Same Study	215
Combining Between-Subjects and Within-Subjects Designs: Mixed Model Designs	218
A Repeated Measures Study of Optimism With Countries as the Unit of Analysis	220
A Mixed Model Study of Implicit Political Attitudes	226
Appendix 8.1: Sample Results of a Study Using a Mixed Model Design	229

Chapter 9: Multiple Regression **231**
    Introduction: Ceteris Paribus   231
    Predictor Variables and Criterion Variables   232
    The Logic of Multiple Regression Analysis   233
    Considering More Data   235
    Checking Your Answers in SPSS   236
    Correlation, Multiple Regression, and Multiple
        Predictor Variables   239
    R-Square, Adjusted R-Square, and Standard Errors
        in Multiple Regression   242
    A Real-World Multiple Regression Application   248
    Logistic Regression: Multiple Regression Analysis
        With Categorical Criterion Variables   251
    Back to Missing Cookies   253
    Logistic Regression Analysis of Cookie Thefts:
        Disentangling Bart and Fred   253
    Understanding Odds Ratios in Logistic Regression   256
    Misunderstanding Odds Ratios in Logistic Regression   257
    Back to Missing Cookies   258
    Confidence Intervals in Logistic Regression   259
    It Sure Is Messy Out There: Multivariate Data Cleaning   259
    Appendix 9.1: Terms for Further Reading or Discussion   261

Chapter 10: Examining Interactions in Multiple
Regression Analysis **263**
    Introduction: Type of Variable Determines Type of Analysis   263
    Moderators and Interactions in Multiple Regression   264
    A More Realistic Example: Centering and Simple Slopes
        Tests in Multiple Regression Analysis   268
    Beyond Median Splits: Isolating and Analyzing Subgroups
        in Multiple Regression   272
    Some Practice With Real Data   273
    More Real Practice Data   275
    Important Moderator Effects Sometimes Add Minimally
        to R-Square Values   276
    Examining Interactions Between Categorical and Continuous
        Predictors in Multiple Regression   279
    Why Does This Technique for Estimating Simple
        Slopes Work?   283
    It's Not Easy Studying Green: Dealing With Interactions
        Involving Categorical Predictors With More
        Than Two Levels   287
    Appendix 10.1: Testing for and Interpreting Three-Way
        Interactions in Multiple Regression   294
    Appendix 10.2: An Example of How to Report the
        Results of a Two-Way Interaction in
        Multiple Regression   300
    Notes   301

Chapter 11: ANCOVA, Covariate-Adjusted Means, and Predicted Scores — **303**
Introduction: Ends to a Mean — 303
The Analysis of Covariance (ANCOVA) — 304
Data Set 1: Gender Differences in Income — 305
The Ghosts in the Machine: Generating Predicted Scores in a Multiple Regression Analysis — 309
Data Set 2: Political Party Affiliation and Attitudes — 312
Data Set 3: A Survey of Smoking and Well-Being — 312

Chapter 12: Suppressor Variables — **315**
Introduction: Multiple Regression and Suppression — 315
Uncovering Causes: Attribution Theory and Suppression — 317
A Practice Example of Suppression: Running and Squatting — 318
Practice With Suppression: Three Data Sets to Analyze — 320
Data Set 1: Anagram Difficulty and Self-Pay — 321
Data Set 2: Predicting Voting Behavior — 322
Data Set 3: Predicting Homicide Rates From Country-Level Statistics — 323
A Cautionary Note Regarding Multicollinearity — 323
Coda: Why Suppression? — 324
Note — 324

Chapter 13: Mediation and Path Analysis — **325**
Introduction: Disentangling Competing Causes — 325
Third Variables Versus Causal Starting Points — 326
Causal Plausibility — 328
Empirical Plausibility — 331
Moderation in All Things—Except for Mediation — 333
A Mediational Model of How Frustration Leads to Aggression — 334
Formal Testing for the Significance of Mediation Requires Knowledge of Standard Errors — 337
What Mediates the Association Between Self-Esteem and Relationship Satisfaction? — 339
Mediation Analysis as a Specific Case of Path Analysis — 341
The Logic of Path Analysis — 343
A Hypothetical Path Model Involving Positive Beliefs and Health — 347
For Further Reading — 349
Useful Web Pages — 350
Appendix 13.1: An Analysis of Teasing From Kruger, Gordon, and Kuban (2006) — 351
Notes — 353

Chapter 14: Data Cleaning — **355**
Introduction: Data Cleaning — 355
Missing Data — 357
That's Not Normal: Outliers — 368
Identifying and Dealing With Univariate Outliers — 371

    Identifying and Dealing With Multivariate Outliers 371
    Putting Your Data-Cleaning Skills to Work 380
    A Final Worry: Multicollinearity 383
    For Further Reading 383
    Appendix 14.1: An Illustration of Multicollinearity 384

Chapter 15: Data Merging and Data Management   **387**

Chapter 16: Avoiding Bias: Characterizing Without Capitalizing   **395**
    Introduction: Some Common Errors and Biases in
        Human Thinking 395
    Confirmatory Biases + Human Statisticians = Statistical Bias 399
    Phineas and Ferb Are Just the Tip of the Iceberg 401
    Four Simple Rules for Avoiding Bias in Data Analysis 405

References   **407**

Author Index   **413**

Subject Index   **417**

Preface

Because the author of this text has (a) a good friend who is a statistician and (b) a good brother who is a middle school math teacher, he has heard his share of sayings and stories about people who either hate or are clueless about mathematics. First, as the quasi-statistician Hart Blanton likes to tell people, "There are three kinds of people in the world—those who are good at math and those who are not." Second, as Jason Pelham often teaches people, there is a right way and a wrong way to expand a simple algebraic equation. I'll let you guess whether I'm sharing the right way or the wrong way (presumably borrowed from an actual exam, but who can know for sure).

$$\text{Question: Expand } (a + b)^2$$
$$\text{Answer: } (\boldsymbol{a} + \boldsymbol{b})^2$$
$$(\ \boldsymbol{a}\ +\ \boldsymbol{b}\ )^2$$
$$(\ \ \boldsymbol{a}\ \ +\ \ \boldsymbol{b}\ \ )^2$$
$$(\ \ \ \boldsymbol{a}\ \ \ +\ \ \ \boldsymbol{b}\ \ \ )^2$$

Although I firmly believe that anyone who can speak can be good at mathematics or statistics, I realize that there is a great deal of math and statistics *phobia* in the world. As a statistics lover, I find this frustrating because it's like having a fantastic recipe for taralli and realizing that very few people in the world even know what a tarallo is (for starters, it's the singular of the plural Italian *taralli*). If only more people could put their statistical phobias aside, the world could be a more wonderful (i.e., statistically sophisticated) place. There is a sense in which this book is part of my personal mission to demystify and popularize that beleaguered field known as statistics.

But that's all personal. From a more pragmatic perspective, this book is meant to serve mainly as a student resource for an intermediate (often a second) course in statistics. In this book, that is, I have often assumed that readers of the text will have already had a prior course in statistics. Making this assumption has freed me up to focus on conceptual issues that are crucial to understanding *why* statistics work the way they do. Unlike many other intermediate statistics textbooks, then, this book does not belabor the details of cumulants or discuss the computational underpinnings of the central limit theorem. This text is highly conceptual rather than highly computational. It focuses, for example, on the beauty and utility of how statistics can take a sea of millions of observations and convert them into one or two useful numbers.

But statistics do not merely allow us to summarize; they also allows us to generalize—that is, to draw broad inferences from a limited number of observations. Sometimes statistics can even allow us to predict the future. Because people constantly draw inferences in their daily lives and frequently try to predict their own futures, I attempt to capitalize on your expertise as a human inference-making expert and help you apply this informal expertise to the formal mastery of statistics.

Although a few purists and math fanatics may find this highly conceptual approach simplistic, if not annoying, I hope most users of the book will appreciate my emphasis on what is logical and intuitive about statistics. I have found that the large majority of students learn more about something by *understanding how and why it works as it does* rather than by reducing it to its mathematical essence.

From this perspective, I also believe that, with a little support from an instructor, there is no reason why this text couldn't be used in a *first* course in statistics, especially for highly pragmatic or highly motivated students. This brings me to a second point. I believe people learn best by doing. Reading about and understanding the neural and cognitive mechanisms that allow people to juggle (e.g., understanding the cerebellum, chaining, or fractional anticipatory goal responses) can be intellectually stimulating, but it's not the same as being *able to juggle*. With this idea in mind, this textbook focuses on how to analyze and communicate about real data. The textbook thus includes about 100 IBM SPSS Statistics* data sets (available at www.sagepub.com/pelham) that have all been carefully selected to illustrate important statistical principles as they apply to either a real or a realistic research question. I thus hope that readers who complete this book, and carefully answer the questions that are interspersed throughout each chapter, will be in a very good position to analyze real data.

On the basis of my desire to help readers understand real data, I have also tried to integrate the SPSS statistical software package seamlessly into this text. This means that in addition to learning about statistical concepts, you will also learn how to use the popular statistical package *SPSS for Windows*. If you are already highly familiar with SPSS, this will give you an advantage, but it won't necessarily be a large one. This is because SPSS is a fantastic tool for doing statistical calculations, but SPSS cannot tell you how to think about or interpret your data. For example, SPSS is just as happy to analyze data from a poorly designed study as from a well-designed study, and it is not yet sophisticated enough to identify confounds or limitations in a research design. Just knowing which buttons to

---

*IBM SPSS® Statistics was formerly called PASW® Statistics.

push is good, but knowing what you're doing conceptually when you push the buttons is much more important. The conceptual knowledge I hope to impart in this book should transfer well to other statistical packages should you ever have the opportunity to learn them.

Statisticians do not simply need to understand data. They also need to communicate about them. Writing about statistics is a different business than simply getting one's statistics correct. For this reason, I have included appendixes in about half the chapters in this text that provide real writing samples, taken from published research papers. These samples demonstrate important aspects of how to write about statistics. I hope you find this unique feature of the text particularly useful when you are lucky enough to need to communicate your own provocative research findings to a scientific audience.

Having emphasized that this book is all about concepts rather than computation and hands-on experiences rather than hypothetical musings, I hasten to add that I have nothing against mathematical proofs and formulas. What I am opposed to, instead, is rotely memorizing formulas without knowing what purpose they serve, or blindly plugging in values to an equation merely to see what number comes out at the end. My purpose in making this text so conceptual is to teach basic statistical concepts, in a highly user-friendly way, so that the formulas you *will* need to use actually mean something.

Because I am a psychologist, this text focuses disproportionately on psychological data. However, most of the analytic techniques covered in this course are highly relevant to research across all of the social sciences. Thus, I hope this textbook is just as useful to readers studying urban planning, sociology, or cultural anthropology as it will be to social, clinical, or health psychologists.

Because of my fervent belief in the power of humor to transform the painful into the slightly less painful, I have tried to make this textbook as entertaining as possible. Even my dumbass brothers all said they found parts of the book highly entertaining—and this was before they were all promised a tiny sliver of the miniscule royalties. Oh yeah, the smart brother and one of the smart sisters liked the often irreverent humor, too. I hope you agree with them.

Finally, I should note that in every part of this book except this Preface, I communicate to you as if there is more than one of me. That is, I use the pronoun *we* rather than *I* and refer to myself individually as your "primary author." If you are wondering whether I think there are two of me, the answer is no. In fact, there are three or four of me. By this, I mean that I did not write this book by myself. Instead, virtually everything I wrote in this text was heavily influenced by what I learned from three people. The first person is Karl Pearson, who invented the most useful, elegant, and flexible mathematical tool known to modern science—namely, the correlation coefficient (not to mention the chi-square test). Second, I was inspired repeatedly by R. A. Fisher, who almost single-handedly invented modern agricultural and social scientific research methods. Last, and clearly least, I was constantly inspired by the insanely creative and sometimes merely insane Curtis Hardin. Hardin taught me more about the fair and proper use of statistics than anyone else I have ever known. In fairness to the other amazing mentors and colleagues I've had over the years, I hasten to add that in the 4 years Curtis and I overlapped at UCLA, I spent at least four times as much *time* with him as I have with anyone else who has ever been my mentor or colleague. It was thus inevitable that I'd learn a few things about statistics from him. If I had ever spent 4 intense years with a real statistician like Dave Kenny or Hart Blanton, who knows how much better this text might be. At any rate, although I would never pretend to speak for any of the three quirky geniuses I consider my unofficial coauthors, I must note that their voices constantly haunt and inspire me. With apologies to the poet e. e. cummings, "I am never without them." I hope that, for the next 16 chapters, you are never without them either.

Acknowledgments

As a graduate student in the late 1980s at UT Austin, I learned a great deal about how to make statistics highly accessible. I did so merely by being lucky enough to serve as a teaching assistant in a statistics class taught by the brilliant psychologist Dan Gilbert. A few years later, I was lucky enough to have the equally brilliant John Hetts as my own teaching assistant (TA) in a statistics class I taught at UCLA. The hardest thing about teaching this demanding class was having a TA who was much, much smarter than I was. Teaching my first statistics class with John ignited my passion for making statistics accessible and entertaining. John enthusiastically encouraged me to teach statistics from a conceptual point of view, and that was when I first began to think about writing a statistics book. The book really began to take shape, however, when I began to teach intermediate statistics at the University at Buffalo (UB) in 2002. This book has thus benefitted tremendously from the insightful questions posed by the numerous graduate and undergraduate students at UB who took my intermediate statistics classes.

I was also lucky enough to be able to try out sample chapters of the book on the many gifted and motivated students who took part in the APA's ASTP (Advanced Statistical Training Program) between about 2005 and 2008. Finally, as proof that luck does sometimes follow people around, I was lucky enough to be able to teach these delightful summer classes with Mauricio Carvallo and Keith Maddox, who are both outstanding models of how to make statistics highly rigorous, highly accessible, and most of all highly entertaining. It wasn't until the ASTP courses were done, however, that I truly got serious about writing an intermediate statistics book. Thus, the person I must thank the most for the evolution of this book is my wife, LJ Fletcher, who tolerated many a lonely evening while I was shaping and molding this book between 2009 and 2011.

Finally, preliminary drafts of this book benefitted enormously from the expert editorial guidance of Vicki Knight and the careful reviews she solicited from

Rebecca Brooks, Ohio Northern University

Tracy DeHart, Loyola Marymount University, Chicago

Andrew S. Fullerton, Oklahoma State University

Sanda Kaufman, Cleveland State University

Michelle Meyer Lueck, Colorado State University

Keith B. Maddox, Tufts University

Jessy Minney, University of Alabama

Jessica W. Pardee, University of Central Florida

Michael J. Poulin, University at Buffalo

Janett M. Naylor, Fort Hays State University

Charlie L. Reeve, University of North Carolina, Charlotte

Yanyan Sheng, Southern Illinois University, Carbondale

N. Clayton Silver, University of Nevada, Las Vegas

About the Author

Brett W. Pelham grew up as the second of six children near the small town of Rossville, Georgia. Brett received his BS from Berry College in 1983 and received his PhD from the University of Texas at Austin in 1989. He began thinking about this book while teaching statistics at UCLA but got serious about doing so while teaching statistics at UB (in Buffalo, NY). After leaving UB, Brett worked for a year as a visiting professor in marketing at Georgetown University. He was then lucky enough to work for about 2 years as a senior research analyst at Gallup. This was followed by 2 not-so-lucky years as a Program Director at the National Science Foundation. At the time of this writing (in January 2012), Brett had just taken a position promoting graduate and postgraduate education in the Education Directorate at the American Psychological Association. With apologies to Lenny Bruce, you can see that Brett has had some difficulties holding down a regular job recently.

The bulk of Brett's research focuses on automatic social judgment and self-evaluation. Over the past two decades, he has taught courses in social psychology, research methods, statistics, social cognition, and the self-concept. In his spare time, he enjoys juggling, sculpting, listening to alternative rock music, cooking, and traveling. His two favorite activities while completing this textbook (in late 2011) were spending time with his 3-year-old daughter, Brooklyn, and his 8-year-old son, Lincoln. Along with his wife, LJ Pelham, Brett is coinventor of the recently released card game *PRIME*. Along with his son, Lincoln, he is coinventor of the soon-to-be-released card game *Cliffhanger*. Along with his daughter, Brooklyn, he is coinventor of the never-to-be-released card game *It's a Hat. You Like It?* His most recent writing project is a novel tentatively titled *Elvis 2.0*, which focuses on scientific and religious problems associated with the apparent resurrection of Elvis Presley.

A Review of Basic Statistical Concepts

The record of a month's roulette playing at Monte Carlo can afford us material for discussing the foundations of knowledge.

—Karl Pearson

I know too well that these arguments from probabilities are imposters, and unless great caution is observed in the use of them, they are apt to be deceptive.

—Plato (in *Phaedo*)

Introduction

It is hard to find two quotations from famous thinkers that reflect more divergent views of probability and statistics. The eminent statistician Karl Pearson (the guy who invented the correlation coefficient) was so enthralled with probability and statistics that he seems to have believed that understanding probability and statistics is a cornerstone of human understanding. Pearson argued that statistical methods can offer us deep insights into the nature of reality. The famous Greek philosopher Plato also had quite a bit to say about the nature of reality. In contrast to Pearson, though, Plato was skeptical of the "fuzzy logic" of probabilities and central tendencies. From Plato's viewpoint, we should only trust what we can know with absolute certainty. Plato probably preferred deduction (e.g., If B then C) to induction (In my experience, bees seem to like flowers).

Even Plato seemed to agree, though, that if we observe "great caution," arguments from probabilities may be pretty useful. In contrast, some modern nonstatisticians might agree with what the first author's father, Bill Pelham, used to say about statistics and probability theory: "Figures can't lie, but liars sure can figure." His hunch, and his fear, was that "you can prove anything with statistics." To put this a little differently, a surprising number of thoughtful, intelligent students are thumbs-down on statistics. In fact, some students only take statistics because they *have* to (e.g., to graduate with a major in psychology, to earn a second or third PhD). If you fall into this category, our dream for you is that you enjoy this book so much that you will someday talk about the next time that you *get* to take—or teach—a statistics class.

One purpose of this first chapter, then, is to convince you that Karl Pearson's rosy view of statistics is closer to the truth than is Bill Pelham's jaded view. It is possible, though, that you fully agree with Pearson, but you just don't like memorizing all those formulas Pearson and company came up with. In that case, the purpose of this chapter is to serve as a quick refresher course that will make the rest of this book more useful. In either event, no part of this book requires you to memorize a lot of complex statistical formulas. Instead, the approach emphasized here is heavily conceptual rather than heavily computational. The approach emphasized here is also hands-on. If you can count on your fingers, you can count your blessings because you are fully capable of doing at least some of the important calculations that lie at the very heart of statistics. The hands-on approach of this book emphasizes logic over rote calculation, capitalizes on your knowledge of everyday events, and attempts to pique your innate curiosity with realistic research problems that can best be solved by understanding statistics. If you know whether there is any connection between rain and umbrellas, if you love or hate weather forecasters, and if you find games of chance interesting, we hope that you enjoy at least some of the demonstrations and data analysis activities that are contained in this book.

Before we jump into a detailed discussion of statistics, however, we would like to briefly remind you that (a) statistics is a branch of mathematics and (b) statistics is its own very precise language. This is very fitting because we can trace numbers and, ultimately, statistics back to the beginning of human language and thus to the beginning of human written history. To appreciate fully the power and elegance of statistics, we need to go back to the ancient Middle East.

How Numbers and Language Revolutionized Human History

About 5,000 years ago, once human beings had began to master agriculture, live in large city states, and make deals with one another, an unknown Sumerian trader or traders invented the **cuneiform** writing system to keep track of economic transactions. Because we live in a world surrounded by numbers and written language, it is difficult for us

Chapter 1   A Review of Basic Statistical Concepts

to appreciate how ingenious it was for someone to realize that *writing things down* solves a myriad of social and economic problems. When Basam and Gabor got into their semimonthly fistfight about whether Gabor owed Basam *five* more or *six* more geese to pay for a newly weaned goat, our pet theory is that it was an exasperated neighbor who finally got sick of all the fighting and thus proposed the cuneiform writing system. The cuneiform system involved making marks with a stylus in wet clay that was then dried and fired as a permanent record of economic transactions. This system initially focused almost exclusively on who had traded what with whom—and, most important, in what quantity. Thus, some Sumerian traders made the impressive leap of impressing important things in clay. This early cuneiform writing system was about as sophisticated as the scribbles of your 4-year-old niece, but it quickly caught on because it was *way* better than spoken language alone.

For example, it apparently wasn't too long before the great-great-great-grandchild of that original irate neighbor got a fantastically brilliant idea. Instead of drawing a stylized duck, duck, duck, duck to represent four ducks, this person realized that *four-ness* itself (like two-ness and thirty-seven-ness) was a concept. He or she thus created *abstract characters* for numbers that saved ancient Sumerians a *lot* of clay. We won't insult you by belaboring how much easier it is to write and verify the cuneiform version of "17 goats" than to write "goat, goat, goat, goat, goat, goat, goat, goat, goat, goat, goat, goat, goat, goat, goat . . ." oh yeah ". . . goat," but we can summarize a few thousand years of human technological and scientific development by reminding you that incredibly useful concepts such as zero, fractions, π (pi), and logarithms, which make possible great things such as penicillin, the Sistine Chapel, and iPhones, would have never come about were it not for the development of abstract numbers and language.

It is probably a bit more fascinating to textbook authors than to textbook readers to recount in great detail what happened over the course of the next 5,000 years, but suffice it to say that written language, numbers, and mathematics revolutionized—and sometimes limited—human scientific and technological development. For example, one of the biggest ruts that brilliant human beings ever got stuck into has to do with numbers. If you have ever given much thought to Roman numerals, it may have dawned on you that they are an inefficient pain in the butt. Who thought it was a great idea to represent 1,000 as M while representing 18 as XVIII? And why the big emphasis on five (V, that is) in a base-10 number system? The short answer to these questions is that whoever formalized Roman numbers got a little too obsessed with counting on his or her fingers and never fully got over it. For example, we hope it's obvious that the Roman numerals I and II are stand-ins for human fingers. It is probably less obvious that the Roman V ("5") is a stand-in for the "V" that is made by your thumb and first finger when you hold up a single hand and tilt it outward a bit (sort of the way you would to give someone a "high five"). If you do this with both of your hands and move your thumbs together until they cross in front of you, you'll see that the X in Roman numerals is, essentially, V + V. Once you're done making shadow puppets, we'd like to tell you that, as it turns out, there are some major drawbacks to Roman numbers because the Roman system does not perfectly preserve place (the way we write numbers in the ones column, the tens column, the hundreds column, etc.).

If you try to do subtraction, long division, or any other procedure that requires "carrying" in Roman numerals, you quickly run into serious problems, problems that, according to at least some scholars, sharply limited the development of mathematics and perhaps technology in ancient Rome. We can certainly say with great confidence that, labels for popes and Super Bowls notwithstanding, there is a good reason why Roman numerals have fallen by the wayside in favor of the nearly universal use of the familiar Arabic base-10 numbers. In our familiar system of representing numbers, a 5-digit number can never be smaller than a 1-digit number because a numeral's *position* is even more important than its shape. A bank in New Zealand (NZ) got a painful reminder of this fact in May 2009 when it accidentally deposited $10,000,000.00 (yes, ten *million*) NZ dollars rather than $10,000.00 (ten *thousand*) NZ dollars in the account of a couple who had applied for an overdraft. The couple quickly fled the country with the money (all three extra zeros of it).[1] To everyone but the unscrupulous couple, this mistake may seem tragic, but we can assure you that bank errors of this kind would be more common, rather than less common, if we still had to rely on Roman numerals.

If you are wondering how we got from ancient Sumer to modern New Zealand—or why—the main point of this foray into numbers is that life as we know and love it depends heavily on numbers, mathematics, and even statistics. In fact, we would argue that to an ever increasing degree in the modern world, sophisticated thinking requires us to be able understand statistics. If you have ever read the influential book *Freakonomics*, you know that the authors of this book created quite a stir by using statistical analysis (often multiple regression) to make some very interesting points about human behavior (Do real estate agents work as hard for you as they claim? Do Sumo wrestlers always try to win? Does cracking down on crime in conventional ways reduce it? The respective answers appear to be no, no, and no, by the way.) So statistics are important. It is impossible to be a sophisticated, knowledgeable modern person without having at least a passing knowledge of modern statistical methods. Barack Obama appears to have appreciated this fact prior to his election in 2008 when he assembled a dream team of behavioral economists to help him get elected—and then to tackle the economic meltdown. This dream team relied not on classical economic models of what people *ought* to do but on empirical studies of what people actually do under different conditions. For example, based heavily on the work of psychologist Robert Cialdini, the team knew that one of the best ways to get people to vote on election day is to remind them that many, many other people plan to vote (Can you say "baaa"?).[2]

So if you want a cushy job advising some future president, or a more secure retirement, you would be wise to increase your knowledge of statistics. As it turns out, however, there are two distinct branches of statistics, and people usually learn about the first branch before they learn about the second. The first branch is descriptive statistics, and the second branch is inferential statistics.

Descriptive Statistics

Statistics are a set of mathematical procedures for summarizing and interpreting observations. These observations are typically numerical or categorical facts about specific people or things, and they are usually referred to as **data.** The most fundamental branch of statistics is **descriptive statistics,** that is, statistics used to summarize or describe a set of observations.

The branch of statistics used to interpret or draw inferences about a set of observations is fittingly referred to as **inferential statistics.** Inferential statistics are discussed in the second part of this chapter. Another way of distinguishing descriptive and inferential statistics is that descriptive statistics are the *easy* ones. Almost all the members of modern, industrialized societies are familiar with at least some descriptive statistics. Descriptive statistics include things such as means, medians, modes, and percentages, and they are everywhere. You can scarcely pick up a newspaper or listen to a newscast without being exposed to heavy doses of descriptive statistics. You might hear that LeBron James made 78% of his free throws in 2008–2009 or that the Atlanta Braves have won 95% of their games this season when they were leading after the eighth inning (and 100% of their games when they outscored their opponents). Alternately, you might hear the results of a shocking new medical study showing that, as people age, women's brains shrink 67% less than men's brains do. You might hear a meteorologist report that the average high temperature for the past 7 days has been over 100 °F. The reason that descriptive statistics are so widely used is that they are so useful. They take what could be an extremely large and cumbersome set of observations and boil them down to one or two highly representative numbers.

In fact, we're convinced that if we had to live in a world without descriptive statistics, much of our existence would be reduced to a hellish nightmare. Imagine a sportscaster trying to tell us exactly how well LeBron James has been scoring this season without using any descriptive statistics. Instead of simply telling us that James is averaging nearly 30 points per game, the sportscaster might begin by saying, "Well, he made his first shot of the season but missed his next two. He then made the next shot, the next, and the next, while missing the one after that." That's about as efficient as "goat, goat, goat, goat. . . ." By the time the announcer had documented all of the shots James took this season (without even mentioning *last* season), the game we hoped to watch would be over, and we would never have even heard the score. Worse yet, we probably wouldn't have a very good idea of how well James is scoring this season. A sea of specific numbers just doesn't tell people very much. A simple mean puts a sea of numbers in a nutshell.

CENTRAL TENDENCY AND DISPERSION

Although descriptive statistics are everywhere, the descriptive statistics used by laypeople are typically incomplete in a very important respect. Laypeople make frequent use of descriptive statistics that summarize the **central tendency** (loosely speaking, the average) of a set of observations ("But my old pal Michael Jordan once averaged 32 points in a season"; "A follow-up study revealed that women also happen to be exactly 67% less likely than men to spend their weekends watching football and drinking beer"). However, most laypeople are relatively unaware of an equally useful and important category of descriptive statistics. This second category of descriptive statistics consists of statistics that summarize the **dispersion,** or **variability,** of a set of scores. Measures of dispersion are not only important in their own (descriptive) right, but as you will see later, they are also important because they play a very important role in inferential statistics.

One common and relatively familiar measure of dispersion is the **range** of a set of scores. The range of a set of scores is simply the difference between the highest and the lowest value in the entire set of scores. ("The follow-up study also revealed that virtually *all* men showed the same amount of shrinkage. The smallest amount of shrinkage observed in all the male brains studied was 10.0 cc, and the largest amount observed was 11.3 cc. That's a range of only 1.3 cc. In contrast, many of the women in the study showed no shrinkage whatsoever, and the largest amount of shrinkage observed was 7.2 cc. That's a range of 7.2 cc.") Another very common, but less intuitive, descriptive measure of dispersion is the **standard deviation.** It's a special kind of average itself—namely, an average

Chapter 1 A Review of Basic Statistical Concepts

measure of how much each of the scores in the sample *differs* from the sample mean. More specifically, it's the square root of the average squared deviation of each score from the sample mean, or

$$S = \sqrt{\frac{\Sigma(x-m)^2}{n}}.$$

Σ (sigma) is a summation sign, a symbol that tells us to perform the functions that follow it for all the scores in a sample and then to add them all together. That is, this symbol tells us to take each individual score in our sample (represented by x), to subtract the mean (m) from it, and to square this difference. Once we have done this for all our scores, sigma tells us to add all these squared difference scores together. We then divide these summed scores by the number of observations in our sample and take the square root of this final value.

For example, suppose we had a small sample of only four scores: 2, 2, 4, and 4. Using the formula above, the standard deviation turns out to be

$$\frac{(2-3)^2 + (2-3)^2 + (4-3)^2 + (4-3)^2}{4},$$

which is simply

$$\frac{1+1+1+1}{4},$$

which is exactly 1.

That's it. The standard deviation in this sample of scores is exactly 1. If you look back at the scores, you'll see that this is pretty intuitive. The mean of the set of scores is 3.0, and every single score deviates from this mean by exactly 1 point. There is a computational form of this formula that is much easier to deal with than the definitional formula shown here (especially if you have a lot of numbers in your sample). However, we included the definitional formula so that you could get a sense of what the standard deviation means. Loosely speaking, it's the average ("standard") amount by which all the scores in a distribution differ (deviate) from the mean of that same set of scores. Finally, we should add that the specific formula we presented here requires an adjustment if you hope to use a sample of scores to estimate the standard deviation in the population of scores from which these sample scores were drawn. It is this adjusted standard deviation that researchers are most likely to use in actual research (e.g., to make inferences about the population standard deviation). Conceptually, however, the adjusted formula (which requires you to divide by $n-1$ rather than *n*) does *exactly* what the unadjusted formula does: It gives you an idea of how much a set of scores varies around a mean.

Why are measures of dispersion so useful? Like measures of central tendency, measures of dispersion summarize a very important property of a set of scores. For example, consider the two groups of four men whose heights are listed as follows:

	Group 1	**Group 2**
Tallest guy	6′2″	6′9″
Tall guy	6′1″	6′5″
Short guy	5′11″	5′10″
Shortest guy	5′10″	5′0″

A couple of quick calculations will reveal that the mean height of the men in both groups is exactly 6 feet. Now suppose you were a heterosexual woman of average height and needed to choose a blind date by drawing names from one of two hats. One hat contains the names of the four men in Group 1, and the other hat contains the names of the four men in Group 2. From which hat would you prefer to choose your date? If you followed social conventions regarding dating and height, you would probably prefer to choose your date from Group 1. Now suppose you were choosing four teammates for an intramural basketball team and had to choose one of the two *groups* (in its entirety). In this case, we assume that you would choose Group 2 (and try to get the ball to the big guy when he posts up under the basket). Your preferences reveal that *dispersion* is a very important statistical property because the only way in which the two groups of men differ is in the dispersion (i.e., the variability) of their heights. In Group 1, the standard deviation is 1.58 inches. In Group 2, it's 7.97 inches.[3]

Another example of the utility of measures of dispersion comes from a 1997 study of parking meters in Berkeley, California. The study's author, Ellie Lamm, strongly suspected that some of the meters in her hometown had been shortchanging people. To put her suspicions to the test, she conducted an elegantly simple study in which she randomly sampled 50 parking meters, inserted two nickels in each (enough to pay for 8 minutes), and timed with a stopwatch the actual amount of time each meter delivered. Lamm's study showed that, on average, the amount of time delivered was indeed very close to 8 minutes. The *central tendency* of the 50 meters was to give people what they were paying for.

However, a shocking 94% of the meters (47 of 50) were off one way or the other by at least 20 seconds. In fact, the *range* of delivered time was about 12 minutes! The low value was just under *2 minutes*, and the high was about *14 minutes*. Needless to say, a substantial percentage of the meters were giving people way less time than they paid for. It didn't matter much that other meters were giving people *too much* time. There's an obvious asymmetry in the way tickets work. When multiplied across the city's then 3,600 parking meters, this undoubtedly created a lot of undeserved parking tickets.

Lamm's study got so much attention that she appeared to discuss it on the *David Letterman Show*. Furthermore, the city of Berkeley responded to the study by replacing their old, inaccurate mechanical parking meters with much more accurate electronic meters. Many thousands of people who had once gotten undeserved tickets were presumably spared tickets after the intervention, and vandalism against parking meters in Berkeley was sharply reduced. So this goes to prove that dispersion is sometimes more important than central tendency. Of course, it also goes to prove that research doesn't have to be expensive or complicated to yield important societal benefits. Lamm's study presumably cost her only $5 in nickels and perhaps a little bit for travel. That's good because Lamm conducted this study as part of her science fair project—when she was 11 years old.[4] We certainly hope she won a blue ribbon.

A more formal way of thinking about dispersion is that measures of dispersion complement measures of central tendency by telling you something about how *well* a measure of central tendency represents all the scores in a distribution. When the dispersion or variability in a set of scores is low, the mean of a set of scores does a great job of describing most of the scores in the sample. When the dispersion or the variability in a set of scores is high, however, the mean of a set of scores does *not* do such a great job of describing most of the scores in the sample (the mean is still the best available summary of the set of scores, but there will be a lot of people in the sample whose scores lie far away from the mean). When you are dealing with descriptions of people, measures of central tendency—such as the mean—tell you what the *typical* person is like. Measures of dispersion—such as the standard deviation—tell you how much you can expect specific people to differ from this typical person.

THE SHAPE OF DISTRIBUTIONS

A third statistical property of a set of observations is a little more difficult to quantify than measures of central tendency or dispersion. This third statistical property is the *shape* of a distribution of scores. One useful way to get a feel for a set of scores is to arrange them in order from the lowest to the highest and to graph them pictorially so that taller parts of the graph represent more frequently occurring scores (or, in the case of a theoretical or ideal distribution, more probable scores). Figure 1.1 depicts three different kinds of distributions: a rectangular distribution, a bimodal distribution, and a normal distribution. The scores in a **rectangular distribution** are all about equally frequent or probable. An example of a rectangular distribution is the theoretical distribution representing the six possible scores that can be obtained by rolling a single six-sided die. In the case of a **bimodal distribution,** two distinct ranges of scores are more common than any other. A likely example of a bimodal distribution would be the heights of the athletes attending the annual sports banquet for a very large high school that has only two sports teams: women's gymnastics and men's basketball. If this example seems a little contrived, it should. Bimodal distributions are relatively rare, and they usually reflect the fact that a sample is composed of two meaningful subsamples. The third distribution depicted in Figure 1.1 is the most important. This is a **normal distribution:** a symmetrical, bell-shaped distribution in which most scores cluster near the mean and in which scores become increasingly rare as they become increasingly divergent from this mean. Many things that can be quantified are normally distributed. Distributions of height, weight, extroversion, self-esteem, and the age at which infants begin to walk are all examples of approximately normal distributions.

The nice thing about the normal distribution is that if you know that a set of observations is normally distributed, this further improves your ability to describe the entire set of scores in the sample. More specifically, you can make some very good guesses about the exact proportion of scores that fall within any given number of standard deviations (or fractions of a standard deviation) from the mean. As illustrated in Figure 1.2, about 68% of a set of normally distributed scores will fall within one standard deviation of the mean. About 95% of a set of normally distributed scores will fall within two standard deviations of the mean, and well over 99% of a set of normally distributed scores (99.8% to be exact) will fall within three standard deviations of the mean. For example, scores on modern intelligence tests (such as the Wechsler Adult Intelligence Scale) are normally distributed, have a mean of 100, and have a standard deviation of 15. This means that about 68% of all people have IQs that fall between 85 and 115. Similarly, more than 99% of all people (again, 99.8% of all people, to be more exact) should have IQs that fall between 55 and 145.

This kind of analysis can also be used to put a particular score or observation into perspective (which is a first step toward making *inferences* from particular observations). For instance, if you know that a set of 400 scores on an astronomy midterm (a) approximates a normal distribution, (b) has a mean of 70, and (c) has a standard deviation of exactly 6, you should have a very good picture of what this entire set of scores is like. And you should know exactly how impressed to be when you learn that your friend Amanda earned an 84 on the exam. She scored 2.33 standard deviations above the mean, which means that she probably scored in the top 1% of the class. How could you tell this? By

Figure 1.1 A Rectangular Distribution, a Bimodal Distribution, and a Normal Distribution

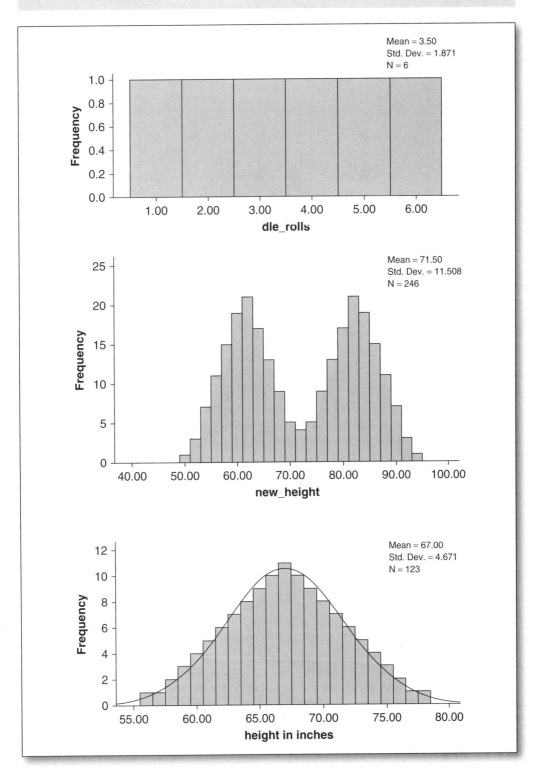

Figure 1.2 Percentage of Scores in a Perfectly Normal Distribution Falling Within 1, 2, and 3 Standard Deviations From the Mean

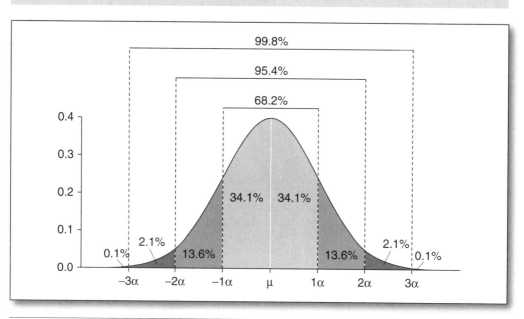

Source: Image courtesy of Wikipedia.

consulting a detailed table based on the normal distribution. Such a table would tell you that only about 2% of a set of scores are 2.33 standard deviations or more from the mean. And because the normal distribution is symmetrical, half of the scores that are 2.33 standard deviations or more from the mean will be 2.33 standard deviations or more *below* the mean. Amanda's score was in the half of that 2% that was well above the mean. Translation: Amanda kicked butt on the exam.

As you know if you have had any formal training in statistics, there is much more to descriptive statistics than what we have covered here. For instance, we skipped many of the specific measures of central tendency and dispersion, and we didn't describe all the possible kinds of distributions of scores. However, this overview should make it clear that descriptive statistics provide researchers with an enormously powerful tool for organizing and simplifying data. At the same time, descriptive statistics are only half of the picture. In addition to simplifying and organizing the data they collect, researchers also need to draw conclusions about populations from their sample data. That is, they need to move beyond the data themselves in the hopes of drawing general inferences about people. To do this, researchers rely on inferential statistics.

Inferential Statistics

The basic idea behind inferential statistical testing is that decisions about what to conclude from a set of research findings need to be made in a logical, unbiased fashion. One of the most highly developed forms of logic is mathematics, and statistical testing involves the use of objective, mathematical decision rules to determine whether an observed set of research findings is "real." The logic of statistical testing is largely a reflection of the skepticism and empiricism that are crucial to the scientific method. When conducting a statistical test to aid in the interpretation of a set of experimental findings, researchers begin by assuming that the **null hypothesis** is true. That is, they begin by assuming that their own predictions are *wrong*. In a simple, two-groups experiment, this would mean assuming that the experimental group and the control group are not really different after the manipulation—and that any apparent difference between the two groups is simply due to luck (i.e., to a failure of random assignment). After all, random assignment is good, but it is rarely perfect. It is always *possible* that any difference an experimenter observes between the behavior of participants in the experimental and control groups is simply due to chance. In the context of an experiment, the main thing statistical hypothesis testing tells us is exactly *how* possible it is (i.e., how likely it is) that someone would get results as impressive as, or more impressive than, those actually observed in an experiment if chance alone (and not an effective manipulation) were at work in the experiment.

The same logic applies, by the way, to the findings of *all* kinds of research (e.g., survey or interview research). If a researcher correlates a person's height with that person's level of education and observes a modest positive correlation (such that taller people tend to be better educated), it is always possible—out of dumb luck—that the tall people in this specific sample just happen to have been more educated than the short people. Statistical testing tells researchers exactly how likely it is that a given research finding would occur on the basis of luck alone (if nothing interesting is really going on). Researchers conclude that there is a true association between the variables they have manipulated or measured only if the observed association would rarely have occurred on the basis of chance.

Because people are not in the habit of conducting tests of statistical significance to decide whether they should believe what a salesperson is telling them about a new line of athletic shoes, whether there is intelligent life on other planets, or whether their friend's taste in movies is "significantly different" from their own, the concept of statistical testing is pretty foreign to most laypeople. However, anyone who has ever given much thought to how American courtrooms work should be extremely familiar with the logic of statistical testing. This is because the logic of statistical testing is almost identical to the logic of what happens in an ideal courtroom. With this in mind, our discussion of statistical testing will focus on the simile of what happens in the courtroom. If you understand courtrooms, you should have little difficulty understanding statistical testing.

As mentioned previously, researchers performing statistical tests begin by assuming that the *null hypothesis* is correct—that is, that the researcher's findings reflect chance variation and are not real. The opposite of the null hypothesis is the **alternative hypothesis.** This is the hypothesis that any observed difference between the experimental and the control group *is* real. The null hypothesis is very much like the *presumption of innocence* in the courtroom. Jurors in a courtroom are instructed to assume that they are in court because an innocent person had the bad luck of being falsely accused of a crime. That is, they are instructed to be extremely skeptical of the prosecuting attorney's claim that the defendant is guilty. Just as defendants are considered "innocent until proven guilty," researchers' claims about the relation between the variables they have examined are considered incorrect unless the results of the study strongly suggest otherwise ("null until proven alternative," you might say). After beginning with the presumption of innocence, jurors are instructed to examine all the evidence presented in a completely rational, unbiased fashion. The statistical equivalent of this is to examine all the evidence collected in a study on a purely objective, *mathematical* basis. After examining the evidence against the defendant in a careful, unbiased fashion, jurors are further instructed to reject the presumption of innocence (to vote guilty) only if the evidence suggests *beyond a reasonable doubt* that the defendant committed the crime in question. The statistical equivalent of the principle of reasonable doubt is the **alpha level** agreed upon by most statisticians as the reasonable standard for rejecting the null hypothesis. In most cases, the accepted probability value at which alpha is set is .05. That is, researchers may reject the null hypothesis and conclude that their hypothesis is correct only when findings as extreme as those observed in the study (or more extreme) would have occurred by chance alone less than 5% of the time.

If prosecuting attorneys were statisticians, we could imagine them asking the statistical equivalent of the same kinds of questions they often ask in the courtroom: "Now, I'll ask you, the jury, to assume, as the defense claims, that temperature has no effect on aggression. If this is so, doesn't it seem like an *incredible coincidence* that in a random sample of 40 college students, the 20 students who just happened to be randomly assigned to the experimental group—that is, the 20 people who just happened to be placed in the uncomfortably hot room instead of the nice, comfortable, cool room—would give the stooge almost *three times* the amount of shock that was given by the people in the control group? Remember, Mr. Heat would have you believe that in comparison with the 20 participants in the control group, participants number 1, 4, 7, 9, 10, 11, 15, 17, 18, 21, 22, 24,

25, 26, 29, 33, 35, 36, 38, and 40, as a group, just *happened* to be the kind of people who are inherently predisposed to deliver extremely high levels of shock. Well, in case you're tempted to *believe* this load of bullsh—." "I object, your Honor! The question is highly inflammatory," the defense attorney interrupts. "Objection overruled," the judge retorts. "As I was saying, in case any one of you on the jury is tempted to take this claim seriously, I remind you that we asked Dr. R. A. Fisher, an eminent mathematician and manurist, to calculate the *exact probability* that something this unusual could happen due to a simple failure of random assignment. His careful calculations show that if we ran this experiment *thousands of times* without varying the way the experimental and control groups were treated, we would expect to observe results as unusual as these less than *one time in a thousand if the manipulation truly has no effect!* Don't you think the defense is asking you to accept a pretty incredible coincidence?"

A final parallel between the courtroom and the psychological laboratory is particularly appropriate in a theoretical field such as psychology. In most court cases, especially serious cases such as murder trials, successful prosecuting attorneys will usually need to do one more thing in addition to presenting a body of logical arguments and evidence pointing to the defendant. They will need to identify a plausible *motive,* a good reason why the defendant might have wanted to commit the crime. It is difficult to convict people solely on the basis of circumstantial evidence. A similar state of affairs exists in psychology. No matter how "statistically significant" a set of research findings is, most psychologists will place very little stock in it unless the researcher can come up with a plausible reason why one might expect to observe those findings. In psychology, these plausible reasons are called *theories*. It is quite difficult to publish a set of significant empirical findings unless you can generate a plausible theoretical explanation for them.

Having made this "friendly pass" through a highly technical subject, we will now try to enrich your understanding of inferential statistics by using inferential statistics to solve a couple of problems. In an effort to keep formulas and calculations as simple as possible, we have chosen some very simple problems. Analyzing and interpreting the data from most real empirical investigations require more extensive calculations than those you will see here, but of course these labor-intensive calculations are usually carried out by computers. In fact, a great deal of your training in this text will involve getting a computer to crunch numbers for you using the statistical software package SPSS. Regardless of how extensive the calculations are, however, the basic logic underlying inferential statistical tests is almost always the same—no matter which specific inferential test is being conducted and no matter who, or what, is doing the calculations.

PROBABILITY THEORY

As suggested in the thought experiment with American courtrooms, all inferential statistics are grounded firmly in the logic of probability theory. Probability theory deals with the mathematical rules and procedures used to predict and understand chance events. For example, the important statistical principle of regression toward the mean (the idea that extreme scores or performances are usually followed by less extreme scores or performances from the same person or group of people) can easily be derived from probability theory. Similarly, the odds in casinos and predictions about the weather can be derived from straightforward considerations of probabilities. What is a probability? From the classical perspective, the **probability** of an event is a very simple thing: It is (a) the number of all specific outcomes that qualify as the event in question divided by (b) the total number of all possible outcomes. The probability of rolling a 3 on a single roll with a standard six-sided die is 1/6, or .167, because there is (a) one and only one roll that qualifies as a 3 and (b) exactly six (equally likely) possible outcomes. For the same reason, the probability of rolling an odd number on the same die is 1/2 or .50—because three of the six possible outcomes qualify as odd numbers. It is important to remember that the probability of *any* event (or complex set of events, such as the observed results of an experiment) is the number of ways to observe that event divided by the total number of possible events.

With this in mind, suppose the Great Pumpkini told you that he had telekinetic powers that allow him to influence the outcome of otherwise fair coin tosses. How could you test his claim? One way would be to ask him to predict some coin tosses and to check up on the accuracy of his predictions. Imagine that you pulled out a coin, tossed it in the air, and asked Pumpkini to call it before it landed. He calls heads. Heads it is! Do you believe in Pumpkini's self-proclaimed telekinetic abilities? Of course not. You realize that this event could easily have occurred by chance. How easily? Fully half the time we performed the test. With this concern in mind, suppose Pumpkini agreed to predict exactly 10 coin tosses. Let's stop and consider a number of possible outcomes of this hypothetical coin-tossing test. To simplify things, let's assume that Pumpkini always predicts heads on every toss.

One pretty unremarkable outcome is that he'd make 5 of 10 correct predictions. Should you conclude that he does, indeed, have telekinetic abilities? Or that he is *half* telekinetic (perhaps on his mother's side)? Again, of course not. Making 5 of 10 correct predictions is no better than chance. To phrase this in terms of the results of the test, the number of heads we observed was no different than the *expected frequency* (the average, over the long run) of a random series of 10 coin tosses. In this case, the expected frequency is the probability of a head on a single toss (.50) multiplied by the total number of tosses (10). But what if Pumpkini made six or seven correct predictions instead of only five? Our guess is that you still wouldn't be very impressed and would still conclude that Pumpkini does not have telekinetic abilities (in statistical terms, you would fail to reject the null hypothesis). OK, so what if he made a slightly more impressive eight correct predictions? What about nine? You should bear in mind that Pumpkini never said his telekinetic powers were absolutely flawless. Pumpkini can't *always* carry a glass of water across a room without spilling it, but his friends usually allow him to carry glasses of water unassisted. Despite your firmly entrenched (and justifiable) skepticism concerning psychic phenomena, we hope you can see that as our observations (i.e., the results of our coin-tossing test) depart further and further from chance expectations, you would start to become more and more convinced that something unusual is going on. At a certain point, you'd practically be forced to agree that Pumpkini is doing *something* to influence the outcome of the coin tosses.

The problem with casual analysis is that it's hard to know exactly *where* that certain point is. Some people might be easygoing enough to say they'd accept eight or more heads as compelling evidence of Pumpkini's telekinetic abilities. Other people might ask to see a perfect score of 10 (and still insist that they're not convinced). After all, extraordinary claims require extraordinary evidence. That's where inferential statistics come in. By making use of (a) some basic concepts in probability theory, along with (b) our knowledge of what a distribution of scores should look like when nothing funny is going on (e.g., when we are merely flipping a fair coin 10 times at random, when we are simply randomly assigning 20 people to either an experimental or a control condition), we can use inferential statistics to figure out exactly how likely it is that a given set of usual or not-so-usual observations would have been observed by chance. Unless it is pretty darn *un*likely that a set of findings would have been observed by chance, the logic of statistical hypothesis testing requires us to conclude that the set of findings represents a chance outcome.

To return to our coin-tossing demonstration, just how likely *is* it that a person would toss 9 or more heads by chance alone? One way to figure this out is to use our definition of probability and to figure out (a) all the specific ways there are to observe 9 or more heads in a string of 10 coin tosses and (b) all the specific outcomes (of any kind) that are possible for a string of 10 coin tosses. If we divide (a) by (b), we should have our answer. Let's begin with the number of ways there are to toss 9 or more heads. At the risk of sounding like the announcer who was describing LeBron James's scoring history without using statistics, notice that one way to do it would be to toss a tail on the first trial, followed by 9 straight heads. A second way to do it would be to toss a head on the first trial and a tail on the second trial, followed by 8 straight heads. If you follow this approach to its logical conclusion, you should see that there are exactly 10 specific ways to observe exactly 9 heads in a string of 10 coin tosses. And in case you actually want to see the 10 ways right in front of you, they appear in Table 1.1—along with all of the unique ways there are to observe exactly *10* heads. As you already knew, there is only one of them. However, it's important to include this one in our list because we were interested in all of the specific ways to observe 9 *or more* heads in a series of 10 coin tosses.[5] So there are 11 ways.

But how many total unique outcomes are there for a series of 10 coin tosses? To count all of these would be quite a headache. So we'll resort to a less painful headache and figure it out logically. How many possible ways are there for 1 toss to come out? Two: heads or tails—which turns out to be 2^1 (2 to the first power). How about 2 tosses? Now we can observe 2^2 (2×2) or four possible ways—namely,

HH, HT, TH, or TT.

What about three tosses? Now we have 2^3 ($2 \times 2 \times 2$), or eight possible ways:

HHH, HHT, HTH, THH, HTT, THT, TTH, or TTT.

Notice that our answers always turn out to be 2 (the number of unique outcomes for an individual toss) raised to some *power*. The power to which 2 is raised is the number of trials or specific observations we are making. So the answer is 2^{10} ($2 \times 2 \times 2 \times 2 \times 2 \times 2 \times 2 \times 2 \times 2 \times 2$) or 1,024 possible unique outcomes for a series of 10 coin tosses. This value of 1,024 includes every possible number of heads (from 0 to 10) and every possible order or position (1st through 10th) for all of these possible numbers of heads. So now we have our probability. The probability of observing 9 or more heads in a series of 10 truly random coin tosses is thus 11/1,024, or .011. So for every hundred times we conducted our coin-tossing study, you'd expect to see 9 or more heads only about once. That's only 1% of the time, and it's pretty impressive. (In fact, it's exactly as impressive as Amanda's score on the astronomy midterm, and we, for what it's worth, were very impressed with Amanda.) So if we had treated the study like a real experiment, if we had set alpha at .05, and if we had observed 9 heads, we would have had to conclude that Pumpkini does, in fact, possess the ability to influence the outcomes of otherwise fair coin tosses.

Now perhaps you're the literal type who is saying, "But wait a minute. I still wouldn't believe Pumpkini has telekinetic abilities, and I certainly don't think most scientists would, either." You are correct, of course, because the theory that you have been asked to accept flies in the face of everything you know about psychology and physics. A much more reasonable explanation for the observed findings is that Pumpkini has engaged in some form of trickery, such as using a biased coin. However, this simply means that, like any scientific practice, the practice of conducting statistical tests must be carried out using a little common sense. If someone is

Table 1.1 All the Possible Ways to Toss Nine or More Heads in 10 Tosses of a Fair Coin: A Single Tail Can Come on Any of the 10 Trials, or It Can Never Come at All

1. THHHHHHHHH
2. HTHHHHHHHH
3. HHTHHHHHHH
4. HHHTHHHHHH
5. HHHHTHHHHH
6. HHHHHTHHHH
7. HHHHHHTHHH
8. HHHHHHHTHH
9. HHHHHHHHTH
10. HHHHHHHHHT
11. HHHHHHHHHH

making a truly extraordinary claim, we might want to set alpha at .001, or even .0001, instead of .05. Of course, setting alpha at a very low value might require us to design a test with a much greater number of coin tosses (after all, 10 out of 10 tosses—the *best* you can possibly do—has a probability higher than .0001; it's 1/1,024, which is closer to .001), but the point is that we could easily design the test to have plenty of power to see what is going on. The exact design of our study is up to us (and, to some extent, to our critics). If people are sufficiently skeptical of a claim, they might also want to see a *replication* of a questionable or counterintuitive finding. If Pumpkini replicated his demonstration several times by correctly predicting 9 or more heads, and if we enacted some careful control procedures to prevent him from cheating (e.g., we let a group of skeptics choose and handle the coins), even the most ardent anti-telekinetician should eventually be persuaded. And if he or she weren't, we would argue that this person wasn't being very scientific.

The logic of the coin-tossing experiment is the same as the logic underlying virtually all inferential statistical tests. First, a researcher makes a set of observations. Second, these observations are compared with what we would expect to observe if nothing unusual were happening in the experiment (i.e., if the researcher's hypothesis were incorrect). This comparison is ultimately converted into a *probability*—namely, the probability that the researcher would have observed a set of results at least this consistent with his or her hypothesis if the hypothesis were incorrect. Finally, if this probability is sufficiently low, we conclude that the researcher's hypothesis is probably correct. Because inferential statistics are a very important part of the research process, let's look at another highly contrived but informative question that could be answered only with the use of inferential statistics.

A STUDY OF CHEATING

Suppose we offered a group of exactly 50 students the chance to win a very attractive prize (say, a large amount of cash, or an autographed copy of this textbook) by randomly drawing a lucky orange ping-pong ball out of a large paper bag. Assume that each student gets to draw only one ball from the bag, that students return the drawn balls to the bag after each drawing, and that the bag contains exactly 10 balls, only 1 of which is orange. Because our university is trying to teach students the values of honesty and integrity, university regulations require us to administer the drawing on an honor system. Specifically, the bag of ping-pong balls is kept behind a black curtain, and students walk behind the curtain—one at a time, in complete privacy—to draw their balls at random from the bag. After drawing a ball, each student holds it up above the curtain for everyone else to see. Anyone who holds up an orange ball is a winner.

Suppose that we're the curious types who want to find out if there was a significant amount of cheating (peeking) during the drawing. At first blush, it would seem like there's nothing we could do. Unless we engage in a little cheating ourselves (e.g., by secretly videotaping the drawings), how can we figure out whether people were peeking as they selected their balls? We're at a complete loss to observe the unobservable—*unless* we rely on inferential statistics. By using inferential statistics, we could simply calculate the number of winners we'd expect to observe if *no one* was cheating. By making a comparison between this expected frequency and the number of winners we actually observed in our drawing, we could calculate the exact probability (based on chance alone) of obtaining a result as extreme as, or more extreme than, the result of our actual drawing. If the probability of having so many winners were sufficiently low, we might reluctantly reject the null hypothesis (our initial assumption that the students were all innocent until proven guilty) and conclude that a significant amount of cheating was happening during the drawing.

Let's find out. To begin with, we need to assume that our suspicions about cheating are completely unfounded and that no one peeked (as usual, we begin by assuming the null hypothesis). Assuming that no one was peeking, what's your best guess about how many of the 50 students should have selected a winning ball? If you are a little fuzzy on your probability theory, remember that you can figure out the expected frequency of an event by multiplying (a) the probability of the event on a single trial by (b) the total number of trials in the series of events. This is how we knew that 5 was the expected number of observed heads in a series of 10 coin tosses. It was $.50 \times 10$. The answer here is also 5 (it's $.10 \times 50$). Now imagine that we had 6 winners. Or 9 winners, or 15—or 50. Hopefully, you can see, as you did in the coin-tossing study, that as our observed frequencies depart further and further from the frequency we'd expect by chance, we become more and more strongly convinced that our observed frequencies are *not* the product of chance.

For the purposes of actually seeing some inferential statistics in action, let's assume that we had exactly 10 winners in our drawing. Because our outcome was a categorical outcome ("success" or "failure" at the draw), and because we had a pretty large sample, we'd probably want to conduct a χ^2 (chi-square) test on these data. The formula for this test appears as follows:

$$\chi^2 = \Sigma \frac{(f_o - f_e)^2}{f_e}.$$

Recall that Σ (sigma) is a *summation sign* that tells you to add together all the appropriate examples of the basic calculation.

f_o refers to the *observed frequencies* of each of the events you care about (successes and failures when it comes to sampling a lucky orange ball).

f_e refers to the *expected frequencies* for each of these same events.

You could think of a χ^2 statistic as a "surprise index." Notice that the most important thing the formula does is to *compare* expected and observed frequencies. Specifically, expected frequencies are compared with (i.e., subtracted from) observed frequencies, and then a couple of simple transformations are made on these difference scores. The more our observed frequencies depart from what you'd expect if chance alone were operating (i.e., the more surprising our results are), the bigger our χ^2 statistic becomes. And as our χ^2 statistic grows, it tells us that it's less and less likely that we're observing a chance process (and, in this case, more and more likely that we're observing cheating).

The χ^2 value for 10 winners (out of 50) when only 5 were expected is computed as follows:

$$\chi^2 = \frac{(10 - 5)^2}{5} + \frac{(40 - 45)^2}{45}.$$

The 10 in the first half of the equation is the *observed* frequency of successes, and the two 5s both refer to the *expected* frequency of successes.

The 40 in the second half of the equation is the *observed* frequency of failures, and the two 45s both refer to the *expected* frequency of failures (this has to be the sample size, which is 50, minus the expected number of successes). When we do the math, we get 25/5 + 25/45, which works out to 5.55. Notice that this *isn't* a probability. The way most inferential statistics work is that you generate both the statistic itself (e.g., a correlation coefficient, a *t* value, an *F* ratio) and then use the exact value of the observed statistic to determine a probability value (one that corresponds to the value of your statistic). If you are doing your calculations on a computer, the software program you are using will always do this for you. That is, it will give you the exact *p* value (i.e., the exact probability) that corresponds to your results after they have been converted to the unambiguous language of your statistic. However, if you are doing your calculations by hand, as we have here, you will need to consult some kind of statistical table to see what the *critical values* are for your statistic. In the case of our study of cheating, the critical χ^2 value that corresponds to an alpha level of .05 is 3.841. Any χ^2 value that exceeds this score will have an associated *p* value that is lower than .05 and will thus be significant when alpha is set at .05. If we were a little bit more stringent, we might set alpha at .02 or .01. Our χ^2 table happens to include critical values for each of these levels of alpha (i.e., for each of these probability values). In a study such as ours, the critical χ^2 value for an alpha of .02 is 5.412, and the critical χ^2 value for an alpha of .01 is 6.635. By these criteria, our result is still significant even if alpha is set at .02. However, if we move to the still more stringent alpha level of .01, the number of winners we observed would no longer be significant (because we're effectively saying that it'd take more than 10 winners to convince us).

Suppose we followed standard practice and set alpha at .05. We'd have to conclude that some people cheated. Notice, however, that we couldn't draw any safe inferences about exactly *who* cheated. Presumably *about* 5 of our 10 winners just got lucky, and *about* 5 cheated. Realizing that *only* about 5 people cheated provides a different sort of perspective on our findings. Specifically, it highlights the fact that there is often more than one way to look at a set of observations. Notice also that an alternate, and equally correct, perspective on our observation is that people are significantly honest! It appears to be the case that about 45 of our 50 students were completely honest—even in a situation that allowed rampant cheating. Why did we say 45? Because we just decided that only about 5 people are likely to have cheated. In light of how hard it is to win the game by playing fairly, these 5 or so cheaters led to a significant amount of cheating. However, if we had started out with the hypothesis that 49 of 50, or 98%, of all people should be expected to cheat under these conditions, and if we had taken 49 (nearly absolute dishonesty) as our standard of comparison rather than 5 (absolute honesty), we would have obtained an *extremely* large χ^2 value:

$$\chi^2 = \frac{(10-49)^2}{49} + \frac{(40-1)^2}{1},$$

which is 1552.04, and which corresponds to an infinitesimally small *p* value. Even if we set alpha at a very, very, very low level (say one in a billion, or .000000001), this would still be significant. In other words, it's important to keep in mind that we appear to have observed a lot more honesty than cheating.[6]

A final aspect of this exercise about drawing ping-pong balls from a bag is that it provides a useful metaphor for thinking about what researchers do when they draw inferences about people in their research. Notice that in the lottery involving ping-pong balls, we could not directly observe the phenomenon in which we were interested. The activities we cared about were shrouded behind a black curtain—just as the activities that psychologists often care about (e.g., dissonance reduction, feelings of passionate love, parallel distributed representations of language) are hidden inside the black box of people's minds. Inferential statistics work hand in hand with things like operational definitions to allow us to make scientific inferences. Operational definitions allow us to draw inferences about *processes* that we cannot observe (those that occur inside the person), and inferential statistics allow us to draw inferences about *people* we can't observe (those we didn't sample in our study). When we conclude that a research finding is significant, we are concluding that it is real and thus that it applies to people who did not take part in our study. This is one sense in which the ping-pong ball demonstration is a little different from most significance tests. Although it would probably be safe to generalize our findings about cheating to other college students, what we really cared about most in this particular test was finding out what was going on in *our particular* sample.

Virtually every inferential statistic that you will ever come across will be based on the logic that was explicated here. Of course, the particular distributions of responses that researchers examine vary enormously from one study to the next, and this, among other things, influences the particular statistics that researchers use to summarize and draw inferences about their data. Moreover, once a researcher has chosen a particular statistic, the specific calculations that she or he will have to carry out (or get a computer to carry out) will typically be a good bit more involved than those you have seen here. For example, in a two-way analysis of variance (ANOVA), there are separate calculations (and separate *degrees of freedom*) for each of the two possible main effects as well as for the two-way interaction. No matter what statistics they are computing, however, researchers will always rely on the logic of probability theory to help them make their case that something significant is at the root of their empirical observations.

Things That Go Bump in the Light: Factors That Influence the Results of Significance Tests

ALPHA LEVELS AND TYPE I AND II ERRORS

Now that you have a better feel for what it means for a research finding to be statistically significant, we feel that it is our duty to warn you that when we look at significance testing in the cold, hard light of day, it has a couple of limitations. In other words, there are a few things that can go wrong when people are conducting statistical significance testing. First of all, it is important to remember that when a researcher conducts a statistical test and obtains a significant result, this does not *always* mean that his or her hypothesis is correct. Even if a study is perfectly executed with no systematic design flaws, it is always possible that the researcher's results *were* due to chance. In fact, the *p* value we observe in an experiment tells us exactly how likely it is that we would have obtained results like ours even if nothing but dumb luck were operating in our study. Statisticians refer to this worrisome possibility—incorrectly rejecting the null hypothesis when it is, in fact, correct—as a **Type I error**. The likelihood of making a Type I error is a direct function of where we set our alpha level. As suggested earlier, if we think it would be a practical or scientific disaster to reject the null hypothesis in error, we might want to set alpha at a very conservative level, such as .001. Then we would be taking only one chance in a thousand of falsely rejecting the null hypothesis.

So why not set alpha at .001 (or even lower) all the time? Because we have to strike a balance between being cautious and being so cautious that we become downright foolish. In statistical terms, if we always set alpha at an extraordinarily low level, we would decrease the likelihood of committing a Type I error at the expense of increasing the likelihood of committing a **Type II error**. A Type II error occurs when we fail to reject an incorrect null hypothesis—that is, when we fail to realize that our study has revealed something meaningful (usually that our hypothesis is correct). The reason it is useful to know about Type I and Type II errors is that there are things we can do to minimize our chances of making both of these troublesome mistakes. As suggested previously, one of the easiest ways to minimize Type I errors is to set alpha at a pretty low level. Over the years, most researchers have pretty well agreed that .05 is a reasonable level for alpha (i.e., a reasonable risk for making a Type I error). And of course, if we want to be a little more cautious, but we don't want to ask anyone to adjust any alpha levels, we can always insist on seeing a replication. In the grand scheme of things, replications are what tell us whether an effect is real.

EFFECT SIZE AND SIGNIFICANCE TESTING

Although no one wants to make a Type I error, no one really wants to make a Type II error either. Several things influence the likelihood that a researcher will make a Type II error (and fail to detect a real effect). Some of these are things over which researchers have little or no control, and some of them are things over which researchers have almost complete control. One thing that researchers can't do too much about is their "effect size," the magnitude of the effect in which they happen to be interested. If you collected a sample of 20 people and measured their heights and their foot sizes, you could probably expect to observe a statistically significant correlation between height and foot size, even though your sample was pretty small. This is because there is a pretty robust tendency for big people to have big feet. Of course there are exceptions, but they are relatively rare. We doubt that you will ever meet a gymnast who squeezes into a size 14 (or an NBA center who slips comfortably into a size 9). On the other hand, if you gave a sample of 20 people a measure of extraversion and a measure of self-esteem, you might not necessarily observe a significant correlation. Although self-esteem and extraversion do tend to go hand in hand, this correlation is much more modest than the substantial correlation between height and foot size. To return to our example about peeking and ping-pong balls, it would have been much easier to detect an effect of cheating if cheating had been rampant. In fact, notice that in this study, it was quite easy to detect an effect of honesty—precisely because honesty was so rampant.

MEASUREMENT ERROR AND SIGNIFICANCE TESTING

Although it's obviously impossible to change the true size of an effect, one thing that researchers can sometimes do to maximize their chances of detecting a small effect is to conduct a within-subjects or repeated measures study. As we argue later in this text, within-subjects designs are usually more sensitive than between-subjects designs. One of the reasons this is the case is that within-subjects designs cut down on extraneous sources of variability that can mask an effect. A person in a cool room might deliver high levels of shock to a confederate just because this person happens to be an unusually aggressive person. However, if we could observe the behavior of the *same* person in both a hot and a cool room (and if we could make sure the person didn't know that she or he was being studied), we would presumably see that the person would deliver even higher levels of shock when the temperature was cranked up a bit. Of course, another reason why within-subjects designs are more powerful than between-subjects designs is that they simply increase the number of observations in a study. If we measure the aggressive behavior of each of our 20 participants in both a hot and a cool room, it is almost as if we had 40 participants in our study rather than 20 (see Pelham, 1993, for a further discussion of the advantages of within-subjects designs).

SAMPLE SIZE AND SIGNIFICANCE TESTING

When researchers are unable to make use of within-subjects designs, they can still do a couple of things to maximize their chances of detecting a real effect. One simple, albeit potentially expensive, thing that researchers can do is to conduct studies with a lot of participants. Increasing your sample size in a study (whether it be an experiment, a quasi-experiment, a survey, or an archival study) can greatly increase the chances that you will detect a real effect. For example, suppose that the true correlation between extroversion and self-esteem among American adults is exactly .32. And suppose that you conducted a survey of 27 randomly sampled American adults and observed a correlation of exactly .32 in your study. Would this be statistically significant? Unfortunately not. In a sample of only 27 people, a correlation of .32 would have a p value slightly greater than .10—at best a marginally significant value. On the other hand, if you had sampled 102 people rather than 27, and if you happened to hit the nail on the head again by observing another correlation of exactly .32, this result would be significant even if you had set alpha at .001. That's because when you have a sample as large as the second, it's quite unusual to observe a correlation as large as .32 when the two variables in question are actually unrelated. If this doesn't quite seem right to you, consider your own intuitive conclusions when we asked you earlier what you'd think if Pumpkini were able to correctly predict 6 heads in 10 coin tosses. If he produced exactly the same proportion of heads (600) in 1,000 tosses, you should be much more impressed.

RESTRICTION OF RANGE AND SIGNIFICANCE TESTING

Limits in the *range or variability* of the variables you are measuring or manipulating (i.e., restriction of range) can also limit your ability to detect a true effect. Wording your dependent measures carefully, choosing the right population, and making sure that your independent variable is as potent and meaningful as possible (which means not shooting yourself in the foot by artificially *diminishing* your real effect size) are all potential solutions to the problem of restriction of range, and thus, they are all potential solutions to the problem of avoiding Type II errors. The particular statistical analysis that you conduct can also play an important role in whether your research findings are significant. When you have a choice between conducting a powerful test (one that can detect even relatively small effects) and a less powerful test, you should always perform the more powerful of the two. For example, performing a correlation between two continuous variables (e.g., self-esteem and the number of minutes people spent reading positive feedback about themselves) is usually more powerful than performing a median split (e.g., on self-esteem) and then conducting an ANOVA or t test to see if the mean difference between the low group and the high group is significant. Similarly, making use of continuous ("How much did you like your partner?") rather than dichotomous ("Did you like your partner?") dependent measures in an experiment usually allows for more powerful statistical tests. As a second example, when you have a choice of conducting more than one separate between-subjects analysis (e.g., three different between-subjects ANOVAs, one on each of your three different dependent measures) versus a single within-subjects or mixed-model analysis on the same set of research findings (e.g., because you asked people to rate a target person on positive, neutral, and negative traits), you will usually be better served by the analysis that incorporates the within-subjects aspect of your design.

The issues discussed here can help you to conduct better research studies. Just as important, they can also help you to better interpret the findings of other people's studies. For example, if a team of experimenters claims that they failed to replicate an important effect, you would do well to ask a few questions about the nature of their manipulation, the nature of their sample, the wording of their dependent measures, and the number of participants they included in their between- or within-subjects study before you abandon your own research on the same topic. If Dr. Snittle noted that he failed to replicate Phillips's archival research on suicide by noting that none of the 23 people in his small Nebraska farming community committed suicide after reading about a front-page suicide, this wouldn't be much cause for concern. However, if Dr. Snittle learned to speak fluent Mandarin, traveled to China, gained access to media and suicide records in several very large Chinese provinces, duplicated Phillips's analytical strategies perfectly, and failed to replicate some aspect of Phillips's findings, we'd want to figure out why. Perhaps some aspect of Chinese culture (or Chinese media coverage) is responsible for the difference. This way of thinking about how to interpret statistics is consistent not only with common sense but also with the logic of the scientific method. It is important to remember that statistics are simply a tool. When effectively applied to an appropriate problem, statistics can be incredibly powerful and effective. However, when misapplied or misinterpreted, statistics—like real tools—can be useless or even dangerous.

The Changing State of the Art: Alternate Perspectives on Statistical Hypothesis Testing

During the past three quarters of a century, statistical hypothesis testing has become a methodological touchstone for evaluating specific research findings. When a provocative research finding proves to be statistically significant, it is considered scientifically meaningful. When an equally provocative research finding proves to be nonsignificant, it usually is not taken seriously in scientific circles. As we have just seen, however, an absolute reliance on significance testing—when divorced from basic considerations involving things such as effect size or sample size—can often lead researchers to inappropriate conclusions. Another way of putting this is that there is more to hypothesis testing than simple significance testing. Critics of significance testing have pointed out, for example, that even when a study is well designed, basing a decision about whether an effect is real solely on the basis of statistical "significance" is not always advisable. In actual practice, for example, when a researcher conducts a study whose results are promising but not significant, the researcher will often run additional participants—or modify the design and run the study again—rather than concluding that the original hypothesis is incorrect. In fact, some researchers have argued that the traditional use of significance testing is an inherently misleading process that should be abandoned in favor of other approaches (J. Cohen, 1994).

Although it seems unlikely that significance testing will be abandoned any time in the near future, most researchers would probably agree that it is often useful to complement significance testing with other indicators of the validity, meaningfulness, or repeatability of an effect. A complete review of the pros and cons of alternate approaches to significance testing is beyond the scope of this book. However, it is probably worth noting that researchers have recently begun to complement significance testing by making use of special statistics to assess the practical or theoretical meaningfulness of research findings. One way in which researchers have done this is to compute estimates of **effect sizes,** that is, indicators of the *strength* or magnitude of their effects. A second way is to compute estimates of (a) the overall amount of existing support for an effect or (b) the consistency or repeatability of the effect. The statistical approach most suited to this second category of questions is referred to as **meta-analysis.**

ESTIMATES OF EFFECT SIZE

When researchers want to assess the practical or theoretical rather than the statistical significance of a specific research finding—that is, when they want to know how big or meaningful an effect is—they typically calculate an **effect size.** Although there are many useful indicators of effect size, the two most commonly reported indicators of effect size are probably r and d. The statistic r is the familiar correlation coefficient, and thus, you probably have had some practice interpreting this frequently used indicator of effect size. Psychological effects that are considered small, medium, and large correspond respectively to correlations of about .10, .30, and .50. The less familiar statistic d is more likely to be used to describe effect sizes from experiments or quasi-experiments because d is based on the difference between two treatment means. Specifically,

$$d = (\text{mean 1} - \text{mean 2})/D,$$

where D is simply the overall standard deviation of the dependent measure in the sample being studied (see Rosenthal & Rosnow, 1991, p. 302). Thus, d tells us *how different two means are in standard deviation units* (or fractions thereof). Because two means in a study can sometimes be more than one standard deviation apart, this means that d, unlike r, can sometimes be larger than 1. Otherwise, the interpretation of d is pretty similar to the interpretation of r. The respective values of d that correspond to small, medium, and large effects are about .20, .50, and .80 (see Rosenthal & Rosnow, 1991, p. 444).

Notice that we used the word *about* when we listed the specific values of r and d that correspond to different effect sizes. The reason we did so is that what makes an effect big or small is partly a judgment call. Moreover, how "big" an effect must be to qualify as meaningful varies quite a bit from one research area to another. If a cheap and easy-to-administer treatment (e.g., a daily vitamin C tablet) could reduce the risk of cancer and turned out to have a "small" effect size (e.g., $r = .10$ or less), this could easily translate into millions of saved dollars in medical expenses (and thousands of saved lives). Moreover, as we just noted, the size of an effect that researchers observe in a particular study is as much a function of how carefully the study is crafted as it is a function of the state of the world. Thus, considerations of effect size, like considerations of statistical significance, should reflect the theoretical or the practical significance of a given finding—regardless of its absolute magnitude. If our easy-to-administer experimental treatment gets blood from only 10% of the turnips that we treat, we will have to consider the relative value of blood and turnips before deciding how meaningful the treatment is.

For many years, when researchers wanted to know how strongly two variables were related, they would compute a **coefficient of determination** by squaring the correlation associated with a particular effect. So if researchers learned, for example, that people's attitudes about a politician correlated .40 with whether people voted for that politician, the researchers might note that attitudes about candidates account for only 16% of the variance in voting behavior ($.40 \times .40 = .16$, or 16%). Although this is a technically accurate way of summarizing the association between two variables, some researchers have noted that it provides a misleading picture of the true strength of the relation between two variables. In particular, Rosenthal and Rubin (1982) developed the **binomial effect size display** as a more intuitive way to illustrate the magnitude and practical importance of a correlation. The binomial effect size display is referred to as binomial because it makes use of variables that can take on only two values (success or failure, survival or death, male or female) to illustrate effect sizes. As matters of convenience and simplicity, Rosenthal and Rubin demonstrate effect sizes using two dichotomous variables whose two values are equally likely. To simplify matters further, they express binomial effect sizes using samples in which exactly 100 people take on each of the two values of each of the two dichotomous variables.

Consider a hypothetical example involving attendance at a review session and performance on a difficult exam. Assume (a) that exactly 100 of 200 students attended the review session and (b) that exactly 100 of 200 students passed the exam. If we told you that the correlation between attending the review session and passing the exam was .20 (meaning that attendance at the review session accounts for only 4% of the variance in exam performance), you might not bother to attend the review session. However, if you examine the binomial effect size display that appears in Table 1.2, you can see that a correlation of .20 corresponds to 20 more people passing than failing the exam in the group of attendees (and 20 more people failing than passing the exam in the group of nonattendees). *More generally, when summarized using a binomial effect size display, a correlation coefficient corresponds to the difference in success rates that exists between two groups of interest on a dichotomous outcome.* If the correlation summarized in Table 1.2 had been .40, we would have seen that 70% of those attending the review (and only 30% of those failing to attend) passed the exam (70 – 30 = 40). Similarly, if we had observed a potential cookie thief for 200 days, if the person had been present in the kitchen for exactly 100 of the 200 days, and if cookies had disappeared on exactly 100 of the 200 days, then a correlation of .66 would mean that when the potential thief visited the kitchen, cookies disappeared on 83 out of 100 days (83 – 17 = 66). Even though the presence of this person accounts for only about 44% of the variance in cookie thefts ($.66^2 = .436$), notice that cookies are almost five times more likely to disappear when the person is present than when the person is absent ($.17 \times 5 = .85$).

Regardless of what format researchers use to illustrate effect sizes, reporting effect sizes provides a very useful complement to traditional significance testing. For example, suppose we know that the effect size for a specific research finding corresponds to a *d* of .43. If a researcher claims that he failed to replicate this finding, it would be useful to consider the effect size the researcher observed (rather than focusing solely on his observed *p* value) before concluding that his finding is different from the original (see Rosenthal & Rosnow, 1991, for a much more extensive discussion). In some cases, researchers have claimed that they failed to

Table 1.2 Performance on an Exam as a Function of Attendance at a Review Session

Attendance at review	Exam Performance		
	Passed	Failed	Total
Attended	60	40	100
Did not attend	40	60	100
Total	100	100	200

replicate findings when they observed effects that were just as large as, or larger than, those observed by previous researchers (e.g., when the second group of researchers had a much smaller sample than the first).

META-ANALYSIS

Estimates of effect size, such as r or d, provide a useful metric for describing and evaluating the magnitude of specific research findings. Regardless of how big a specific finding is, however, researchers are often interested in questions that have to do with the consistency or repeatability of the finding. Questions about the repeatability of a finding almost always have to do with a *group* of studies (and perhaps even an entire literature) rather than a single specific study. How many failed studies would have to exist to indicate that a set of findings is a statistical fluke rather than a bona fide phenomenon? If a phenomenon is bona fide, how consistently has it been observed from study to study? Even more important, what are the limiting conditions of the effect? That is, when is the effect most and least likely to be observed? Questions such as these can rarely be answered by any single study. Instead, researchers need systematic ways to summarize the findings of a *large number of studies*.

Fortunately, researchers have developed a special set of statistical techniques to summarize and evaluate entire sets of research findings. Not surprisingly, R. A. Fisher (1938), the person who popularized modern statistical testing, was one of the first researchers to address the question of how to combine the results of multiple studies. In the days since Fisher offered his preliminary suggestions, researchers have developed a wide array of techniques for summarizing and evaluating the results of multiple studies (see Rosenthal & Rosnow, 1991, for an excellent conceptual and computational review of such techniques). Statistical techniques that are designed for this purpose are typically referred to as *meta-analytic* techniques. The more commonly used term **meta-analysis** thus refers to the use of such techniques to analyze the results of *studies* rather than the responses of individual *participants*. From this perspective, meta-analyses are to groups of *studies* what traditional statistical analyses are to groups of specific *participants*. Literally, meta-analysis refers to the analysis of analyses.

Prior to the development of meta-analysis, the only way researchers could summarize the results of a large group of studies was to logically analyze and verbally summarize all the studies. Meta-analyses complement such potentially imprecise analyses by providing precise mathematical summaries of different aspects of a set of research findings. For example, a meta-analysis of effect sizes can provide a good estimate of the average effect size that has been observed in all the published studies on a specific topic. Other meta-analytic techniques can be used to indicate how much *variability* in effect sizes has been observed from study to study on a specific topic (see Hedges, 1987). Finally, meta-analysis can be used to determine the kinds of studies that tend to yield especially large or small effect sizes (e.g., studies that did or did not make use of a particular control technique, studies conducted during a particular historical era, studies conducted in a particular part of the country). This final kind of meta-analysis can provide very useful theoretical and methodological information about the nature of a specific research finding.

As an example of this third approach, consider a couple of meta-analyses conducted by Alice Eagly. Eagly (1978) analyzed findings from a large number of studies of the effects of gender on conformity and social influence. Many researchers had argued that women are more easily influenced than men are. When Eagly looked at studies published prior to 1970 (i.e., prior to the beginning of the women's movement), this is exactly what she found. However, when she focused on studies published during the heyday of the women's movement (during the early to mid-1970s), Eagly observed very little evidence that women were more easily influenced than men. Furthermore, in a second meta-analysis, Eagly and Carli (1981) found that (a) the gender of the researcher conducting the study and (b) the specific topic of influence under investigation were good predictors of whether women were more conforming than were men. When studies were conducted by men or when the topic of influence was one with which women were likely to be unfamiliar (e.g., football), most studies showed that women were more conforming than were men. However, when studies were conducted by women or when the topic of influence was one with which men were likely to be unfamiliar (e.g., fashion), men often proved to be more conforming than were women.

Although meta-analysis may be used for many different purposes, the biggest contribution of meta-analysis to psychological research is probably an indirect one. The growing popularity of meta-analytic techniques has encouraged researchers to think about research findings in more sophisticated ways. Specifically, instead of treating alpha as an infallible litmus test for whether an effect is real, contemporary researchers are beginning to pay careful attention to the question of *when* a given effect is most (and least) likely to be observed. Ideally, when a meta-analysis suggests that an effect is magnified or diluted under certain conditions, researchers should conduct a study in which they directly manipulate these conditions. Doing so boils down to designing *factorial* studies in which at least two independent variables are completely crossed. An example would be a single experiment on persuasion that randomly assigned half of all men and half of all women to read about a stereotypically masculine topic before seeing how much they conform to others' opinions on this topic. Of course, the other half of men and the other half of women would be randomly assigned to read about a stereotypically feminine topic before the researchers assessed conformity on this topic. If the results of the experiment confirmed the results of the meta-analysis, researchers could be even more confident of the conclusion suggested by the meta-analysis.

Summary

This chapter provided a brief review of statistics. We noted that statistical procedures can be broken down into *descriptive statistics* and *inferential statistics*. As the name suggests, descriptive statistics simply describe (i.e., illustrate, summarize) the basic properties of a set of data. Along these lines, measures of central tendency describe the typical or expected score in a given data set. In contrast, measures of dispersion reveal how much the entire set of scores varies around the typical score. The most common measures of central tendency are the mean, the median, and the mode, and the most common measures of dispersion are the range and standard deviation. Of course, psychologists interested in testing psychological theories are typically interested in inferential statistics as well as descriptive statistics. This second branch of statistics applies probability theory to determine whether and to what degree an observed data pattern truly differs from a chance pattern. Inferential statistics thus provide a basis for determining whether an observed research finding reveals a systematic association that is likely to be true in a population of interest or whether it merely reflects noise or error. For instance, if a treatment group differs from a control group to a degree that would not be expected by chance alone, then researchers will view this as evidence that the treatment is actually causing changes in the outcome. As another example, if people who tend to score high in self-esteem also tend to score high on a measure of aggression, inferential statistics can tell us whether this tendency for the scores to go hand in hand could have happened easily by chance or whether the tendency is strong and consistent enough that it probably reflects a true association between these two variables in the general population (or at least the population that most closely resembles the researcher's sample). Both of these examples reveal the logic of significance testing. More recently, statisticians have begun to complement traditional statistical tests with indicators of effect size. Whereas statistical significance tells you whether an effect is likely to exist, estimates of effect size tell you how large an effect is likely to be.

Appendix 1.1: Some Common Statistical Tests and Their Uses

In Cervantes's classic novel *Don Quijote*, there is a point at which Don Quijote expresses tremendous self-satisfaction when he learns that he has been speaking *prose* his entire life. Unlike Don Quijote, most statisticians are more impressed with proofs than with prose. Thus, many statistics texts offer readers flowcharts, formulae, or decision trees to help them decide what kind of statistical analysis to perform on different kinds of data. Because of our abiding love of prose as well as our pathological fear of decision trees, we offer an alternative in this appendix. That is, instead of a decision tree, we offer a series of definitions and concrete examples that are much richer than a decision tree. If you spend a little while reading over the list of analytic techniques covered in this appendix, we hope that you'll have a good sense of how to analyze most basic data sets while also gaining a good sense of all of the major topics we cover in this textbook. The one way in which this list does vaguely resemble a decision tree is that it is loosely organized in increasing order of the complexity or sophistication of the research question. It thus begins with a couple of simple descriptive statistics and progresses through a series of increasingly complex inferential statistical tests. Readers who are interested in a true decision tree can find an excellent one in Chapter 2 of Tabachnick and Fidell (2007, pp. 28–31).

Mean, median, and mode: Very often, researchers who have collected data on a continuous (i.e., interval or ratio) scale simply want to summarize what the typical score is like. When the scores are normally distributed with very little skew (and modest to high kurtosis), the mean is an excellent indicator of the typical score. Height, SAT scores, and the highway mileage for one's car are all normally distributed without too much skew or kurtosis (the Hummer and Prius notwithstanding). In contrast, variables such as personal income, number of criminal convictions, and number of depressive symptoms only approximate a normal distribution because they are typically highly skewed. In such cases, the median and/or mode are often better indicators of central tendency because the median and the mode are influenced very little by extreme outliers. When researches wish to make inferences about populations rather than merely summarize a set of scores, they will want to report the standard error of the mean and/or a 95% or 99% confidence interval for the mean—to give others some idea of how far from the observed sample mean the true population mean is likely to fall. The standard error of the mean is a function of the observed sample standard deviation (see below) and the sample size, and it grows smaller as the sample size gets larger. All else being equal, this means, for example, that a researcher who randomly samples 1,000 people will be able to make a more precise statement about the likely range of the population mean than will a

Chapter 1 A Review of Basic Statistical Concepts

researcher who randomly samples only 100 people. We discuss measures of central tendency in great detail in Chapter 2. In that same chapter, we also discuss the important topic of variability (especially the **standard deviation**) while also covering important topics such as skewness and kurtosis. These last two topics set the stage for subsequent chapters on inferential statistics and data cleaning.

Standard deviation: The standard deviation is an indicator of the variability of a set of continuous scores around the mean. Whereas central tendency tells us what the typical score is like, the standard deviation tells us how well that typical score describes *all* of the scores in the distribution—because the standard deviation is an indicator of the average amount by which all of the scores in a distribution vary around the mean. We discuss both traditional and nontraditional (creative) uses of the standard deviation in Chapter 2.

Variance: The variance of a set of scores is simply the **standard deviation** of that set of scores squared.

Pearson's *r*: describes the strength and direction of the *linear* association between two continuous (interval or ratio) variables. An *r* of zero means that there is no linear association whatsoever between two variables. Absolute values of *r* closer to 1.0 indicate a stronger association. If *r* is negative, it means that as scores on one variable (*X*) increase, scores on the other variable (*Y*) decrease. If *r* is positive, it means that as scores on one variable (*X*) increase, scores on the other variable (*Y*) also increase. The concept of correlation is closely linked to prediction. Thus, for example, if height and weight are correlated $r = .70$, one can minimize errors of prediction by predicting that a person who is exactly one standard deviation above the mean in height is 0.70 standard deviation units above the mean in weight. We discuss the correlation coefficient in Chapter 3, including a brief discussion of how to assess curvilinear as well as linear associations between continuous variables.

Phi coefficient (φ): The phi coefficient is very similar to *r* except it has no sign because it is used to describe the strength of association between two nominal or categorical variables (variables that do not indicate quantity or amount). It thus ranges between zero and 1.0. We discuss both the **phi coefficient** and the **chi-square test of association** in Chapter 4.

Chi-square test of association: A chi-square (χ^2) test of association is conceptually identical to a phi coefficient (φ) because, like phi, this test of association indicates whether two categorical variables are related. In fact, it is very easy to convert a chi-square test of association to a phi coefficient using the simple formula $\varphi = \sqrt{(\chi^2_{obt} / N)}$.

Single-sample *t* test: This very simple test is designed to test see whether the mean score for a continuous, normally distributed variable (e.g., IQ score, height) in a specific sample differs from some known or hypothesized (e.g., theoretically meaningful) population value. For example, the SAT scores for the students at a particular high school might be compared with the known U.S. population mean for SAT scores to see if the students at that high school tend to be more academically prepared than the average American high school student. We discuss this test in the beginning of Chapter 6.

Independent samples *t* test: This common test is used to assess the reliability (statistical significance) of mean differences on a continuous variable between two independent groups or categories of people. Some examples of the use of this test are (a) drawing inferences about the results of a lab experiment that has one experimental and one control group, (b) assessing gender differences on a continuously measured emotional performance test, and (c) comparing people with versus without a disease on a suspected health consequence of the disease. The main assumption of the test is that the dependent measure is normally distributed (although the test is pretty robust to many, but not all, violations of this assumption). We discuss this test in some detail in the latter portion of Chapter 6.

One-way analysis of variance (ANOVA): This test is used to assess the reliability (statistical significance) of mean differences between three or more groups. This test is very similar to an independent samples *t* test (and shares the same assumption of a normally distributed dependent measure). However, the difference is that a one-way ANOVA controls for the experiment-wise error rate that occurs as researchers consider all of the many possible comparisons that can be made between specific groups when there are *multiple* experimental or naturally existing groups (three or more levels of the independent variable). Some examples of the use of this statistic are (a) outcomes in a lab experiment that has three different conditions (e.g., three dosage levels of a drug), (b) comparing kids in four different grades on an intellectual outcome, and (c) comparing the attitudes of soldiers from five different military ranks. When the researcher has a clear a priori reason to expect the various conditions to yield results that follow a specific pattern, the researcher can greatly increase statistical power by conducting a **planned comparison** based on this specific expectation. We discuss one-way ANOVA, including planned comparisons, in the first half of Chapter 7.

Factorial ANOVA: This technique is used to assess joint effects of two or more fully crossed categorical independent variables on a continuously scored outcome. It allows for the statistical separation of main effects of all independent variables and, if desired, interactions between two or more independent variables. This technique is used frequently to draw conclusions about the results of laboratory experiments. It also assumes a normally distributed dependent measure. We discuss factorial ANOVA, including follow-up comparisons such as simple effects tests, in the second half of Chapter 7.

Analysis of covariance (ANCOVA): Sometimes researchers may wish to control for a confound or nuisance variable in an ANOVA and—rather than reporting the raw, between-groups means—hold the different, naturally occurring or experimental groups constant on that nuisance variable. This is both conceptually and mathematically identical to a simultaneous multiple regression analysis with at least one categorical variable (the independent variable or variables) and at least one continuous variable (the covariates in an ANCOVA). The main advantage of ANCOVA over a regression analysis is the fact that ANCOVA readily yields covariate-adjusted means, which look very much like regular means and thus are very easy to interpret.

One of the crucial assumptions of ANCOVA is **homogeneity of covariance,** meaning that the covariate for which the analysis makes a statistical adjustment should have roughly the same association with the dependent measure in all of the various experimental conditions. For example, a researcher studying gender differences in aggressive behavior in the lab might wish to control statistically for the fact that people who more strongly believe in the concept of defending one's honor (reported, say, on a 9-point scale) behave more aggressively than people who do not believe in the concept of defending one's honor (e.g., see D. Cohen & Nisbett, 1994). So long as the association between beliefs about honor and laboratory aggressive behavior was about equally strong for women and men, it would be appropriate to reduce the noise associated with this belief to see if a significant gender difference remained after controlling for the belief. The test for a gender difference could thus be more powerful than it would have been otherwise after controlling for any effects of this nuisance variable that is more or less independent of gender. In Chapter 11, we compare and contrast ANCOVA with ANOVA and with multiple regression analysis. We emphasize that whereas ANCOVA is mathematically identical to a simultaneous multiple regression analysis, the two techniques yield very different kinds of outputs. For example, it is often much easier for people to understand covariate-adjusted means (because they look just like traditional means) than to understand standardized regression coefficients.

Reliability analysis: A reliability analysis is used to determine the degree to which the multiple items in a scale all behave in the same fashion (i.e., are positively correlated with one another). Cronbach's alpha (α) is a very useful and easy-to-calculate statistic. However, a very high alpha statistic for a scale does not always guarantee that the items in the scale form a single factor. On the other hand, low corrected item-total *r*s for specific items in a scale are a useful indicator that the specific items are not correlated with the other scale items (and thus should probably be excluded from the scale). Treating individual raters of a specific behavior or judgment as if they were specific items in a scale can allow an assessment of the reliability of individual raters. A rater whose item-total correlation is low is in disagreement with the average of all of the other raters and can either be retrained, if possible, or dropped. We discuss reliability analysis (along with principal components analysis and factor analysis) in Chapter 5.

Multiple regression analysis: Multiple regression analysis is used to assess the strength and direction of the unique linear association between multiple, continuous predictors of a continuous outcome (a criterion) and that outcome. Because each predictor is controlled statistically for the association between that predictor and all other predictors, an assessment of the relative strength of each predictor is possible. Each predictor is thus assessed controlling for the natural *confounds* between that predictor and all other predictors. The primary statistical indicator is a B or b (an unstandardized regression coefficient) or beta (β), a standardized regression coefficient that is conceptually very similar to r. Each coefficient has its own associated p value that indicates the reliability (statistical significance) of that unique association. Although both univariate (single-variable) normality and multivariate normality of the continuously measured predictors are assumed, this analysis is usually quite robust to the inclusion of one or more categorical variables, especially when these variables are not highly skewed. Gender, for example, is often dummy-coded without any problems in a multiple regression analysis that also includes several continuous, normally distributed predictors. We discuss the basics of multiple regression analysis, with a conceptual emphasis on how multiple regression identifies the unique association between different predictor variables and a criterion variable, in the first part of Chapter 9.

In most cases of multiple regression analysis, researchers expect the zero-order association between a predictor variable and the criterion of interest to grow smaller (and sometimes even disappear altogether) when all the other predictor variables are statistically held constant. However, a multiple regression analysis can also reveal that a predictor variable that appeared to be unrelated (at the simple or zero-order level) to a criterion variable is actually associated with the criterion variable once statistical adjustments are made for the effects of one or more additional predictors. This unusual situation is referred to as **suppression,** and it is discussed in great detail in Chapter 12.

Moderator analysis (multiple regression analysis of statistical interactions): If a researcher also wishes to know whether the association between a predictor and the criterion *differs* at different levels of some other predictor, it is possible to conduct a **moderator** analysis by examining the cross-product(s) of the two or more predictors of interest and following up a significant interaction with simple slopes tests (analogous to simple main effect tests in ANOVA). A moderator analysis in multiple regression could thus be conducted to see if the association between the number of times people moved as children and their physical health as adults is stronger for introverts than for extraverts. (Introverts who moved a lot as children often have poorer than average adult health, but extraverts seem to show no such association.) A moderator variable can also be categorical while the other predictor and the criterion variable are both continuous. For example, the association between implicit self-esteem and explicit self-esteem (both continuous variables) seems to be stronger (more positive) for women than for men (Pelham, Koole, et al., 2005). We discuss moderator analyses in multiple regression, including simple slopes tests to elucidate the exact nature of a significant interaction, in Chapter 10.

Logistic regression: Logistic regression analysis is conceptually identical to a standard multiple regression analysis except that the criterion variable (and sometimes one or more of the predictors) is categorical rather than continuous. The primary output statistic is a predictor-adjusted odds ratio that is the rough conceptual equivalent of a B or a β. Unlike a simple odds ratio, however, the odds ratio from a logistic regression analysis refers to the association between one categorical variable and another while holding all other predictor variables in the model constant. We discuss the basics of logistic regression in the last section of Chapter 9.

Principal components analysis and factor analysis: These two closely related techniques are designed to uncover underlying dimensions along which a set of many separate responses vary. These numerous individual responses might be answers to individual personality questions, specific political or social attitudes, or self-reported liking for many different kinds of foods or many different specific types of music. For example, contemporary research in human personality suggests that hundreds of individual personality questions all boil down to five core personality dimensions: openness to experience, conscientiousness, extraversion, agreeableness, and neuroticism (see Goldberg, 1990). A factor analysis of dozens of specific personality traits might thus reveal, for example, that trait terms such as *energetic, friendly, outgoing, outspoken,* and *loud* would all load heavily on the basic dimension of **extraversion**. In contrast, specific trait terms such as *reliable, punctual, obedient, organized,* and *honest* might all load heavily on the core dimension of **conscientiousness**. One key difference between principal components analysis and factor analysis is that principal components analysis is usually a bottom-up, purely empirical way to distill a large set of observations into a smaller number of dimensions. In contrast, factor analysis is more likely to be used when the researcher has a clear a priori theory about how the different items ought to load together and how many factors there ought to be in a data set. We discuss principal components analysis and factor analysis in Chapter 5, including a discussion of how these methods are related to the concept of reliability (specifically, the internal consistency of a multiple-item scale).

Paired-samples *t* test: When two measures come from the same organism (or similar organisms), the two different measures are likely to be highly correlated with one another. For example, a specific child who excels in reading is also likely to excel in spelling. A similar lack of independence in specific observations often occurs when genetically related or experimentally yoked members of a pair are tested on the same outcome. When exactly two such repeated measures are obtained, a paired-samples *t* test can be used to assess the statistical significance of any observed behavioral or performance difference between the two measures. This test is also very useful when the same person fills out the same measure or task under different (manipulated) experimental conditions. For example, a child might be given the same intellectual measure by two different experimenters, one of whom expresses a positive expectancy about her performance and one of whom expresses no expectations. If appropriate experimental controls are used (e.g., counterbalancing the order of the two expectancies across participants), a paired-samples *t* test can reveal whether performance varies reliably with experimental condition, with a very high level of statistical power. Like the independent samples *t* test, this test assumes that the dependent measure (which in this case is the *difference* between two scores) is normally distributed. Importantly, the increased power that usually comes with this test comes precisely to the degree that the two measures of interest are strongly correlated with one another. It is this strong correlation between two related measures that effectively reduces the variance that serves as the error term for this analysis. We discuss the paired-samples *t* test in Chapter 8, with a special emphasis on how the correlation between two repeated measurements plays a crucial role in the power of a repeated measures *t* test to reveal differences between paired means.

Repeated measures ANOVA: If three or more within-subjects conditions (or measurement periods) rather than two are collected from the same (or related) participants, a repeated measures ANOVA is the appropriate statistical test for differences between the means. Just as a one-way ANOVA replaces an independent samples *t* test once you graduate from two to three or more groups, the one-way repeated measures ANOVA replaces the paired-samples *t* test once you graduate from two to three or more within-subjects conditions. For example, if children are exposed to three different expectancies rather than two, a repeated measures ANOVA could be conducted on the mean performance scores in the three within-subjects conditions. Repeated measures studies can also involve complete factorial designs. For example, a completely within-subjects experiment might separately study reactions to sexist and nonsexist jokes that also vary independently in how funny the jokes are pretested to be. In its simplest form, this study would be a 2 (Level of Sexism: High vs. Low) × 2 (Level of Humor: High vs. Low) completely within-subjects study, analyzed using a within-subjects ANOVA. We discuss repeated measures ANOVAs in Chapter 8.

Mixed model ANOVA: If a study includes at least one within-subjects variable and at least one between-subjects variable (whether measured or manipulated), a mixed model ANOVA can test simultaneously for both between-subjects and within-subjects effects. Mixed model ANOVAs can also test for statistical interactions between one or more between-subjects variables and one or more within-subjects variables. For example, a cognitive psychologist might manipulate cognitive load on a between-subjects basis while assessing both implicit and explicit memory for studied material. She might predict, for example, that the cognitive load manipulation (e.g., rehearsing a 7-digit number) will have a large effect on explicit memory (recall memory) while having little or no effect on implicit memory (e.g., based on performance on a word fragment completion task). Thus, the researcher would expect to observe a Load × Memory–type interaction in this mixed model design. We discuss mixed model ANOVAs in Chapter 8.

Mediation analysis: In both laboratory experiments and passive observational studies, researchers often wish to know *why* one variable is related to another. For example, research on frustration and aggression suggests that one of the main reasons why frustration often leads to aggression is because frustration leads to anger, which then leads to aggression. In the language of mediation, this is to say that anger mediates the simple association between frustration and aggression. Mediation analyses are merely variations on a multiple regression analysis with an emphasis on assessing the degree to which the association between the original independent variable and the dependent variable disappears or gets weaker once you statistically control for the significant effects of the mediator on the dependent measure. Prototypically, if anger fully mediates the association between frustration and aggression, (a) frustration should predict aggression, (b) frustration should predict anger, (c) anger should predict aggression, and finally (d) the association between anger and aggression should completely disappear once you statistically control for the effects of anger on aggression (because frustration affects aggression indirectly through the route of increased anger). We discuss both **mediation analysis** and **path analysis** in Chapter 13.

Path analysis: Path analysis is a special version of multiple regression analysis that is designed to assess the plausibility of a proposed causal chain leading from one or more source variables to an ultimate ("downstream") outcome variable. In fact, the simplest possible kind of path analysis is a three-variable mediation model with the mediator representing the causal step between just one source variable and just one outcome variable. In most cases, however, path analyses involve four or more variables, ideally measured at different carefully selected time points (e.g., in a longitudinal or prospective design). Moreover, researchers do not always expect every variable in the middle of a causal chain to mediate the associations between the source variables and the ultimate outcome variable. Instead, some of the source variables might be expected to have a direct (nonmediated) as well as an indirect (mediated) connection to the downstream variable. Path analysis is the historical and conceptual precursor of modern **structural equation modeling,** which can be thought of as a hybrid combination of path analysis and factor analysis. In fact, some researchers refer to structural equation modeling as confirmatory (aka "theory-driven") factor analysis. A detailed discussion of structural equation modeling is beyond the scope of this intermediate text.

Notes

1. For more details, see http://www.news.com.au/business/story/0,27753,25515799-462,00.html.

2. See the *Time* magazine story at http://www.time.com/time/magazine/article/0,9171,1889153,00.html.

3. We adapted this example of men of varying heights from an illuminating statistics lecture by Daniel Gilbert, who probably adapted it from a lecture by Plato.

4. http://imgs.sfgate.com/cgi-bin/article.cgi?f=/c/a/1998/12/28/MN9307.DTL&type=printable

5. Computing the probability of an event as extreme as *or more extreme than* an observed event (or set of events) is standard practice for most statistical tests. At first blush, paying attention to events even more extreme than an observed event may seem a little odd. However, if we care about events *as unusual as or more unusual than* our observed event—which we almost always do—it makes a lot of sense. If you think of the unusualness of a set of observations (e.g., a lot of heads tossed, a pair of means that are noticeably different) as a standard of experimental performance that a researcher hopes to meet or exceed, this may help make sense of this practice. If we set a high-jump bar at exactly 6 feet and Amanda clears it, the set of outcomes that Amanda, the judges, and the fans all care about is jumps of exactly 6 feet *or higher.* Furthermore, if we tried to calculate the probability of a specific observation or event, probabilities would almost always be pretty low—because the probability of any specific event is always quite low. For example, the probability of tossing a fair coin 20 times and observing *exactly* 10 heads is .176, even though this is the *most* likely of all the possible outcomes. Once we move to continuous rather than discrete events, this is even truer. The probability that a particular high jump would be *exactly* 6 feet—even for a very good jumper who was trying to jump exactly 6 feet—is extremely low.

6. Speaking of cheating, we cheated. Unless we increased our sample size to about 250 people, we couldn't actually conduct this second χ^2 analysis. That's because we're allowed to use the χ^2 statistic only in situations in which all our expected frequencies have a value of at least 5.0. With values lower than 5, the χ^2 values that are generated can be pretty unstable and pretty inaccurate. In an extreme case such as this one, however, it's safe to say that people were significantly honest. If nothing else, we could always choose to make a very conservative comparison and set 90% (instead of 98%) dishonesty as our standard of comparison. This would yield 5 rather than 1 as the expected number of nonwinners. In case you want to practice your calculations, the value you should get if you do the analysis this more conservative (but legal) way is $\chi^2 (1, N = 50) = 272.22$. The 1 in the parentheses indicates the *degrees of freedom* you'd report in an actual research report in which you conducted this analysis. We come back to this in the section on reporting commonly used statistics.

Descriptive Statistics 2

Before we explore inferential statistics, it should be instructive to spend a little time exploring the kind of statistics that often dominate newspaper headlines, barbershop conversations, and public debates about highly controversial topics such as health care reform, gay marriage, and whether the Boston Red Sox are better than the New York Yankees. Of course, we are talking about simple *descriptive* statistics such as percentages, rankings, means, and standard deviations. After we have explored these familiar descriptive statistics in some detail, we conclude the chapter with a discussion of a couple of more esoteric descriptive statistics—namely, skewness and kurtosis. This discussion serves as a bridge to the rest of the text because, as it turns out, skewness and kurtosis can have a very powerful impact on the results of inferential statistical tests (tests designed to allow us to draw inferences about populations).

At the same time that you learn about statistics, you will also learn how to use the SPSS software package. People sometimes liken learning statistics to learning a foreign language, and you could say the same thing about learning to use SPSS. As you know if you have ever studied a foreign language, the best way to increase your fluency is via immersion. In this text, then, we are going to immerse you in the related but distinct languages of statistics and of SPSS, with a particular emphasis in this chapter on descriptive statistics. After we have discussed the most well-known descriptive statistics in some detail, we begin, in the last section of the chapter, to explore some of the more esoteric descriptive statistics, with an emphasis on how these statistics can influence the results of inferential statistical tests.

In this discussion of descriptive statistics, let's make sure we start off on the right foot. But let's begin by taking some very small steps. In other words, let's just get our feet wet before we try to take any giant steps. To put this yet another way, before we expect you to think on your feet or take a walk on the wild side, we want to remind you that a journey of a thousand steps begins when you just put one foot in front of the other. As we hope you have guessed by now (because we have run out of clichés involving feet), our first topic is feet.

The Very Small Survey of Moderately Large Shoe Sizes

If you don't think shoe sizes are a very interesting, you should know that there is a pretty popular website (www.averageshoesize.com/) devoted exclusively to the meaning of shoe sizes. Although we cannot agree with some of the sweeping conclusions the author of the webpage is willing to draw based on shoe size (e.g., " the ideal shoe size for a man, in terms of mobility, being able to ride a horse, speed, ability to kick and the general ability to fight, wrestle and kill in warfare ... [is] ... size 11."), we *do* appreciate the fact that this webpage is one of the few sources we could find that seems to have collected survey data on average shoe size. According to a survey of more than 200 men, the average shoe size for men (in American sizes) is a size 10. Although most statisticians would probably have some serious concerns about the sampling techniques that yielded this mean of 10.0, we are going to accept the value of 10.0 so that we have *some* kind of standard with which to work. Needless to say, if you were doing serious scientific work on foot size, you would want to make sure you used a representative sample of Americans rather than a convenience sample. Moreover, you'd probably want to measure people's feet yourself, to avoid any reporting biases. Unfortunately, we don't yet have a budget for that lofty scientific research project.

So based on this starting assumption, the question we want to address is very simple: Do the grown men in the first author's immediate family have big feet? Here are the shoe size data listed in order of the men's ages (oldest to youngest):

Bill: 11½, Stacy: 10½, Brett: 13, Jason: 10½, Barry: 13½

Because the entire adult male population of my immediate family includes only five people, it should be pretty easy for you to eyeball the data and see how we compare with the presumed population mean of 10.0. But what score best summarizes where we stand?

Here are the data again—ranked from lowest to highest shoe size, with fractions converted to decimals (the way SPSS likes to see them). *Notice how ranking the scores makes it easier to evaluate them.*

Jason: 10.5, Stacy: 10.5, **Bill: 11.5**, Brett: 13, Barry: 13.5

Based on these ranked scores, we hope you can now see that the **mode** is 10.5 and that the **median** is 11.5. This very small, tremendously trivial data file provides us with an ideal introduction to data files in SPSS because you can fully understand all of the data points and focus on some important aspects of how SPSS data files work. The file you'll be working with is called [**Pelham men shoe size data.sav**], and once you have started the SPSS software package, you can open this data file by clicking the File button you see in Image 1 and going to the web page, flash drive, or other location where all the SPSS files for this text exist. From "File," just move your cursor to "Open" and then click on "Data. . . ." as illustrated in Image 1.

Chapter 2 Descriptive Statistics

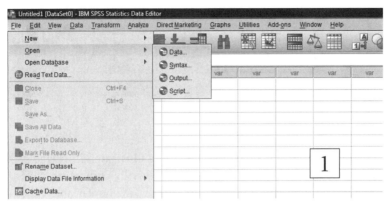

You may need to browse around a bit to find the file you want. Once you have found it and have double clicked on it, your formerly blank SPSS data screen should now look very much like the screen capture you see in Image 2.

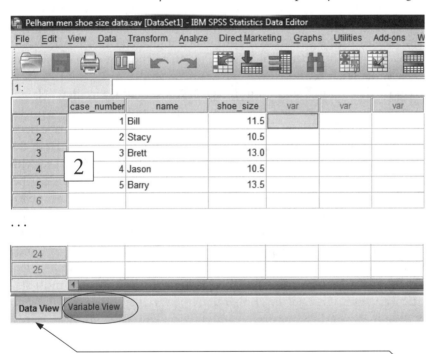

You should notice, first of all, that this screen capture shows you the Data View screen, which is the version of an SPSS data file that allows you to see the **actual data.** In this case, the file contains only three variables for five respondents: (1) an arbitrary case number, (2) each person's name, and (3) his shoe size. However, there is a lot more you should know about this very simple SPSS data file. If you click on the Variable View button, which is circled for you in Image 2, you will toggle into the Variable View mode. This mode allows you to find out (or specify) things you need to know about the nature of each variable. This is very important because the nature of a variable (e.g., whether it is a word or a number) determines the kind of *analysis* you can perform on the variable.

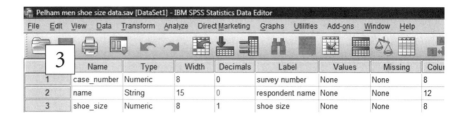

Notice that in the Variable View mode illustrated in Image 3, SPSS tells you what *kind* of variable each variable is by indicating each variable's "Type" (see the second column of Image 3). The variable called "Name," for example, is a "String" or alphanumeric variable meaning that it has no numeric value. You could never calculate a mean or standard deviation for this kind of nominal variable. Stacy may be twice as strong and half as good a poker player as Jason (both of which are roughly true), but those facts can't be expressed using letters. Names, on the other hand, *must* be expressed as strings of letters. If your data file included country names or Social Security numbers, you'd want to make those nominal variables *String* variables, too.

Under the "Width" column in Image 3, you should notice that "Name" is 15 characters wide, that "case_number" is 8 characters wide (with no decimals), and that "shoe_size" is 8 characters wide—but allows one decimal. Notice also that SPSS allows you to label your variables. Here the *variable names* and the *labels* are obvious, but if you've ever worked with data sets collected by others, you'll appreciate how useful a "Label" can be. The first author has worked with data files in which the variable **Name** for gender was "sex" or "gender," but he has worked with others in which, for good reasons, the gender variable was named "dem3," "sc7," or "wp1219." If variable names are not patently obvious, labels really come in handy because in more recent versions of SPSS, you can search unfamiliar data files for words that are part of a label. Moreover, when you run most statistical analyses, it is the *labels* rather than the variable names that show up by default in your SPSS output (results) files.

Getting back to the data, recall that the *median* was 11.5 and that the *mode* was 10.5.

QUESTION 2.1a. Which statistic, the median or the mode, is a better indicator of central tendency in these data?

QUESTION 2.1b. If the data more closely approximated a normal distribution, would the two values (the mode and the median) be more likely to be in perfect agreement? If you're not sure, you should examine an alternate data file that *does* strongly resemble a perfectly normal (nonskewed) distribution and take a close look at it. How about 8.5, 9.5, **10.0**, 10.5, 11.5? Or what about 13, 16, 18, **19**, 20, 22, 25?

Chapter 2 Descriptive Statistics

Returning to the Pelham men shoe size data (10.5, 10.5, **11.5**, 13.0, 13.5), we hope you can see that the *mean* has to be at least a *little* larger than the median of 11.5 (and thus much larger than the mode of 10.5). Notice that the two scores *below* the median (10.5 and 10.5) are each exactly one point below it. In contrast, the two scores above the median of 11.5 (13.0 and 13.5) are each *more* than one point above it. Let's confirm this logic about the mean electronically—by asking SPSS to calculate the three most common measures of central tendency in these data.

To do so, go back to the Data View mode (using the toggle button in the lower left-hand corner of Image 2). This will take you to a screen that resembles Image 4, where you can click on "Analyze." After clicking, move your arrow to "Descriptive Statistics" and then move to, and click your mouse on, "Frequencies. . . ." This will generate a dialog box like the one you see in Image 5. This box will allow you to choose a variable or a set of variables for analysis.

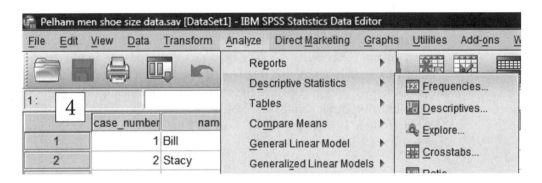

From this Frequencies dialog box, click on the variable you wish to analyze (which will highlight it, as in Image 5). Then click the arrow circled in Image 5 to send that desired variable to the "Variable(s):" list.

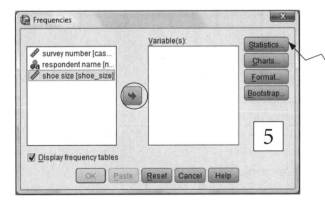

Once you've sent "shoe_size" to the Variable(s): list (which we didn't yet do in Image 5), you can ask for each of the three statistics we need.

Do this by clicking the "Statistics. . ." button, and you will see another dialog box (shown in Image 6) that will allow you to select your desired statistics.

Be sure to check the "Mean," the "Median," and the "Mode," as you can see we did in Image 6.

Once you have selected your three descriptive statistics, click on "Continue" to return to the *first* dialog box.

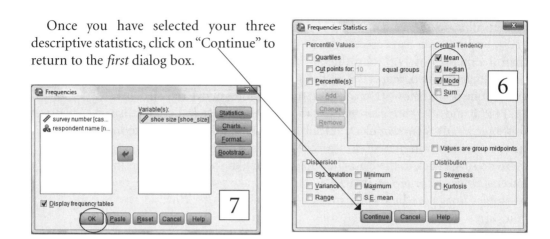

Notice in Image 7 that the "OK" and "Paste" buttons have now been activated (prior to this—in Image 5—they appeared only faintly and could not yet be clicked). This means that if you click on "OK," SPSS will *run the frequency analysis*. If all goes well, you'll generate a second kind of SPSS file, an **output** (i.e., results) file. The heading for the file will simply remind you of which analysis you just ran. Unless you happen to be using the first author's laptop, the part of the file that begins "[DataSet1]" won't look exactly like this, but it should resemble it. The really important part of the output, though, is what *follows* the heading labeled "Frequencies" that you see in Image 8.

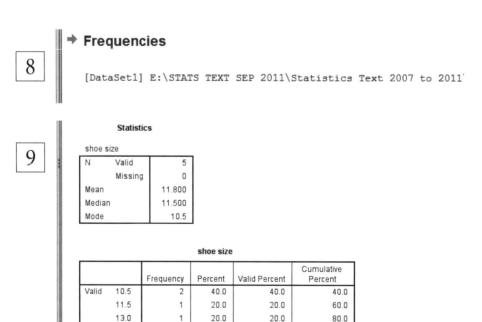

Chapter 2 Descriptive Statistics

If you focus on the top portion of the output in Image 9 (where the heading is "Statistics"), you should be able to see that (a) we had a sample size of 5, (b) there were no missing data for shoe size, and (c) the mean shoe size in our sample was 11.8. You should also be able to see (d & e) that SPSS agrees perfectly with our observations regarding the median (11.5) and the mode (10.5).

In the bottom portion of the output—beneath "shoe size"—you should see that the "Frequencies" command also gives you, *in ranked format*, the frequency of every score for the selected variable. We hope you can see that if we had collected data from 5,000 people rather than 5, this frequency output would be *very* handy. Assuming that shoe size is normally distributed, this output would likely have included a few very large and a few very small shoe sizes (at very low frequencies). Less extreme shoe sizes would presumably have increased in frequency until we reached a very frequent score at or very near the sample mean of about 10.0. This very frequent score, of course, would represent the peak of a normal distribution. You'll have at least one more chance to see how useful the Frequency command is before this activity is done. For now, though, let's return to the question at hand.

QUESTION 2.2a. Report the mean value provided in your SPSS output. In addition, confirm that the median and mode reported in your SPSS output match perfectly what you observed by simply eyeballing the ranked data.

QUESTION 2.2b. Which of the *three* measures of central tendency best summarizes these data? Why?

QUESTION 2.2c. What other population statistic in addition to the mean would be very useful if we wanted to make a more precise statement about whether the men in my family have big feet? A clue: All we can say now is that, as a group, we are 1.8 sizes above the presumed population mean. What statistic, loved by statisticians yet rarely discussed by laypeople, would also be useful to know? (A clue: You need this statistic to create *percentile* scores.) Would there be any statistical problems estimating this population statistic from the 5 cases we have at our disposal? (A clue: Do these five data points satisfy the important statistical assumption of the independence of individual observations?)

Estimating Spending in the U.S. Population

Now that you've had a little practice with SPSS, let's move on to a more serious topic, one that U.S. economists and politicians would love to understand better. How much money is the typical American spending these days? Because of the tremendous importance of daily spending to the U.S. economy, you should not be surprised to learn that government agencies and polling firms sometimes take surveys of daily spending. For the purposes of this activity, we are going to provide you with a small file containing hypothetical data from 200 respondents. You may assume not only that the respondents were *sampled randomly* (which is almost always the case in soundly designed surveys) but also that the large majority of people who were invited to take part in the survey agreed (which is rarely true, even in the best surveys). On the other hand, whether everyone was willing to tell the interviewers exactly how much money they spent yesterday is a question you'll want to answer based on the data.

The hypothetical data file for this activity is called [**hypothetical US daily spending data.sav**]. Your open SPSS data file should strongly resemble the screen capture that appears in Image 10:

Missing Data and Variable Values

If you look closely at the spending scores in your data file, you might wonder why so many Americans seem to have reported that they spent exactly $9,999.00 yesterday! The answer is that, in these data, a score of 9999 does *not* indicate $9,999.00. Instead, the researcher who designed this survey decided to use this very large value as an indication that respondents *did not answer* this particular question. This could happen for a variety of reasons. Perhaps the absentminded interviewee cannot remember what he spent yesterday. Or perhaps he remembers exactly what he spent but isn't comfortable admitting to a complete stranger that he blew $8,473 at the casino. A problem that plagues people who study money (e.g., income, spending, charitable contributions) is that a substantial minority of people don't like to tell interviewers the intimate details of their financial lives. Interviewers have devised a wide range of imperfect solutions to this vexing problem (e.g., asking people to report which of many spending or income *categories* best describes them). For now, we will just take it as a given that we will have some missing data for spending. Of course, part of your job in this activity will be to determine just how big a problem this is—by assessing the percentage of people who did not answer the spending question.

But to address this missing data problem at any level, researchers must begin by deciding *how to represent* missing data in a data file. The simplest solution to this problem is usually to choose *an arbitrary but sensible value that indicates that data are missing*. Doing so allows SPSS to ignore the missing data value (or *values,* if there is more than one kind of missing data) when performing any statistical calculations. This isn't quite as simple as it sounds. For example, researchers sometimes make the mistake of choosing a value that could easily be a valid response, which can lead to major errors if someone forgets to tell SPSS to ignore this missing value score.

It is usually up to the person who creates an SPSS data file to let SPSS know exactly what values indicate missing data for *every* variable in the data file (or at least for every variable for which there are any missing data). To see exactly how this hypothetical researcher chose to deal with missing data, you need to switch from the Data View mode to the Variable View mode. Once you do so, your screen should strongly resemble the one depicted in Image 11:

If you click on the far right-hand portion of the cell in the "Values" column of the "daily_spend" variable (see the rectangle in Image 11), this will open a dialog box (not shown here) that shows you exactly what each value of "daily_spend" means. In the case of daily spending, this researcher decided to remind users of the data file that a score of zero means no spending at all, that a score of 9998 means someone spent any amount *equal to or greater than* $9,998, and that a score of 9999 means that no data were available. It is useful to have this information duly noted in the "Values" column. However, this is *not* how SPSS knows how to interpret scores of 9999. Instead, *missing value scores must be specified in the "Missing" column*. To get a peek at how missing values are indicated, just click the far right-hand portion of the "Missing" column for daily spending (where you see the oval in Image 11). Doing so will open a dialog box that strongly resembles the one in Image 12.

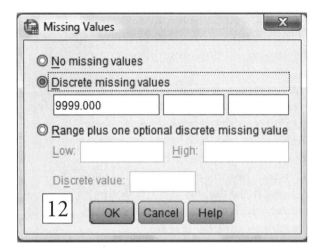

Notice that this researcher chose only *one* (discrete) missing value. Sometimes researchers may need to specify a whole range of missing values. For example, a researcher who administers pencil-and-paper surveys (rather than telephone interviews) without the assistance of a computer might find that people occasionally give answers that extend beyond the limits of a scale. For a 5-point scale, then, a researcher might need to specify that all answers between 6 and 9999 be treated as missing data.

It is worth noting that in the case of public opinion polls, researchers usually treat answers such as "don't know" or "refused" as valid answers. Where public opinion is concerned, it's highly informative to know the percentage of people who are unaware of an issue, and responses such as "don't know" are simply coded as one of several possible answer options (and *not* treated as missing). Research on human memory provides another example of an area in which a researcher would need to have a valid code for "don't remember." In this case, a separate missing data code might still be useful but only to indicate that a research participant *was never asked* an optional question.

Getting back to our spending data, we should note that beginning in January 2008, the polling and consulting firm Gallup began assessing U.S. spending on a daily basis. More specifically, Gallup began *randomly sampling* 1,000 Americans *every day* and asking them to report exactly how much money, in U.S. dollars, they spent yesterday on discretionary items (such as food, clothing, or entertainment). For the purposes of this activity, you should pretend that a statistically sophisticated but overworked economist has asked you to save her some much-needed time by analyzing these spending numbers.

So how much money *did* these hypothetical Americans typically spend on a daily basis in 2008? To see how you might find out, you can use the same commands you just used to analyze the shoe size data. However, we recommend that *before you do this*, you scroll through the entire SPSS data file and try to get a general feel for how much money the typical American is spending. OK, now that you have ignored this sage advice and are about to jump right into the analysis, remember that you'll start by clicking on the "Analyze" button you see in Image 13. Once you make it to the "Statistics" box, don't forget to ask for all three measures of central tendency.

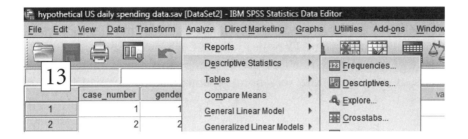

QUESTION 2.3a. What did you find? That is, report the three measures of central tendency, along with the statistical pros and cons of each measure. Cap off your analysis by discussing the tricky issue of whether it is really possible to characterize the spending habits of the typical American using only one measure of central tendency. By the way, make sure that you describe your findings using common language rather than focusing exclusively on list of statistics. A statement such as "The average American reported having spent $315 million on the day prior to being surveyed" is better than "The mean was $315 million."

QUESTION 2.3b. What percentage of survey respondents failed to answer the spending question? What concerns should this figure raise? Consider how those who preferred not to answer the question may have differed from those who did answer it. More specifically, try to come up with at least one reason why **nonresponse bias** might make our estimate too high. That is, why might the typical American spend less than the average person who was willing to answer this question? Now take the opposite view and try to think of a reason why nonresponse bias could conceivably make our estimate too low.

Describing the Ethnic Diversity of U.S. States

For the third activity in this chapter, let's focus on a very simple question that turns out to have a pretty complex answer. Answering this question requires you to go beyond measures of central tendency to make creative use of the standard deviation. Furthermore, because this question requires you to compare all 50 U.S. states, you will eventually want to *rank* the states as a way of organizing the data. The simplest version of the question is "What are the most and least ethnically diverse U.S. states?" A more thorough version of the question is "How can we best summarize and compare the ethnic diversity of all 50 U.S. states?" Before you try to answer the question empirically, write down your two best guesses for both the most and the least ethnically diverse states. When we are done, you can see how close you came to the correct answer (because these are real data, taken from the U.S. Census).

One reasonable indicator of ethnic diversity is simply the **standard deviation** of the four state-level percentage scores for the four most common U.S. ethnic groups. Consider Table 2.1, which contains the percentages of people in each of four ethnic groups in three hypothetical U.S. states, two of which are very homogeneous and one of which is very diverse:

Table 2.1 Ethnic Composition of Three Bogus U.S. States

STATE	Percent White	Percent Latino	Percent Black	Percent Asian	SD of Four Scores
Whitoming	100	0	0	0	50
Diversitania	25	25	25	25	0
New Korea	4	2	3	91	44

Notice that Whitoming, the state with no diversity whatsoever, has a very large standard deviation. New Korea is a *little* more diverse than Whitoming, but the very high percentage of Asians in New Korea is almost as high as the percentage of Whites in Whitoming. In contrast, the state that is the epitome of diversity has a standard deviation of zero! In Diversitania, every group is represented in perfectly equal percentages. The mean percentage of people in the four ethnic groups is 25%, and no group deviates at all from this aggregated percentage. If one ethnic group had been slightly more common than the others, the standard deviation would have necessarily been at least slightly greater than zero.

Your job in the next step of this activity is to use an SPSS syntax file (the third kind of SPSS file) to calculate an indicator of ethnic diversity for all 50 U.S. states. After you're done with these calculations, you'll be ready to rank the 50 states on the diversity indicator and see if the rankings make sense.

Let's begin, though, by opening the data file that we created based on U.S. Census data for 2007 (the most recent data available at the time we wrote this chapter). The SPSS data file is depicted in Image 14. It's called **[US states ethnic percentages 2007.sav]**. We won't show you this here, but if you switch to the Variable View mode, you should be able to see that whereas "states" is a String (nominal) variable, all the other variables (which are percentages) are numeric. By the way, to avoid confusion down the road, we should note that not every possible ethnic group is included in this simplified file. This means that for most if not all states, the percentages for the four ethnic groups will add up to less than 100%.

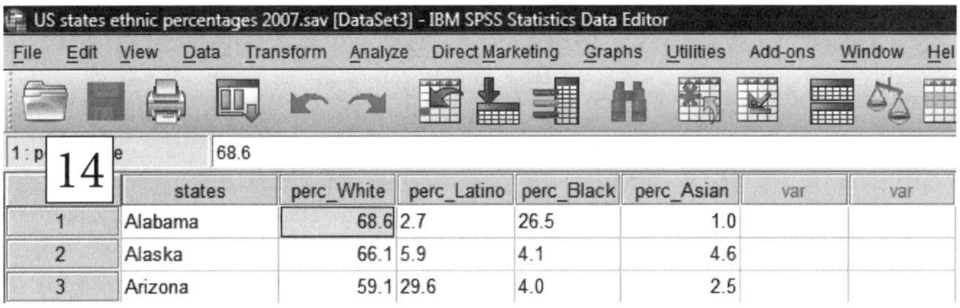

Chapter 2 Descriptive Statistics

[Image 14: Screenshot of SPSS Data Editor showing "US states ethnic percentages 2007.sav" with columns states, perc_White, perc_Latino, perc_Black, perc_Asian. Rows: Alabama 68.6, 2.7, 26.5, 1.0; Alaska 66.1, 5.9, 4.1, 4.6; Arizona 59.1, 29.6, 4.0, 2.5.]

As the data stand now, no single number will tell us which state is the most ethnically diverse. All we can do so far is to identify the states with the highest percentage of people in the four different ethnic groups. To help you get a feel for the data before we begin the important task of creating an ethnic diversity score, we'd like you to rank the states on each of the four ethnicity variables. If you don't need to print out your scores, you can rank the data in any SPSS data file within the data file itself. As shown in Image 15, you do this from the Data View mode by first clicking on the variable on which you wish to *rank your cases*. By the way, recall that in this unusual file, cases are individual states; more often, cases are individual respondents.

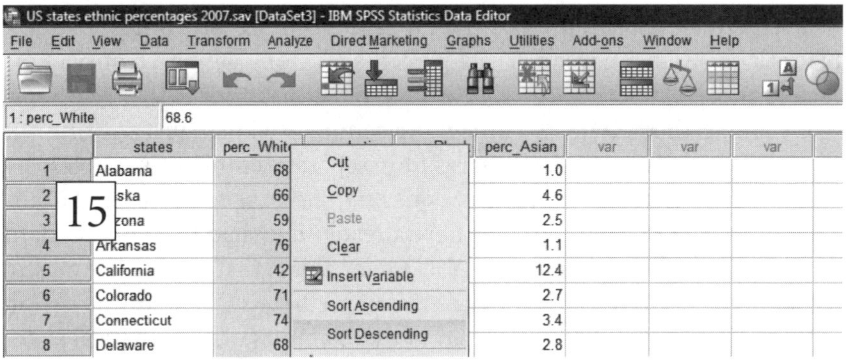

[Image 15: Screenshot of SPSS Data Editor with a right-click context menu showing Cut, Copy, Paste, Clear, Insert Variable, Sort Ascending, Sort Descending options, on the perc_White column. Rows shown: Alabama 68, Alaska 66, Arizona 59, Arkansas 76, California 42, Colorado 71, Connecticut 74, Delaware 68.]

Here we chose to rank cases on the variable "perc_White," which is the percentage of non-Hispanic Whites in each state. After clicking on the variable name, the next step is to move your cursor to either "Sort Ascending" (low to high) or "Sort Descending" (high to low). If you click the "Sort Descending" option, your entire data file should now be ranked on "perc_White," as shown in Image 16. Thus, you should now know that the U.S. state with the highest percentage of non-Hispanic Whites in 2007 was Maine, followed very closely by Vermont. This could have been very useful information if you were helping Barack Obama get out the minority vote in early 2008. You probably wouldn't want to kick off this effort in Maine or Vermont. Of course, Obama must have known this, and many other things, about U.S. state demographics or he would not have been savvy enough to have been elected U.S. president.

	states	perc_White	perc_Latino	perc_Black	perc_Asian	var	var
1	Maine	95.5	1.2	1.0	.9		
2	Vermont	95.3	1.3	.8	1.2		
3	West Virginia	93.6	1.1	3.5	.7		
4	New Hampshire	93.4	2.5	1.2	1.9		
5	Iowa	90.6	4.0	2.6	1.6		
6	North Dakota	90.0	1.9	1.0	.8		

We trust that you can follow these general instructions to re-sort your entire data file on the percentages of Latinos, Blacks, and Asians. However, your instructor may not be quite as trusting as we are. To make sure *your instructor* knows that you know how to do this, we have a few very simple questions for you.

QUESTION 2.4a. Which U.S. states had the highest percentage of Whites in 2007? What about the highest percentage of Latinos, Blacks, and Asians? Which U.S. states had the *lowest* percentage of Whites, Latinos, Blacks, and Asians? Give the exact percentages along with each answer for each group.

Of course, this activity is a little more complicated than just figuring out how the U.S. states rank on these four separate ethnicity variables. By definition, diversity is a question about *multiple groups*. So we need a single indicator of diversity, and we have already suggested that the standard deviation is one such indicator. You might think that you could click a button or two and ask SPSS to generate an ethnicity standard deviation score for each state, much as we did in the examples of hypothetical states a couple of pages back. As far as we know, however, that is not possible without a *lot* of restructuring of your data file. This is because your SPSS file treats *states*, not ethnic groups, as the unit of analysis. SPSS could easily tell you the standard deviation of the 50 "perc_Black" scores, for example, but this is not what we need. Instead, we need to know—*within each individual state*—the standard deviation of the four different ethnic percentage scores. The easiest way to get this is to deal with the third kind of SPSS file, which is a syntax file. As you work your way through this text, we hope that you will become very familiar with SPSS syntax files. However, because this is your first real SPSS activity set, we have created a syntax file for you that contains all the calculations you need. Your job will merely be to open and run that syntax file, which will lead to some important updates in your original SPSS data file.

The syntax file you need is called [**compute std deviation of ethnic percentages from state data.sps**]. By the way, opening an existing syntax

Chapter 2 Descriptive Statistics

file is very much like opening an existing data file, but we are showing you a screen capture in Image 17 just to be sure you don't confuse SPSS by trying to open the syntax file using the button reserved for opening data files. *SPSS is very particular about knowing what kind of file you wish to open before you try to open the file!*

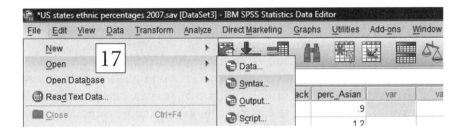

Image 18 provides a glimpse of what the syntax file should look like.

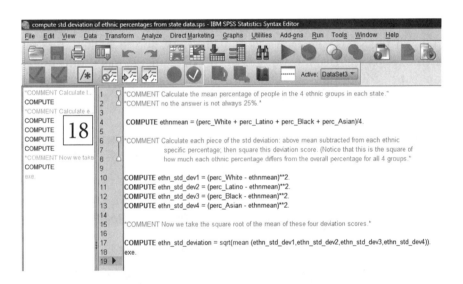

If you carefully examine what's in this syntax file, you will see that we are instructing SPSS to compute a standard deviation for the four ethnicity percentages in each state. We start by calculating an ethnic percentage mean for each state (it won't simply be 25%, and it will vary from state to state). We then take the squared difference between each of the four ethnic percentage scores and the mean score for that state (you square a variable in syntax commands with the symbols "**2"). Finally, we take the square root (sqrt) of the average (mean) of these four deviation scores within each state. (Incidentally, there are a couple of reasons why we used the formula for the population standard deviation rather than the sample standard deviation. One reason is that these *census* data come from *populations*.)

You might also notice that the command that begins a lot of simple arithmetic operations is the "COMPUTE" command. Furthermore, the command that ends each operation is usually a simple period. If you were to erase even one period in one of the COMPUTE commands in your syntax file, you would not get an ethnic diversity standard deviation score for any of your states. Perhaps the most important syntax command of all is the "exe." command, which is short for "EXECUTE." If you didn't add this at the end, SPSS would still *do* all of the calculations but would never add them to your *data file*.

In fact, even if you do have all your computes, periods, and exe commands perfectly in place, this syntax file will not produce any of the scores you need until you *run* all of these syntax commands. The easiest way to run a syntax command is to highlight the command or commands you wish to run (just as you would do if you were highlighting a section of a Word document you wished to cut, move, or copy) and then to click on the run arrow you see circled in the screen capture that follows. In a later chapter, we discuss other ways to get SPSS to run a syntax command. For now, let's be pragmatic and just learn one way. So if you highlight everything in this syntax file, as we did in Image 19, and then click the Run arrow, which is circled for you in Image 19, SPSS should make some pretty dramatic additions to your data file. Every specific variable you computed (beginning with "ethnmean" and ending with "ethn_std_deviation") should now be part of your data file.

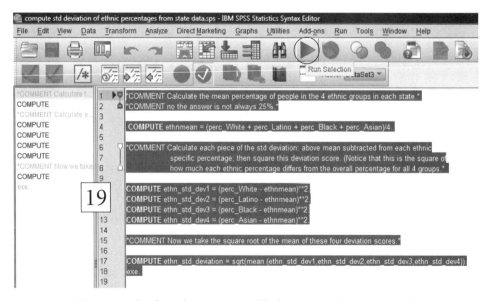

However, the first thing you are likely to see when you run these syntax commands is *not* the revised data file. Instead, you'll probably be taken automatically to an SPSS output file that just tells you that all the compute statements have been run. If they cannot be run (e.g., because you

Chapter 2 Descriptive Statistics

accidentally erased a period), you'll get some kind of error message in your output file. Image 20 contains a section of the kind of output file you should hope to see.

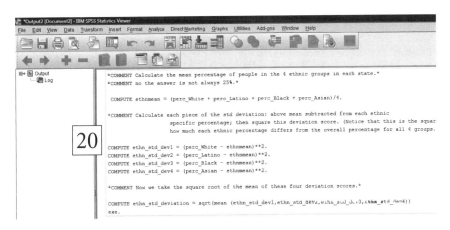

To get from the output file to the data file, by the way, you simply click on the little data file icon at the very bottom of your screen, the same way you'd shuttle between different files, webpages, and so on. This icon is circled for you in Image 21.

If you have a recent version of SPSS, a "thumbnail" of the data file will show up as your move your cursor over the data file icon. Regardless of whether the thumbnail appears, though, just click on the data file icon to go back to the data screen. You will notice that your data file now contains all the variables you just calculated, including the crucial "ethn_std_deviation" variable (see the far right-hand column of Image 22).

You are now ready to rank the 50 states on the newly created ethnic diversity variable. Recall that you can do so by merely right clicking on the variable that is the basis of your sort. Image 23 provides a quick reminder of how to do this.

You want an **ascending** sort, by the way, because a *smaller* standard deviation indicates *greater* ethnic diversity. So the state that begins the rankings once you click on "Sort Ascending" should be the *most* ethnically diverse U.S. state. Before you summarize these results for ethnic diversity, however, it might be useful to learn one more descriptive technique that will allow you to compare different kinds of rankings. If you scroll to the last case in your newly sorted data file, you will see that, by our definition, Maine is the least ethnically diverse U.S. state. You may recall that Maine is also the U.S. state with the highest percentage of non-Hispanic Whites. You may further recall that Hawaii is the state with the lowest percentage of Whites—and Hawaii proved to be the most diverse state based on our measure. Did we go to all this trouble just to produce a standard deviation measure of ethnic diversity when a simple percentage of non-Whites would have yielded exactly the same results?

To find out, at a glance, it would be useful to see the rankings for these two variables (percent non-Whites and ethnicity standard deviation) side by side. To convert *scores* to *rankings*, you use an SPSS command—part of the "Transform" command—that we have not yet explored. Let's explore it now. Image 24 will help guide you through this.

Chapter 2 Descriptive Statistics

After you click on "Transform" and then move down to and click on "Rank Cases...," a dialog box very much like the one in Image 25 will open. In this box, we have already sent "perc_White" and "ethn_std_deviation" to the "Variable(s):" list. You should also notice (see the oval) that the default for these rankings is to assign a rank of 1 to the *smallest* value. That's usually *not* how you rank most things (when higher numbers of the ranked things mean more of something), but this default is good for us because both of the measures we want to rank are smallest when ethnic diversity is highest.

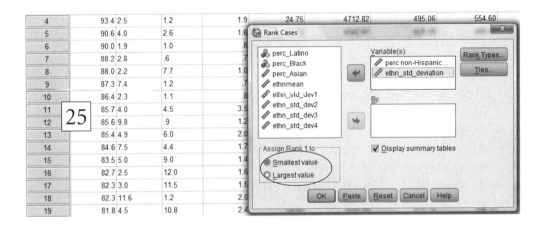

If you simply click the "OK" button, SPSS will create two new variables for you. Moreover, SPSS is so thoughtful that it will even make labels for these new variables—so you don't have to do so yourself. Here is what your data file should look like once these two rankings have become a part of the file. If you move your cursor over the variable "Rethn_st," you can even see (as shown in Image 26) that the label SPSS created for this variable is "Rank of ethn_std_deviation." Recall that it was us, rather than SPSS, that decided to create a variable called "ethn_std_deviation." SPSS labels for ranked variables are always just "Rank of _____" where the blank is always your original variable name.

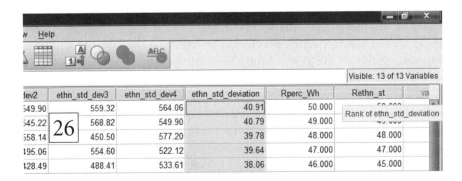

If you now sort these data in ascending order based on "Rethn_st," the least ethnically diverse states (rank1, rank2, rank3, etc.) will now come to the beginning of the data file. And if you examine these rankings for the two variables of interest, you should be able to see, for example, that the state that ranks 10th in ethnic diversity based on the percentage of Whites living in that state ranks 5th based on the more sophisticated indicator of ethnic diversity. So we hope we have answered this critical question.

QUESTION 2.4b. Now based on our fancy measure of ethnic diversity, describe (a) the two most ethnically diverse U.S. states, (b) the two least ethnically diverse states, and (c) the two states whose scores make them the most typical in ethnic diversity (the two states at the median in diversity). How is each member of each two-state group similar to and/or different from the other? Finally, criticize these results in at least one way. For example, identify at least one conceptual limitation or potential source of bias in these results. (A hint: The well-known social psychologists whose surnames are Sekaquaptua and Schwarz might offer conceptually similar critiques of the ethnic diversity measure.)

Descriptive Statistics in Public Opinion Polls

Any discussion of descriptive statistics would be incomplete if we did not make *some* mention of what is probably the most commonly reported category of descriptive statistics—namely, **public opinion polls.** News agencies, government agencies, and marketing researchers, among others, make frequent use of public opinion polls to understand what a large group of people (a state, a country, a market for the popular and delightful card game *PRIME*) think or feel about something. This something might be anything from whether consumers prefer *PRIME* to poker or *UNO* to whether people are in favor of or opposed to the death penalty. Usually, the goal of public opinion polls is to describe populations, and one of the most powerful contributions of statistics to modern life is the insight that we can get a very good idea of what a population of many millions of people think about something if we properly sample the opinions of a tiny subset of that population. As a concrete example, if we randomly sampled 1,000 of the more than one billion people in China, we could get a very good idea of Chinese national opinion by knowing the views of less than one millionth (1/1,000,000) of the Chinese population!

Of course, this is true only to the degree that a public opinion poll makes use of *random sampling* and has a respectable *response rate*. But even if we were to do everything right and sample a respectable number of people in our ideal survey (let's say $n = 1,000$), it is extremely unlikely that the responses of these 1,000 people would be *perfectly* representative of the population we care about (let's say U.S. citizens). Even a well-designed survey of 1,000 carefully selected people leaves room for error. In fact, we have a word for such error. It is called **sampling error,** and it is a fancy way of saying that in any survey based on a random sample of *n* people, there is always an element of luck based on *which **n*** people you happened to sample. It is always possible that, by means of simple bad luck, we happened to sample a group of people who are *not* all that similar to the parent population we wish to describe. This is no different than saying that if we flip a fair coin 100 times, it is *possible* that we will observe quite a few more (or less) than 50 heads.

However, an important rule that applies to coin tosses applies in the same way to random sampling. The bigger our random sample, the more likely it is to resemble the parent population from which it was drawn. It wouldn't be all that unusual to flip a coin 10 times and observe 7 or more heads. However, the chances of observing 700 or more heads in 1,000 tosses of a fair coin are *incredibly* low (in fact, less than one in a million, million, million, million, million million!). Fortunately, statisticians have worked out a pretty simple formula that reliably tells us, based on a result observed in a random sample of size *n*, how well we can expect that sample result (e.g., 31% Republicans, 98% *PRIME* lovers) to match whatever is true in the population.

For nominal (i.e., categorical) questions such as whether people favor or oppose something, have an illness, or own a deck of the intriguing card game *PRIME*, this formula works in two simple steps. First we calculate the **standard error** of our finding based on two things: the result (expressed as a probability, *P*) and how many people we sampled (*n*).

In the second step, we take that standard error and multiply it by ±1.96. If you remember the normal distribution from Chapter 1, you may recall that about 95% of the scores in a normal distribution fall within two standard deviations of the mean. To be more precise, exactly 95% of all scores in a perfectly normal distribution fall between ±1.96 standard deviations from the mean. So when we multiply a standard error (the sampling equivalent of a standard deviation) by this value of ±1.96, we are saying that there is only a 5% chance that the true population score is *outside* this very wide range. Of course, this means that there is a 95% chance that the population mean falls *within* ±1.96 standard deviation units from our observed sample result. This 95% chance corresponds to the 95% **margin of error** for a particular survey result. Let's look at a concrete example.

Suppose we randomly sampled 100 Canadians and asked them if they believe that intelligent life exists elsewhere in the universe. If exactly 48 Canadians reported that they do believe this, this would not *guarantee* that *exactly* 48% of all Canadians believe that intelligent life exists elsewhere in the universe. No matter how carefully we sampled, the true population value could easily be higher or lower than this observed sample result. But *how much* higher or lower? For a dichotomous outcome such as this one where we have already calculated the observed sample percentage (in this case 48%), the only thing we need to know to calculate the margin of error is the sample size.

Here are the calculations, beginning with the **standard error of the proportion** (abbreviated S_p):

$$S_P = \sqrt{P(1-P)/n}\,.$$

So if $P = .48$, note that $(1 - P) = (1 - .48)$, which is $.52$.

So S_p is the square root of $(.48 \times .52)/100$ (proportion *yes* times proportion *not yes*, all divided by n), which rounds to $.05$ (expressed, remember, as a probability). So expressed as a proportion, the 95% margin of error for this result is $P \pm (1.96 \times .05)$, which is $.48 \pm 0.098$.

Converting from proportion to percentage, this means that there is a 95% chance that the real percentage of Canadians who believe there is intelligent life elsewhere in the universe falls between 38.2% (48% − 9.8%) and 57.8% (48% + 9.8%)—that is, within about 10 percentage points of our observed sample percentage. Would things differ if we had sampled many more people? Let's see.

QUESTION 2.5. Calculate by hand and report the margin of error based on the observed sample percentage of 48% and a sample size of $n = 1,000$ (rather than 100). Then do the same thing for a random sample of 1,000 Canadians who were asked whether they think their taxes are too low. Assume that only 4% of Canadians report this belief (taxes in Canada, by the way, are somewhat higher than taxes in the United States, which many, many Americans erroneously think are too high). Comment briefly on whether—and why—this margin of error for 4% yes responses ("my taxes are too low") in a sample of 1,000 people is any smaller or larger than the margin of error for 48% yes responses ("I believe in alien life") in a sample of 1,000 people. If you want a clue about *why* the margins of error are different for questions to which almost everyone responds yes (or no) and questions on which the public is evenly split in their opinion, recall that margins of error are always based on (a) sample size and (b) error (some cousin of the standard deviation). We held sample size constant in the previous example, so it must have something to do with error. Either by doing hand calculations or by getting SPSS to do the calculations for you, calculate the standard deviation in the following two sets of scores:

Set 5: 0, 0, 0, 0, 0, 0, 0, 0, 0, 0, 0, 0, 0, 0, 0, 0, 0, 0, 0, 1 (5% yes = 1 yes out of 20)

Set 50: 0, 1, 0, 1, 0, 1, 0, 1, 0, 1, 0, 1, 0, 1, 0, 1, 0, 1, 0, 1 (50% yes = 10 yeses out of 20)

Be sure to incorporate the insight you get from these two small data sets into your answer about exactly why margins of error differ when different percentages of respondents answer yes to a question with only two answer options.

Shape Matters: The Normal Distribution, Skewness, and Kurtosis

If you ever wondered why statisticians spend so much time thinking about the shape of distributions, remember that there are few things in life that matter more than shape. We are not just saying that it's hard to put a square peg in a round hole. We are also saying that everything from the mileage your car gets to which faces you find attractive depends crucially on shape. To be more specific, the fact that shape matters is one of seven key reasons why a highly aerodynamic Toyota Prius (🚗) gets much better mileage than a highly un-aerodynamic '55 Chevy (🚙). Shape (and how to control it) is also what the Wright brothers spent much of their time perfecting (lift, for example, depends heavily on wing shape). Moving from aerodynamics to interpersonal dynamics, symmetrical faces are almost universally preferred over asymmetrical faces, and hourglass figures are generally preferred over pear-shaped ones. (We said shape was *important*, not that it was *fair*.) Of course, these examples all have to do with physical shape. In contrast, the shapes statisticians spend so much time debating are abstract properties of distributions such as skewness and kurtosis. However, if you care about getting the correct answers to important statistical questions, the shape of distributions can sometimes be just as important as the shape of windshields or prospective spouses. Furthermore, carefully assessing the shape of a distribution generally becomes more important the more sophisticated your statistical question is. The last exercise notwithstanding, you don't usually need to know a lot about the shape of statistical distributions to know that ground beef that is 93% lean is leaner than ground beef that is 80% lean. However, suppose you are an agricultural researcher who wants to conduct multivariate statistical tests to see if a three-phase dietary intervention increases lean meat yield in Beefmaster cattle as well as it does in Pineywoods cattle. In this case, you're going to be spending some time thinking carefully about issues such as skewness and kurtosis, especially as they relate to whether there are any **multivariate outliers** in your data.

We will defer a discussion of multivariate outliers until Chapter 14. For the present purposes, though, we want to lay a foundation for this and other subsequent chapters by reminding you that the normally distributed set of scores you see depicted in Figure 2.1 is pretty unusual. However, it is pretty unusual *not* because of how *weird* the distribution of scores is but because

Figure 2.1 An Extremely Normal Distribution

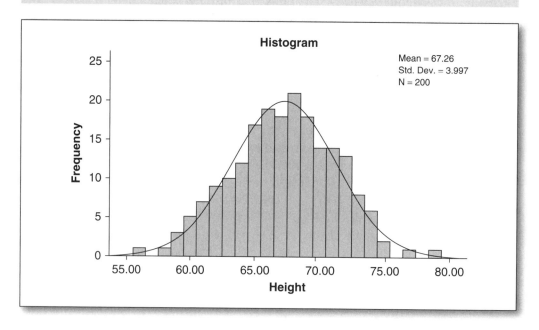

of how decidedly *normal* it is. Thus, for example, the two tails of this distribution are pretty symmetrical. Furthermore, about 70% of the scores lie within one standard deviation of the mean (which is very close to the hypothetical expected normal value of 68%). If real data were always as beautifully normal as are these somewhat idealized height data, statisticians who specialize in distortions such as skew and kurtosis might be out of a job.

So these 200 carefully chosen height scores pretty strongly resemble the idealized, highly symmetric normal curve. Unfortunately, though, many of the variables that social scientists study are only quasi-normally distributed. (If you're not sure what *quasi* means, you may recall that in the second *Austin Powers* movie, Dr. Evil rebukes his son Scott for being merely "quasi-evil" rather than truly and profoundly evil. And if you don't know what *rebukes* means, you'll just have to watch the movie.) Getting back to the point, if you assess continuous scores such as self-esteem, liking for *Austin Powers* movies, IQ, extraversion, or even temperature, you will often find that the scores in your sample only approximate a normal distribution. As it turns out, it is very useful to know about two crucial ways in which a set of continuous scores may deviate from the ideal of a perfectly normal distribution. These ways are **skewness** and **kurtosis.**

Loosely speaking, skewness has to do with whether the bulk of the scores in the distribution are shifted toward the right (referred to as a **negatively skewed** distribution) or shifted toward the left (referred to as a **positively skewed** distribution). Kline (2005) offers one of the most succinct definitions of skew, noting that "skew implies that the shape of a unimodal distribution is asymmetrical about its mean" (p. 49). Along similar lines, a useful rule of thumb for knowing whether a distribution is skewed (and how so) has to do with the relative position of the mean and the median. As Agresti and Finlay (1997) put it, "For skewed distributions, the mean lies toward the direction of skew (the longer tail) relative to the median" (p. 50). Notice that this definition focuses on the **tails** of the distribution. This is a very useful way to think about skewness because introductory statistics students are often puzzled about why a *positively* skewed distribution has all the scores bunched up on the *left* ("negative") side. A positively skewed distribution of $n = 60$ scores (the greatest number of Oreos that participants reported having ever eaten in one sitting) appears in Figure 2.2. In this figure, the median PR (personal record) for number of Oreos eaten (20) lies well to the left of the mean number eaten (27.45). So the mean of about 27 lies toward the (positive) direction of skew relative to the median of 20.

This should help you remember the otherwise confusing way in which skewness is traditionally labeled. The key to remembering the labels *positively* and *negatively* skewed is that statisticians lose a lot of sleep worrying about **outliers** (scores that lie very far away from the typical scores in a distribution). So in a *positively* skewed distribution, there tend to be some potentially worrisome *positive* outliers (like the two scores between 60 and 80 that appear in Figure 2.2). Of course, in a negatively skewed distribution, there tend to be some potentially worrisome negative outliers. Thus, the positive part of positively skewed refers not to the *bulk* of the scores

Figure 2.2 A Positively Skewed Distribution

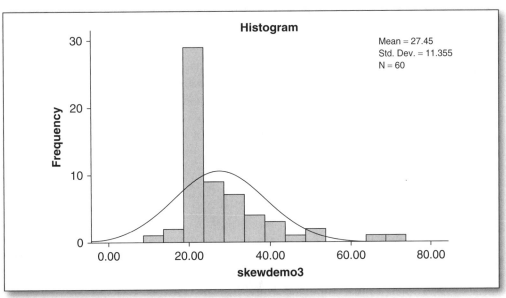

but to the worrisome outliers. If this still isn't all that helpful, by the way, there is an even easier way to remember how skew is labeled. In the case of a very positively skewed distribution, with a lot of scores bunched up on the left, just imagine a skier, gleefully skiing from the peak of the negative scores toward the tail of positive scores. She would be skiing in a positive direction, wouldn't she? So a positively *skewed* distribution is a lot like a positively *skied* distribution.

Although we like Agresti and Finlay's concise definition of skewness, we should note that this definition, like most other textbook definitions of skewness, is based on a rule of thumb (a useful way to think about skewness) rather than the precise statistical formula for calculating skewness. Unfortunately, the formula for skewness is based on a ratio of the third to the second cumulant—with the second cumulant being raised to the 1.5th power. Real statisticians love cumulants, but your primary author always gets cumulants confused with cubits, stimulants, and rainclouds. Fortunately, we do not have to delve into the precise details of how skewness is calculated to see that it is not *always* the case that the mean will lie in the direction of skew (relative to the median). We don't have to, by the way, because statisticians have carefully done this for us. As von Hippel (2005) showed, for example, if the positive outliers in a set of scores are not very extreme, they may not have enough "pull" on the mean to displace it at all in their direction. von Hippel shows, for example, that in the case of household size, there are so many U.S. households with only one or two people in them that the households with three, four, or more people in them simply aren't numerous enough to pull the mean household size upwards past the median (which is 2). In fact, the mean of about 1.8 is a little *less than* the median, but the scores are nonetheless positively skewed.

Having said all this, we hasten to add that in the very large majority of cases of skewness, the mean *will* be pulled past the median in the direction of the outliers. For example, in most negatively skewed distributions, the mean is less than the median. To see this for yourself, take a look at the variables "pos_skew" and "neg_skew" in Image 27. This image shows the data file **[shape of distributions demo (skew, kurtosis, etc) jan 2011.sav]**. By merely inspecting the conveniently sorted variables, we hope you can see that the two variable names having to do with skew are very fitting. For the variable "pos_skew," there are a bunch of 1s, a few 2s, and then a 3, a 4, and a 5. For the variable "neg_skew," the bunching is a little less intense, and it occurs at the high rather than the low end of the scores. To put this a little differently, if you carefully compare the two sets of scores, we hope you can see that the deviation from normality is a bit more serious for the "pos_skew" scores than for the "neg_skew" scores. For instance, whereas 13 extreme scores of 1 are bunched together for "pos_skew," only 9 extreme scores of 11 are bunched together for "neg_skew." To get a better sense of exactly how skewed the two sets of scores are, let's compare both sets of scores with the aptly dubbed "normal" scores.

Chapter 2 Descriptive Statistics

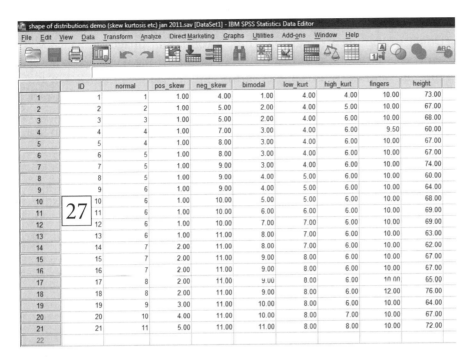

To do so, follow the lead of Image 28 and click on "Analyze." Then select "Descriptive Statistics" and then "Frequencies" (as you did in the previous exercise on daily spending). In this case, however, you are going to want to ask for a statistic we did not bother to ask for earlier. You're right; that statistic is skewness.

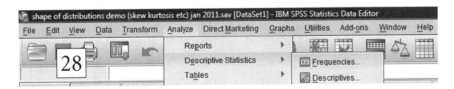

But before you can ask SPSS about skewness, make sure you send the three variables "normal," "pos_skew," and "neg_skew" to the Variables box, to select them for analysis (as shown in Image 29).

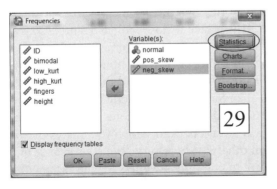
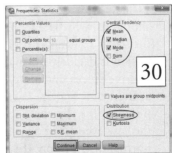

Next, click on the "Statistics..." button (which is circled in Image 29). Doing so will open the dialog window you see in Image 30, which will allow you to request information about the mean, the median, the mode, and the *skewness* by clicking all of the appropriate boxes (circled in two different spots in Image 30). Next, click "Continue" (marked with a rectangle in Image 30), which will return you to the main dialog window. If you now click "Charts..." (just below the circled "Statistics..." button that appears in Image 29), you will open the "Charts..." dialog box depicted in Image 31. From here, first select the "Histograms:" button, and then make sure to check the "Show normal curve on histogram" option. Then click "Continue" to get back, yet again, to the main Frequencies window. Unless you wish to paste the analysis commands into a syntax file, just click the "OK" button. If all goes well, you will generate an SPSS output file, part of which strongly resembles what you see in Image 32.

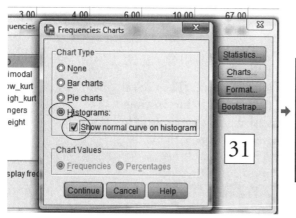

Because you asked for "Charts..." your SPSS output file should contain not only the the table that includes the three skewness scores that appear not only in the next to last row of Image 32 but also in a chart that strongly resembles Figure 2.3 (plus two other charts that we have not bothered to reproduce here, one for each of the other two variables). To get a better feel for skewness, you might begin by noting that the skewness for the "normal" variable is exactly zero (.000). If you take a close look at Figure 2.3, you should see why. The scores for normal are *perfectly* symmetrical. Of course, the scores only take on 11 different values (1–11), but otherwise this distribution is decidedly normal. The mean, the median, and the mode are all 6, and the scores that are *greater* than 6 taper off in exactly the same pattern as the scores that are *less* than 6. If a distribution of scores is perfectly symmetrical, the skewness score will always be zero. Moreover, this would be the case even if the distribution otherwise bore very little resemblance to a normal distribution. If you were to compute skewness for the rectangular distribution 1, 2, 3, 4, 5, for example, the skewness would be 0.00. This tiny rectangular distribution isn't at all normal (in the statistical sense), but it's perfectly symmetrical about the mean of 3.0.

Figure 2.3 A Distribution With No Skew Whatsoever (Skewness = 0)

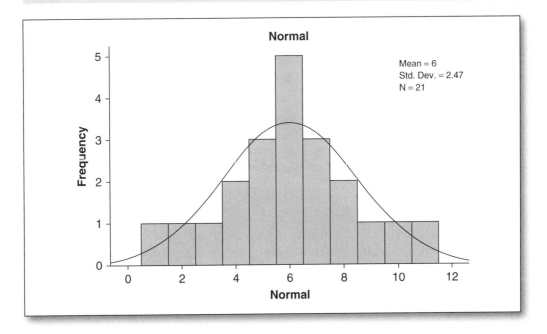

QUESTION 2.6. After confirming (for your own benefit) that you get the same three skewness scores shown in Image 32, run the same frequency analysis we just reviewed for "normal," "pos_skew," and "neg_skew" but replace these three variables with the "bimodal" variable that is part of the same "shape of distributions. . ." data set. Report the skewness score for "bimodal." Then, by consulting both the plotted chart for "bimodal" and the sorted frequency list that is part of your output, explain why your "bimodal" skewness score took on the exact value that it did. Next, add a new variable to your SPSS data file, label it as you please, and enter the scores 1, 1, 1, 1, 1, 1, 1, 1, 1, 1, 1, 1, 1, 1, 1, 1, 1, 1, 1, 1, 7 (that's twenty 1s and a 7). Analyze this variable, too, and explain the sign and strength of its skewness score.

If this analysis of skewness leaves you feeling a little flat, this may actually help you understand our last descriptive statistic. This is because **kurtosis** has to do with how *flat* versus *sharply peaked* a set of scores is. To simplify things, let's assume that we're dealing with a unimodal, roughly normal distribution. Kurtosis has to do with just how tightly packed the scores are at or near the *center* of the distribution. Consider, for example, the $n = 10$ scores 1, 2, 3, 4, 5, 5, 6, 7, 8, 9. If you plotted this set of scores, you'd see that the scores are very flat relative to what you'd expect for a normal distribution. In fact, if you got rid of just one of those two 5s, the distribution would be perfectly flat (i.e., it would be a *rectangular* distribution). Compare this with the n = 20 scores 1, 2, 3, 4, 5, 5, 5, 5, 5, 5, 5, 5, 5, 5, 5, 5, 6, 7, 8, 9. This highly "peaked" distribution has the same mean and range as the first, but it deviates from the prototype of a normal distribution because fully 60% (n = 12) of the 20 scores are now bunched together precisely at the mean, median, and mode (or center) of the distribution. To get a better feel for kurtosis (including how it is different from skew), you should repeat the same frequency analysis you just ran using four variables: "normal," "low_kurt," "high_kurt," and "fingers." In this case, however, after opening the "Statistics…" box (see Images 29–32 for a refresher), you'll want to add "Kurtosis" to the list of requested statistics. (See Image 30.) Then, after making sure you have still requested charts, run the basic frequency analysis. Take a careful look at the chart and the kurtosis for each distribution of scores.

QUESTION 2.7. Report the kurtosis for each of the four variables you analyzed. After explaining briefly why "normal" has a kurtosis score very near zero, explain why each of the other three variables has a kurtosis score that is *not* close to zero. Among other things, make sure you explain (a) why the variable "low_kurt" is not flat at all but still has a very low (negative) kurtosis score and (b) why "fingers" (the number of fingers each participant has) has an even higher kurtosis score than "high_kurt."

Sometimes Shape *Really* Matters

One of the reasons why statisticians worry about skew and kurtosis is that they can sometimes lead to spurious results of statistical tests that are based on the assumption that variables are normally distributed. To see the trouble that kurtosis can cause in data analysis, go back to your "shape of distributions…" SPSS data file one more time and sort the cases in ascending order based on the fingers variable, using the reminder you see in Image 33. Doing this sort should reveal (just as your last frequency analysis did) that the fingers variable consists of 19 scores of 10, 1 score of 9.5 (a person who, like the first author's dad, is missing half a finger), and 1 score of 12 (a highly unusual person who has 12 rather than 10 fingers). To see how problematic kurtosis can be, let's calculate a simple correlation using fingers and height. We'll discuss correlation in much greater detail in Chapter 3, but for now, let us remind you that a correlation reflects the strength and direction of the linear association between two variables. Do tall people tend to have more fingers? That is, are height and fingers positively correlated (in these data)? To find out, click "Analyze," then "Correlate," then "Bivariate..." as shown in Image 34. As shown in Image 35, just select the variables fingers and height, and place them in your "Variables..." box. Then click the "OK" button. Before you interpret the correlation you'll observe, let us bring to your attention some facts about the relation between fingers and height. First, the mean height in this sample is exactly 67 inches, which would be pretty typical in a mixed-gender sample of adults. If you go back to the data file and simply look over all the height scores, you should see, for example, that about half the scores are lower than 67 inches, and about half are higher.

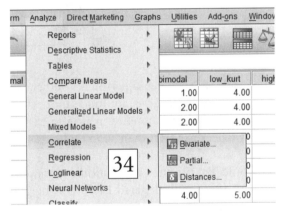
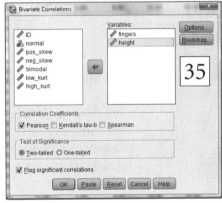

Furthermore, the mean height for the 19 people whose "fingers" score is 10 is exactly 67 inches, which is the mean for the total sample of $n = 21$ people. However, there are two people left in the sample. As highlighted in Image 36, the person with the lowest score on fingers (9.5) turns out to be very short (60 inches or 5 feet tall). And the person with the highest score on fingers turns out to be very tall (76 inches or 6 feet 4 inches). So for 19 of your 21 scores, there cannot be any correlation between height and fingers because there is no variation on fingers. Given this fact, what impact will these two outliers on fingers have on the correlation we just calculated? Before you answer that (by examining your SPSS output file), let us remind you that these two cases are arguably minor outliers for the height variable. (We say "arguably" and "minor" because these scores, although extreme, are still within 3 standard deviations of the mean.) However, when it comes to the fingers variable, it is not so clear whether the two scores are true outliers at all. Each is highly unusual, of course, but to an alien from a 2-32 fingered world, for which number of fingers is truly normally distributed, each person in our sample would be scarcely distinguishable from his or her 10-fingered fellow earthlings.

ID	normal	pos_skew	neg_skew	bimodal	low_kurt	high_kurt	fingers	height
1	4	4	1.00	7.00	3.00	4.00	6.00	9.50	60.00
2	1	1	1.00	4.00	1.00	4.00	4.00	10.00	73.00
3	2	2	1.00	5.00	2.00	4.00	5.00	10.00	67.00
4	3	3	1.00	5.00	2.00	4.00	6.00	10.00	68.00
5	5	4	1.00	8.00	3.00	4.00	6.00	10.00	67.00
6	6	5	1.00	8.00	3.00	4.00	6.00	10.00	67.00
7	7	5	1.00	9.00	3.00	4.00	6.00	10.00	74.00
8	8	5	1.00	9.00	4.00	5.00	6.00	10.00	60.00
9	9	6	1.00	9.00	4.00	5.00	6.00	10.00	64.00
10	10	6	1.00	10.00	5.00	5.00	6.00	10.00	68.00
11	11	6	1.00	10.00	6.00	6.00	6.00	10.00	69.00
12	12	6	1.00	10.00	7.00	7.00	6.00	10.00	69.00
13	13	6	1.00	11.00	8.00	7.00	6.00	10.00	63.00
14	14	7	2.00	11.00	8.00	7.00	6.00	10.00	62.00
15	15	7	2.00	11.00	9.00	8.00	6.00	10.00	67.00
16	16	7	2.00	11.00	9.00	8.00	6.00	10.00	67.00
17	17	8	2.00	11.00	9.00	8.00	6.00	10.00	65.00
18	19	9	3.00	11.00	10.00	8.00	6.00	10.00	64.00
19	20	10	4.00	11.00	10.00	8.00	7.00	10.00	67.00
20	21	11	5.00	11.00	11.00	8.00	8.00	10.00	72.00
21	18	8	2.00	11.00	9.00	8.00	6.00	12.00	76.00

Regardless of whether you consider a 12-fingered person an outlier, we hope you can see that (a) in an absolute sense, 12 is not much more than 10 and (b) there is no compelling physical or theoretical reason to expect people with extra fingers to be any taller or shorter than people with exactly 10 fingers. So one would hope that the dumb luck of sampling just 2 unusual people (out of 21) would *not* produce a *statistically significant* correlation between polydactylism (i.e., "multifingered-ism") and height. To save all of us a little trouble, let us just tell you that the correlation of $r = .55$ ($p = .01$) you should see in your SPSS output file is correct. At least that's the correct figure you get if you ignore the perils of computing a correlation coefficient using a variable that is extremely *leptokurtic* (i.e., extremely high in kurtosis).

QUESTION 2.8a. Explain why the extreme kurtosis of the fingers variable means that we probably cannot trust either the correlation of $r = .55$ or its associated p value of .01. As a hint, the extremely thoughtful statistician Tom Wickens used to argue that one should never trust the results of any statistical significance test if changing only a couple of data points in a sample would alter the basic pattern of results. How does a high level of kurtosis make it more likely that one would have to worry about this problem? As another hint, is there any sense in which the n for this study should best be thought of as 3 rather than 21? Do you think an *r* of .55 would be statistically significant if it were truly based on a sample size of 3?

QUESTION 2.8b. Finally, if the fingers variable were highly *platykurtic* (e.g., low in kurtosis, as would be the case with a flat or rectangular distribution) and the height variable kept its highly normal shape, would you be any more or less worried about observing a spurious correlation based on one or two unusual cases? A hint: Biological sex (1, 2) and rolls of a single die (1, 2, 3, 4, 5, 6) are highly platykurtic.

How Much Skewness or Kurtosis Is Too Much (or Too Little)?

Depending on what you plan to do with your data, you may wish to find out if any observed skewness or kurtosis in your data (positive or negative) is statistically significant. If you sampled your participants in a representative fashion, and if the skewness in your sample is statistically significant, for example, it presumably means that the variable you have measured is skewed in that same direction in the *population*. Tabachnick and Fidell (2007) provide some very simple formulas for testing the statistical significance of skew and kurtosis. However, many researchers worry about skew and/or kurtosis (for good reason) even if they are not statistically significant. Unfortunately, different researchers often use different rules of thumb for how much skewness or kurtosis is problematic. Almost all statisticians agree that skewness with an absolute value of 3.0 or greater is worrisome. Some statisticians worry about scores whose absolute values are 2.0 or greater, and some of the first author's more fidgety colleagues begin to worry about scores whose absolute values are 1.0 or greater. Because kurtosis can take on high absolute values more easily than skewness, few statisticians would worry much about kurtosis at absolute values less than 2 or 3. However, as Kline (2005) put it when comparing skewness to kurtosis, "There is less consensus about the kurtosis index, however—absolute values from about 8.0 to 20.0 have been described as indicating 'extreme' kurtosis" (p. 50). DeCarlo (1997) offers a similar view. Although Tabachnick and Fidell (2007, p. 80) argue that sample sizes of $n = 200$ or larger greatly reduce worries about both positive and negative kurtosis, it is possible in principle for extreme kurtosis to wreak havoc even in very large samples. As discussed in Chapter 14, if you are conducting a multivariate analysis, you can sometimes resolve the issue of whether any skew and/or kurtosis in your data is problematic by identifying (and then trimming or deleting) specific cases in your data that are the likely sources of your problems (e.g., by testing for **multivariate outliers** based on a case-by-case **Mahalanobis** statistic). As discussed in the following, there are also some simple transformations you can perform that will sometimes correct for skew and kurtosis. These transformations can prove equally useful whether you are conducting a complex multivariate analysis or a simple correlation.

Correcting for Skew and Kurtosis

There are usually things you can do to reduce skew or kurtosis in your data. For example, three of the most common things researchers try to reduce positive skew are (a) taking the square root of the raw scores, (b) taking the inverse ($1/x$) of the raw scores, and (c) taking the base-10 logarithm of the raw scores. Consider, for example, the $n = 11$ positively skewed scores 1, 1, 1, 4, 4, 4, 4, 4, 9, 16, 25. Notice that if you take the square root of these 11 scores (1, 1, 1, 2, 2, 2, 2, 2, 3, 4, 5), the three positive outliers aren't such radical outliers any more. Now that we have made this correction business look easy, let us remind you that we said researchers often *try* these three common transformations. The fact is that some skewed distributions are so uncooperative that none of these three transformations does much good. When this happens, researchers often resort to alternate procedures such as **winsorizing** (i.e., trimming) extreme outliers or creating categories (e.g., based on percentiles, so as to produce two or more groups of roughly equal sample size).

If you recall that positive and negative skew are mirror images of one another, we hope you can see that the same three transformations that are commonly used to reduce positive skew can also be used to reduce negative skew. However, the trick is to try these transformations (e.g., the square root transformation) on the mirror image of the negatively skewed distribution (by simply subtracting each of the original scores from the maximum score plus 1; see Kline, 2005, p. 50). The other trick, of course, is to *remember* that you have reverse-coded your original scores. If higher numbers once meant higher weight, for example, higher numbers will now mean lower weight. Furthermore, whether you transformed your raw scores or their mirror images, it is also very important to remember that transformed scores no longer have the same scale that the original scores had. If you are using the logarithm of income rather than raw income to predict long-term divorce rates, for example, you'll need to remember that a 1-unit change in the log of income is not the same as a 1-unit change in the raw scores (e.g., in simple euros or pesos).

If it's not intuitive to you that you can easily convert a negatively skewed distribution into a positively skewed one, consider the following negatively skewed distribution: 1, 10, 17, 22, 22, 22, 22, 22, 25, 25, 25. If you subtract each of these scores from 26 (25 + 1), you'll get 25, 16, 9, 4, 4, 4, 4, 4, 1, 1, 1. If you now sort these 11 numbers from lowest to highest, you'll see that you get 1, 1, 1, 4, 4, 4, 4, 4, 9, 16, 25, which is the same positively skewed distribution we just discussed two paragraphs ago. Anything that reduces skew in that distribution will reduce it in this reverse-coded distribution.

If you need to correct for kurtosis rather than (or in addition to) skew, (a) taking an odd root of the original scores (e.g., the cubed root) or (b) sine transforming the scores can sometimes reduce positive kurtosis (Kline, 2005, pp. 50–51). Winsorizing a set of scores (e.g., "trimming" both positive and negative outliers so that no score is allowed to exceed a z score of ±3.0) can also reduce positive kurtosis (by bringing both tails closer to the mean). Conversely, raising raw scores to a higher power (e.g., x^3) can reduce negative kurtosis. (When this doesn't work, some people suggesting *praying* to a higher power, but your authors have had limited success with this approach.) Speaking of higher powers, an excellent source for addressing a wide range of problems with badly behaved distributions is Erceg-Hurn and Mirosevich (2008). They even provide advice about how to handle not-so-normal distributions using SPSS (as well as SAS and R, two other popular data analysis software packages).

Just as there is no guarantee that you can always correct skew, there is no guarantee that you can always correct kurtosis. Consider the case of number of fingers discussed previously. It wouldn't help much to transform a set of scores when 90% or more of your sample all had the *same* score. Transforming 9.5, 10, 12 into 1, 2, 3 might look clean, but it wouldn't much change any spurious correlations involving polydactylism. This doesn't mean that it's impossible to study polydactylism, by the way. If you had a reason for studying it (perhaps you are a geneticist), you could simply sample a sizable group of people (or cats) with extra digits and compare them with people (or cats) who were genotypically and/or phenotypically normal. In this case, your significance test might be something as simple as an independent samples t test, which we discuss in detail in Chapter 6.

For Further Thought

If you are familiar with the methodological problem of **restriction of range,** it may have occurred to you that *ceiling effects, floor effects,* and what Pelham and Blanton (2013) call *"middle of the room effects"* (e.g., cases in which all the scores in a set are bunched together near the middle of a physical or psychological scale) are related to skewness as well as to kurtosis. These specific forms of restriction of range are usually considered impediments to detecting a real effect. For example, if every single person in your marketing study gives four out of four stars to the amazing card game *Cliffhanger,* it will be hard to show that extraverts rate it more favorably than do introverts or that women like it more than do men. However, as we hope you can see from the height and fingers example, if restriction of range is not absolute, this restriction can sometimes contribute to spurious effects—at least it can when researchers are not careful to check for skewness or kurtosis.

Linear and Curvilinear Correlation 3

Introduction: A Brief Tribute to Karl Pearson

Your primary author has long argued that the world's most undercited scientist is Karl Pearson, the man who, in 1896, proved mathematically that the **product-moment correlation coefficient** provides the best possible (i.e., most accurate) prediction of one set of scores from a statistically related set of scores (e.g., the heights of sons based on the heights of their fathers; the sizes of plants based on whether they received fertilizer). Without belaboring the history of the correlation coefficient too much, suffice it to say that many brilliant people (including Pearson's mentor, Sir Francis Galton) struggled with the basic concept of correlation but never got the formula just right. Galton, for example, used the **median** rather than the **mean** as his measure of central tendency when struggling with his version of the correlation coefficient. As you may recall, the median does not do quite as good a job as the mean, in the long run, of minimizing errors of prediction in a set of scores. In other words, if you take an entire set of scores and predict that every score is the mean, you will do the best possible job, in the long run, of minimizing errors of prediction. Pearson *further* established that the formula that best allows one to predict one set of scores from a related set of scores must make use of the mean and the standard deviation (rather than, for example, the median and the range; see Stanton, 2001). Every time a scientific paper reports a correlation coefficient without citing the person who invented it, the author arguably fails to cite one of the most influential scientists in modern history.

To get back to the correlation coefficient itself, there is no question that the social and physical sciences were changed forever when Pearson derived his scale-free indicator of whether—and in what way—two measures are related. Because of Pearson, you probably have a very good idea of what we mean when we say that there is a correlation of about $r = -.30$ between income and religiosity. Poorer people tend to be, but are not always, more religious than wealthy people. Because of Pearson, you probably appreciate how interesting it is that there is a correlation, across occupations, of about $r = .75$ between how much creativity is required to succeed in an occupation (think actors vs. actuaries) and the percentage of people in that occupation who suffer from any form of mental illness. If accountants and engineers were the actors and rock stars of our culture, tabloids probably wouldn't have very much to write about. Finally, because Pearson's correlation coefficient has become so much a part of modern scientific language, you probably have a good idea of what we mean when we say that the correlation, across countries, between a nation's gross domestic product (GDP) and its infant mortality rate is about $r = -.78$ (children are much less likely to die during the first few years of life in richer as compared with poorer countries).

In this chapter, we explore the correlation coefficient (along with its close cousin, the phi coefficient, and its more distant cousin, the odds ratio) as a way of describing the nature and strength of the association between two variables. This chapter also repeats and reinforces some rules of navigating SPSS (e.g., how to navigate an SPSS data file, how to use a syntax file). To get things started, we are going to focus on some simple but provocative statistical questions. More specifically, in the first section of this chapter, you will learn to use SPSS to compute some simple correlations involving a few familiar variables (e.g., age, height, income). As part of this first activity, you also will have to create your own SPSS data file from scratch.

A Hypothetical Study of How Unfair Life Is

The following screen captures show the top and bottom of a blank SPSS data screen that is awaiting data input. If you were to conduct a lab experiment without the benefit of data collection software such as MediaLab, or if you were to conduct a short pencil-and-paper survey, you would need to enter the data from your survey into an SPSS data file.

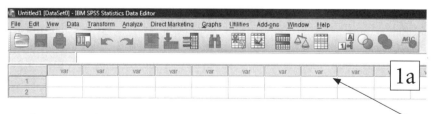

The spots where you enter your variable names are all marked "var." Recall that, at any time, you can toggle between the Data View mode and the Variable View mode.

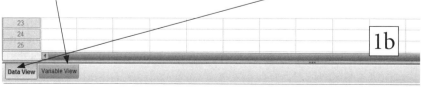

Chapter 3 Linear and Curvilinear Correlation

If you switch to the Variable View mode, you can enter important things users of the data will need to know, including the descriptive **Labels** that will prove to be extremely useful to anyone else who ever wishes to use the data file (and to you as well, after you've had time to forget much of what you once knew about the data file).

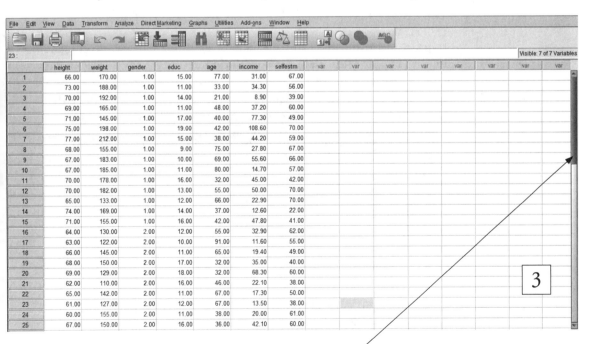

So before you go any further, enter the same seven variable **Names** (e.g., "height") and variable **Labels** (e.g., "height in inches) you can see in Image 2. Remember that you can only do this from the Variable View mode. Once you have named and labeled all seven of your variables, you are ready to switch to the Data View mode and enter all of your data. We recommend entering the data *row by row* because each row corresponds to one individual respondent.

Unfortunately, on the particular laptop monitor we used to create our own version of these data, SPSS only fits about 25 rows (25 cases) in a data file. However, if you move down a little in the data file (either by hitting the down key on your keyboard or by dragging the screen down as you are probably used to doing in other software packages), you'll be able to enter the last five cases in this file of $n = 30$ cases. Those last five cases are shown separately in Image 4.

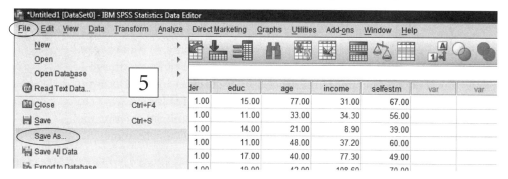

When you have entered all of your data, you should go to the File menu (circled in Image 5) to save all your hard work. From File, choose "Save As . . ." (also circled in Image 5) and find a nice, safe spot to save your data file.

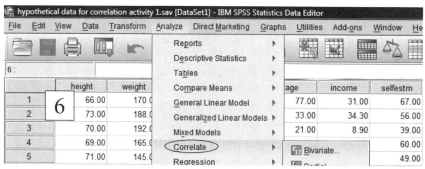

In principle, you could call your data file anything, but we recommend calling it [**hypothetical data for correlation activity 1.sav**]. (Ignore the square brackets, of course.)

Once you have saved your data file, by the way, you should notice that the file now has a name—where it once merely said "Untitled."

Now that you've entered and saved your data, you're ready to run a correlation. To do so, click "Analyze," and then choose the *kind* of analysis you want to run. One of the options from the list is "Correlate," which is circled for you in Image 6. Slide your cursor across this "Correlate" command to select the "Bivariate" option. Clicking the "Bivariate" option will open a new dialog box that will allow you to choose the variables you wish to correlate with one another.

Chapter 3 Linear and Curvilinear Correlation

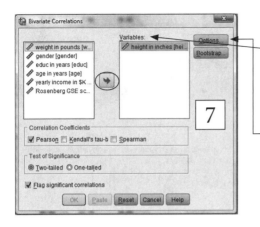

In this case, you want to correlate every variable with every other variable, which you can do by sending all the variables in the left-hand box to the "Variables:" box. To do so, you first select variables from the left box and then click the arrow that is circled in Image 7.

Once you have chosen your variables, clicking on the "Options" button will allow you to ask for some basic descriptive information about all of your variables.

From the Options dialog box, click "Means and standard deviations" (circled in Image 8).

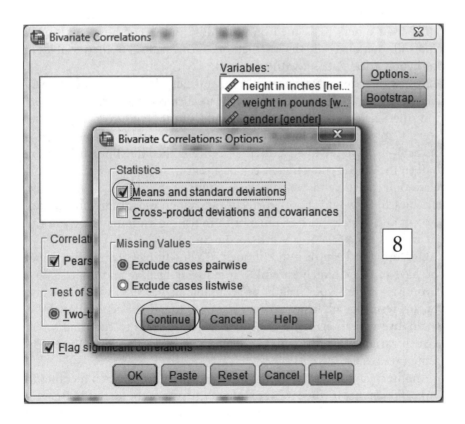

If you then click "Continue" (also circled in Image 8), you will return to the original Bivariate Correlations dialog box. From there, if you click on "Paste," you will generate an SPSS syntax file that should resemble the screen capture in Image 9.

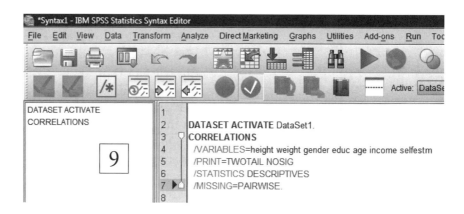

As you may recall, a syntax file is an SPSS file that contains some kind of *instructions*. These instructions, like the ones you see in Image 9, could be instructions to perform a specific statistical analysis on a particular set of variables. However, a syntax file could also contain instructions to create a new variable, sort a data file, or focus on some cases in a data file rather than others (e.g., men only, children only). For now, our simple syntax file contains only the instructions to correlate all of the variables in our hypothetical data set with one another.

Syntax files are *extremely* useful, and you can edit them in any way you like. For example, if you had forgotten to click on the variable "gender," from the Correlate dialog box, you could simply type in "gender" after "weight" in your syntax file and your syntax command line would look—and act—just like the one pictured here. Along the same lines, if you were to open the saved syntax file a week or a year from now—wanting to correlate only height and weight—you could simply delete the other five variables. SPSS would simply correlate height and weight for you. You can also cut and paste syntax commands in much the same way that you can cut and paste material in a Word document.

One thing to keep in mind, though, when editing a syntax file is that you have to be extremely careful to mind your spelling and punctuation. If you were to mistype "heigth" as a variable name, SPSS would not give you any results at all. Instead, when you tried to run the correlation, SPSS would give you an error message—because you tried to correlate a non existent variable (heigth) with the other valid (correctly spelled) variables in the variable list.

Another example of how detail oriented you must be when it comes to syntax files is that if you copied or edited the syntax command in Image 9 but left off the period at the end, you would either get an error message telling you that an expression ended unexpectedly or, worse yet, SPSS would simply do nothing, and you might wonder why it was stubbornly ignoring your instructions to run the correlation. Speaking of running the correlation, you may recall from the last chapter that the most intuitive

Chapter 3 Linear and Curvilinear Correlation

way to run an analysis from a syntax file is to highlight whatever you wish to run in the syntax file and then to click on the "Run" arrow. The screen capture in Image 10 should clarify this.

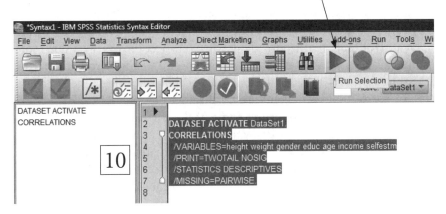

Once you run the correlation command shown above, SPSS should produce an SPSS output file that contains all of the correlations between all of the variables in this data file. Image 11 contains the crucial part of that output file. In this image, the correlation between height and weight is circled for you, as is the correlation between education and income.

Correlations

		height in inches	weight in pounds	gender	educ in years	age in years	yearly income in $K	Rosenberg GSE score
height in inches	Pearson Correlation	1	.769**	-.680**	.424*	-.455*	.514**	-.053
	Sig. (2-tailed)		.000	.000	.020	.012	.004	.781
	N	30	30	30	30	30	30	30
weight in pounds	Pearson Correlation	.769**	1	-.759**	.078	-.244	.285	.085
	Sig. (2-tailed)	.000		.000	.680	.194	.126	.653
	N	30	30	30	30	30	30	30
gender	Pearson Correlation	-.680**	-.759**	1	.012	.081	-.270	-.056
	Sig. (2-tailed)	.000	.000		.950	.670	.149	.770
	N	30	30	30	30	30	30	30
educ in years	Pearson Correlation	.424*	.078	.012	1	-.655**	.623**	-.074
	Sig. (2-tailed)	.020	.680	.950		.000	.000	.699
	N	30	30	30	30	30	30	30
age in years	Pearson Correlation	-.455*	-.244	.081	-.655**	1	-.370*	.194
	Sig. (2-tailed)	.012	.194	.670	.000		.044	.304
	N	30	30	30	30	30	30	30
yearly income in $K	Pearson Correlation	.514**	.285	-.270	.623**	-.370*	1	.396*
	Sig. (2-tailed)	.004	.126	.149	.000	.044		.030
	N	30	30	30	30	30	30	30
Rosenberg GSE score	Pearson Correlation	-.053	.085	-.056	-.074	.194	.396*	1
	Sig. (2-tailed)	.781	.653	.770	.699	.304	.030	
	N	30	30	30	30	30	30	30

**. Correlation is significant at the 0.01 level (2-tailed).
*. Correlation is significant at the 0.05 level (2-tailed).

Thus, you can tell from these hypothetical data that there is a very strong tendency for taller people to be heavier as well as a substantial tendency for wealthier people to be more highly educated. Focusing on the correlation between income and education, if you examine your output file carefully, you should see that the $r(28) = .62$ correlation between income and education seems to be significant at $p = .000$. If this seems a little weird, it should. Nothing can ever *really* be significant at $p = .000$, but whenever the p value that corresponds to a correlation is *less than* .001, it looks like a p value of zero because SPSS only shows you the first three decimal places for most p values. If you wanted to know the exact p value that corresponds to this correlation, you could double click on the value of .000 and SPSS would open a **pivot table** that would allow you to inspect the exact p value more carefully. By the way, if it also seems a little weird that we wrote $r(28) = .62$ when there are $n = 30$ cases, it *shouldn't*. Recall that the degrees of freedom (df) for correlation coefficients are $n - 2$ rather than n.

Before we ask you to interpret some of the correlations in this correlation matrix, we should inform you that there is one variable in the list whose name and label may not be highly informative (unless you happen to study the self-concept). This is the variable "selfestm," which refers to global self-esteem (GSE). Over the past few decades, the most commonly used measure of self-esteem is a 10-item, face-valid measure developed by the sociologist Morris Rosenberg (1965). It contains items such as "I am able to do things as well as most other people" and "All in all I am inclined to think that I am a failure" (which is scored in reverse, so that higher scores mean higher self-esteem). For the purposes of this assignment, all you need to know is that scores on this version of the Rosenberg Self-Esteem Scale could range from 10 to 70, with higher scores indicating higher self-esteem levels.

You may also wish to print this SPSS output file. To do so, you can find your local printer by working from the same File button you would use to save your output file. As soon as you have printed this output file, you would also be wise to save your syntax file. To avoid confusion down the road, we recommend giving SPSS syntax files the same (or much the same) descriptive prefix you give to the data files to which they correspond.

QUESTION 3.1a. Mauritania (gender unknown) is 62.88 inches tall. Based on the correlation between height (x) and weight (y) as well as the other information in your output file (i.e., the means and standard deviations for height and weight), what is your best guess about Mauritania's weight? Show all of your work and be ready to defend your logic. In case you are a little rusty on how information about correlations can be used to predict scores on one variable from scores on a correlated variable, note that we have included an appendix at the end of this chapter on prediction scores. Furthermore, notice that your SPSS output file contains not only the correlation matrix you see in Image 11 but also some descriptive information (means and standard deviations for all variables) that we did not bother to copy from the output.

QUESTION 3.1b. As it turns out, Mauritania happens to weigh 146.8 pounds. What's the statistical term for the difference between the score you predicted (y') and Mauritania's actual weight (y)? Your answer should begin with either the letter e or the letter r.

QUESTION 3.2a. Imagine you just learned that Mauritania is a man. If you had the luxury of doing some additional calculations (e.g., computing a different set of correlations, examining some additional means, etc.), could you predict Mauritania's weight any better now than before? Exactly how would you go about doing this? Well, do it then (starting by running some additional correlations). Be concise, but spell out your logic for the revised predictions and show your hand calculations again.

QUESTION 3.2b. What's the difference now between the score you predicted and the score you observed? Explain why your second prediction was better than, worse than, or about the same as the first prediction. A complete answer should include the words *mean, standard deviation,* and *correlation.*

QUESTION 3.3a. Based on your complete ($n = 30$) correlation matrix, identify what you consider the three most interesting correlations in the set and provide social scientific interpretations for why the correlations exist. Clue 1: At least one of these correlations should involve height, and at least one should involve self-esteem. Clue 2: During the past 80 to 100 years, the taller of the two major candidates in the U.S. presidential race has won the popular vote (although not necessarily the Supreme Court vote) in the overwhelming majority of the races (Bush's 2004 victory over Kerry and Carter's 1976 victory over Ford were the two exceptions).

For at least one these three correlations, describe the importance of the predictor variable by documenting its predictive utility in meaningful units of the criterion (outcome) variable. For example, people sometimes illustrate how important education is to income by reporting how much additional yearly income people tend to get (on average) for each additional year of education they have. In these hypothetical data, for example, each additional year of education earns people about $3,000 extra. This means that the predicted incomes of a person with 16 years of education will be about $12,000 per year higher than the predicted income of a person with only 12 years of education. Do this same kind of analysis for some other variable and show where your numbers come from.

QUESTION 3.3b. Now criticize your original interpretations (e.g., come up with equally or more plausible but less interesting explanations for why the correlations exist). A clue: There are many potential criticisms, but some of them could be supported by the actual data you have at your disposal (e.g., they could be supported by one or more of the correlations we observed in these data).

A Hypothetical Correlational Study of Afrocentrism

Social psychologists such as Irene Blair and Keith Maddox (e.g., see Blair, Judd, & Fallman, 2004; Maddox & Gray, 2002) have examined how people stereotype others on the basis of cues that are correlated with ethnicity, specifically whether people possess physical traits that are associated with being Black or White. Interestingly, both Blair and Maddox have shown that people rely on cues associated with ethnicity even when people are making judgments *within* a specific ethnic group.

As an example, Blair et al. (2004) showed that people make global judgments about targets using "Afrocentric" facial cues. In one study, Blair et al. showed headshots of both Black and White men to a group of raters. They then asked the raters to make a global judgment of the degree to which each man possessed stereotypically Black physical features. Next they showed the same photos to a different group of raters and asked the raters to estimate the likelihood that each depicted person fit a specific personality profile (e.g., a person who was highly athletic and had difficulty keeping up his grades). Blair et al. found, for example, that obviously White targets whose faces were higher as opposed to lower in Afrocentric features were judged to be more likely to fit the description that described a stereotypically Black person.

Although this is an intriguing finding, it does not tell us which specific traits people associate with Afrocentric features. For example, perhaps the results by Blair et al. were driven *completely* by the tendency for people to associate Black features with a lack of intelligence and not at all by any presumed tendency to associate Black features with athleticism. We cannot tell this because Blair et al. presented their stereotypical traits as a complete package (e.g., they never gave participants a description of a person who sounded unintelligent as well as unathletic). Let's assume, then, that we don't yet know exactly *which* specific stereotype-relevant traits are influenced by cues for Afrocentrism. Specifically, let's imagine that a researcher conducted a follow-up study to predict people's tendency to stereotype targets on positive as well as negative traits. Does the effect Blair et al. observed occur because of positive as well as negative stereotypes? For example, do we assume that a person with Afrocentric features is athletic and musical (two positive traits) *as well as* unintelligent and aggressive?

So to address the nature of this stereotype in more detail, imagine that you conducted the study summarized in the following data set. Assume that in this follow-up study, the researcher only used White targets (i.e., White faces that had been rated on their degree of Afrocentrism). The data file for this hypothetical study is called [**Afrocentrism stereotyping study for correlation.sav**].

In this correlational study, there is only one *independent* variable (blind judges' ratings of how Afrocentric the various White faces were that were used in the study). However, there are four separate *dependent* measures,

Chapter 3 Linear and Curvilinear Correlation

all of which are tightly linked to stereotypes of African Americans. Importantly, two of these measures are positive (athletic, musical) and two are negative (aggressive, lazy). You may further assume that for each of these four stereotypic traits, the values in the data file indicate participant ratings of the target on a 1-to-9 scale on that particular trait. If you check out the first line of the screen capture in Image 12, you will see that the first target was judged to be extremely athletic (9), somewhat aggressive (6), only moderately musical (5), and pretty lazy (7).

	afrocentric	athletic	aggressive	musical	lazy	var
1	3	9.00	6.00	5.00	7.00	
2	4	7.00	7.00	7.00	8.00	
3	7	9.00	7.00	9.00	9.00	
4	6	5.00	7.00	4.00	6.00	

12

Your job with regard to this study is to conduct every possible correlation you might want to conduct to learn everything possible from the study—and to relate your new findings to Blair's established findings.

QUESTION 3.4a. What, exactly, were the results of the study? How do these new findings (i.e., in the form of four separate correlations) support or fail to support Blair's original findings?

QUESTION 3.4b. Do the new findings tell you anything about whether positive and negative stereotypes are related? For instance, are targets who are judged to be more musical or athletic judged to be any more (or less) lazy or aggressive? What might this tell you about the nature of stereotypes about African Americans?

QUESTION 3.5. What are some likely limitations to the validity of the new study?

A Study of Freedom of the Press and Perceived Corruption in Europe

Our next topic in this chapter has to do with whether people believe that there is widespread corruption in their country and whether their country enjoys a great deal of freedom of the press. We hope you agree (a) that freedom of the press is to be preferred over unfettered censorship and (b) that a corruption-free world is to be preferred over a world full of corruption. In fact, one of the many reasons why people have historically argued that freedom of the press is important is that it presumably helps to minimize corruption. As former Supreme Court Justice Louis Brandeis reputedly said, "Sunlight is the best disinfectant." It is easy to do shady things in the darkness than in the light of public scrutiny. Brandeis's idea suggests that if we compared different countries, we might expect to see that in countries with greater freedom of the press, people would perceive less corruption (because there *is* less corruption). In the final activity in this chapter, you will examine press freedom data and perceived corruption data from 20 European countries to see whether Europeans who live in countries where there is greater freedom of the press report less widespread corruption.

As a concrete example, according to the human rights organization Freedom House, European countries such as Denmark, Finland, and Belgium frequently rank at or near the top of the world in freedom of the press. At the other end of the spectrum, many former Soviet Union countries, such as Belarus and Russia, frequently rank much closer to the bottom of the world in press freedom. Do the residents of Denmark, for example, collectively report that there is less corruption in their country than do the residents of Belarus? If so, and if this trend extends to other European countries, there should be a *negative correlation* between a country's level of press freedom and a country's average corruption level. Beginning in late 2005, the Gallup Organization began a yearly poll of representative samples of about 1,000 citizens in about 130 countries all over the world (as of 2011, the number of countries polled had increased to more than 150). Two of the many questions Gallup asks people in this World Poll survey are whether they perceive corruption to be widespread in (a) business and (b) government in their country. For the purposes of this activity, we harvested the Gallup data for the 20 European countries for which Gallup had perceived corruption data in 2007 (the most recent year for which we were able to get Freedom House ratings at the time of this original writing). You can view country-level data from the Gallup Poll yourself at https://worldview.gallup.com/signin/login.aspx.

Merging the Freedom House and the Gallup data allows us to test Brandeis's idea. Is Brandeis's idea supported by a negative correlation between press freedom and perceived corruption? Let's find out. The data file you will need for this analysis is called [**Freedom House PF & Gallup World Poll European Corruption 2007.sav**]. In addition to an arbitrary country code, the only variables contained in this data file are (a) freedom of the press ratings and (b) perceived corruption levels based on the Gallup corruption index. Incidentally, for those of you who are familiar with Freedom House, we should note that we *reversed the scaling* of the Freedom House press freedom ratings for the purposes of this activity. For some reason, Freedom House gives countries with *greater* press freedom *lower* scores. We made the scores more intuitive without changing the range of the scores by simply subtracting them all from 100. By the way, in case you ever design your own scale or index, we suggest that it is almost always best to design measures so that *higher* scores on something indicate *more* of that something. SPSS doesn't care which end of a scale means more or less of something, but the human brain does not deal well with numbers that run in the opposite direction from what they indicate. So for this assignment, just as greater scores on the corruption measure indicate higher levels of perceived corruption, greater scores on our recoded Freedom House freedom of the press measure indicate *greater* freedom of the press.

QUESTION 3.6. Report the correlation between press freedom and perceived corruption in these 20 European countries. Based on the magnitude and direction of the correlation as well as the *p* value, do the data support Brandeis's hypothesis?

At the risk of answering Question 3.6 for you, we must say that these results could be taken to suggest that Brandeis was wrong. Was he, though? Whenever a correlation (or a lack thereof) really surprises you, there are a few things you can do to figure out why you failed to see what you expected. One thing is to look for outliers in the data. It is possible, in principle, that one or two extremely unusual countries could be muddying the waters when it comes to the potential connection between press freedom and perceived corruption. If just one country with a very free press yielded very high perceived corruption scores, this might be worth noting. By the way, we are *not* endorsing the corrupt (i.e., unethical) practice of eliminating annoying data points that get in the way of one's pet hypothesis. Instead, we are reminding you that looking for outliers is part of routine data cleaning and that it *can* also be part of the scientific discovery process.

Sometimes outliers are more interesting than data points that are more predictable. For example, social scientists have had a hard time explaining why the United States, which is one of the world's wealthiest countries, (a) is much more religious than most other wealthy countries and (b) has a higher murder rate than most other wealthy countries. From a global perspective, then, the United States is an interesting outlier on two dimensions that are otherwise very highly correlated with GDP (i.e., national wealth).

The Power of Impossible Outliers

Sometimes, when you are working with a new data set, you learn that outliers are simply errors. For example, the first time the first author ever collected data on self-reported depressive symptoms and global self-esteem (two variables that are typically correlated about $r = -.60$), he observed virtually no correlation between the two variables! Moreover, his sample size was well over 500 people, and so sampling error (due to a small sample size) wasn't the culprit. The problem was data cleaning. In this sample of more than 500 people, there were two or three very unusual participants. To simplify a little, they all had self-esteem scores of about 90 and depression scores that greatly exceeded 50. Because the self-esteem scale *stops* at 50 (it was made up of 10 items scored on a 5-point scale) and scores on the depression scale rarely exceed 10, these participants seemed to be about 20 standard deviations above the mean in self-esteem and about 30 standard deviations above the mean in depressive symptoms! This is roughly the equivalent of saying that someone has an IQ of 5 and holds 14 PhDs. To clarify the nature of this problem, we should note that these participants were filling out their surveys using an optically scanned computer answer sheet. In this case, the problem is that these participants were (a) using the wrong scale anchors for the self-esteem questions and (b) skipping a page of the survey—so that they were probably answering questions about their self-perceived skills and talents rather than questions about depressive symptoms. If you have ever accidentally skipped one question on your answer sheet when bubbling in answers on an optically scored multiple-choice exam, you know how disastrous errors like this can be. Needless to say, it would have been a mistake to use these erroneous data. When the first author removed the erroneous cases from the data analysis, the correlation between self-esteem and depression went from roughly zero to roughly $r = -.70$.

It's hard to visualize 500 cases without a lot of overlapping scores, but in the SPSS output scatterplot that appears in the left-hand portion of Figure 3.1, you can see a hypothetical example with 52 cases that is even more extreme than the first author's real example. This scatterplot shows a *positive* correlation of $r = .50$ between scores on the Rosenberg Self-Esteem Scale and scores on the BDI (Beck Depression Inventory). The problem with this substantial positive correlation is that it is tainted by two erroneous (and way off-the-scale) self-esteem and BDI scores. Because there are only 50 cases rather than 500 cases to balance out the two highly erroneous cases, what should have been a negative correlation is actually a positive correlation (rather than no correlation). That's how powerful extreme outliers can be. To fully appreciate this, compare the first scatterplot with the scatterplot that appears immediately to the right in Figure 3.1. This second scatterplot includes the same set of scores after removing the two erroneous cases. The correlation goes from $r = .50$ to $r = -.65$! (If it's hard to see how all but two of the scores are the same in the side-by-side scatterplots, remember that we changed the range of the "bdi" scale once we took out the two impossible outliers.) If this makes you worry that the correlation coefficient is a flawed measure, you shouldn't. This would be like learning that a table saw can do serious physical harm to people and concluding that there is something inherently wrong with table saws. Table saws used carefully are incredibly useful tools. One of the statistical equivalents

Figure 3.1 The Correlation Between Self-Esteem and Depressive Symptoms With and Without the Influence of Two Erroneous and *Extreme* Outliers

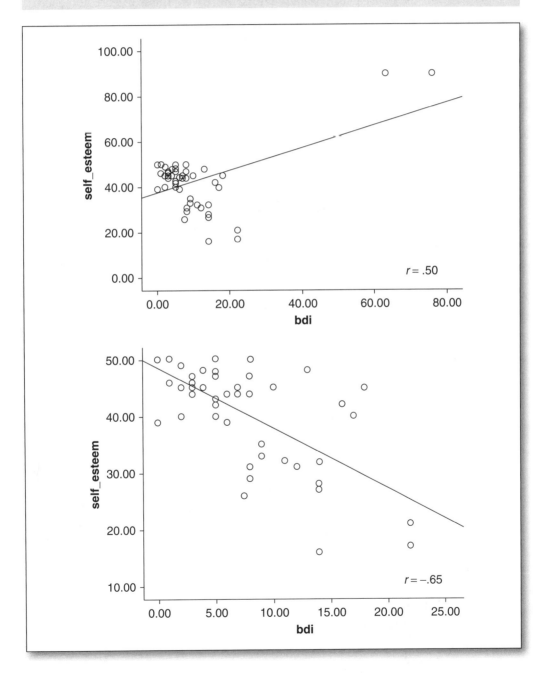

of following appropriate safety precautions with power tools is data cleaning. In fact, this topic is important enough that we devote Chapter 14 to it. For now, we merely wish to warn you that one cannot blindly interpret any statistic without making sure that the scores that go into it are free of errors.

Getting back to our story about press freedom and corruption, let us remind you that the first author carefully cleaned these data for you. So *you don't have to worry about errors*. Moreover, unless you are doing a dissertation on press freedom and corruption, we are going to spare you the trouble of looking for theoretically meaningful outlier countries. However, we would like you to see how a little extra data mining might yield a useful insight you couldn't unearth by simply computing a standard correlation coefficient. We'd like you to begin by entertaining a couple of possibilities that may complicate Brandeis's idea a little. Remember that what Gallup assessed in the World Poll was not corruption per se but *public opinion* on corruption. If at least some corrupt governments impose restrictions on the press precisely because they don't want the public to *know* exactly how corrupt they are, consider what effect this might have on public opinions of corruption in countries with very little press freedom. Furthermore, if, in countries with at least a modicum of press freedom, stories about corruption sell a lot more newspapers than stories about honesty, politeness, and lawfulness, what effect might this press freedom have on public awareness of corruption? Putting all this together, should we really expect a strictly linear association between press freedom and perceived corruption?

A Look at Brandeis's Hypothesis Through a Curved Lens

Let's take a closer look by seeing if perhaps the association between press freedom and corruption is *curvilinear* rather than linear. To do this, we need to move beyond a simple (linear) correlation and choose one of the options available in the SPSS multiple regression analysis. We will have much more to say about linear multiple regression later in this text. For now, we are going to run a multiple regression analysis because only the SPSS multiple regression analysis allows us to test for curvilinear in addition to linear associations. The screen capture in Image 13 is from the same data set you should have used to run your original correlation between press freedom and perceived corruption. From your data screen, go to "Analyze," then "Regression," then "Curve Estimation...." Once you click on "Curve Estimation..." you should see a dialog box much like the one in Image 14.

Chapter 3 Linear and Curvilinear Correlation

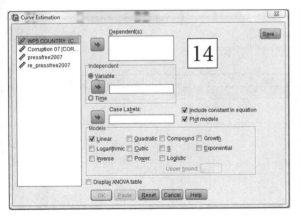

Notice that (a) you need to choose an independent (*x*, aka "predictor") and a dependent (*y*, aka "outcome") variable. Notice also that only the "Linear" box is checked under "Models" in Image 14. This means that, by default, SPSS only tests for a linear association between whatever variables you chose to analyze—even though you specifically asked to do a curvilinear analysis! The reason for this is that there are many different kinds of curvilinear analyses, and SPSS wants you to choose which curvilinear analysis or analyses you want. In this case, the most appropriate analysis is the test for a "quadratic" (U-shaped) effect (although we will have more to say about this in a later chapter).

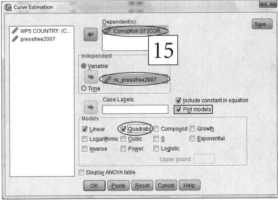

Notice that in the screen capture in Image 15, we have sent corruption to the **dependent** variable box, sent the *recoded* press freedom variable to the **independent** variable box, and clicked on the "Quadratic" model option to ask for a curvilinear analysis. Make sure you do this yourself. Finally, notice that we drew a rectangle around the "Plot models" option in Image 15, even though it comes as a default. Plotting a curvilinear association is always a good idea because it is very hard to appreciate a curvilinear association between two variables unless you see the association plotted. If you now click on Paste at the bottom (just left of center) of the dialog box in Image 15, you'll send the curvilinear analysis command to a syntax file (depicted in Image 16).

Once you have highlighted the commands in your syntax file, you can run the curvilinear regression command by clicking the Run arrow (circled in Image 16). Doing this should generate an SPSS output file that includes two crucial kinds of output, both of which appear in Image 17. The first part of the output file provides crucial statistics (e.g., p values, degrees of freedom), and the second part is a scatterplot (i.e., a picture) of the association in question.

Model Summary and Parameter Estimates

Dependent Variable: Corruption 07

Equation	Model Summary					Parameter Estimates		
	R Square	F	df1	df2	Sig.	Constant	b1	b2
Linear	.045	.842	1	18	.371	78.229	-.182	
Quadratic	.533	9.691	2	17	.002	10.143	3.257	-.032

The independent variable is re_pressfree2007.

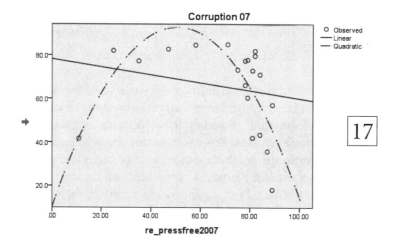

Chapter 3 Linear and Curvilinear Correlation

The first ("Model Summary...") section shows two important things. First, the test for a linear association does not yield a significant effect. In fact, you might notice that the p value for this linear test (Under "Sig." see $p = .371$) is the same as the nonsignificant p value you observed when you ran the correlation earlier (it should be because a correlation is just the simplest possible kind of linear regression analysis—the kind with only one predictor). Along the same lines, the square root of the R-square value (.045) for the linear effect is .21. This value of .21 is the absolute value of the correlation of –.21 you should have observed earlier. By the way, if you are wondering how you could tell that the correlation is *negative* from this output file, it is by looking at the scatterplot in the bottom portion of the output. Notice that the line of best fit for the *linear* effect slopes slightly *downward* rather than upward. But, of course, that is old news, too.

Now for the new news: If you examine the results for the "Quadratic" equation, which appear directly below the results for the "Linear" equation, you will see SPSS provides you with an R-square value and a p value for the quadratic effect—just as it does for the linear effect. Hopefully, you can now appreciate how useful it is to have a scatterplot of the press freedom and corruption scores.

QUESTION 3.7. Provide a description and interpretation of the curvilinear effect. Make sure to indicate whether the effect is statistically significant, describe the nature of the effect (e.g., What happens to perceived corruption scores as press freedom scores increase from low to medium and from medium to high?), and provide some kind of interpretation of the effect. How do these findings qualify what Brandeis said about corruption?

Appendix 3.1: A Primer for Predicting Scores on Y From Scores on X

Some of the information you need to answer Questions 1 and 2 in this chapter is kind of technical. To save you the trouble of a lot of potential Googling for statistical formulae, here is a quick primer on how to use information about correlation to generate predicted scores.

If x and y are correlated, we can predict a person's score on y better than we would otherwise be able to do by knowing his or her score on x. Furthermore, the more strongly the two variables are correlated, the more accurate our prediction of y will be (based on knowing a person's score on x). In the extreme case, if height and weight were perfectly correlated, this would mean that we could predict a person's weight perfectly (exactly) by knowing his or her height. Here are some of the important things you need to recall to understand prediction and predicted scores when dealing with correlations.

The general formula for computing Y' (Y prime or Y predicted) from X is

$$Y' = bx + a,$$

where

b is the **slope** of the regression line that summarizes the association between X and Y (it's a lot like r),

x is merely a specific x score for which you wish to predict a y score, and

a is the **intercept** = the value of Y at the point in a graph where the regression line crosses the Y axis. It's closely related to the mean of Y. (Here it is zero.)

So you can use the formula for Y' to predict Y scores. But you need to know the slope and the intercept in the data with which you are working. Unfortunately, the formula for the slope is almost as complex as the formula for Pearson's r. Fortunately, the slope is a close cousin of Pearon's r and is very easy to compute when you already know r (as you do, because SPSS told you what it was).

Thus, the easy formula is $b = r\, S_y/S_x$, where r is the correlation between X and Y, and S_y and S_x are the standard deviations for y and x, respectively (also available in your output file).

Once you know this, it's pretty easy to compute the slope. And it's easy to compute the intercept (a) from the slope.

$a = \overline{Y} - (b)(\overline{X})$ (the mean of Y minus the product of b and the mean of X).

Notice that your SPSS printout gives you everything you need to calculate any predicted score you'd want to calculate—because it gives you all of your correlations as well as all of the means and standard deviations for all of the variables you examined.

This should help you work through some of the early questions in this chapter.

4
Nonparametric Statistics (Tests Involving Nominal Variables)

Introduction: The Correlation Coefficient's Nominal Cousins

In the days before high-speed computers, researchers often had to do a *lot* of painstaking hand calculations to generate the correlation coefficients SPSS can now generate for you in a few seconds. It is probably for this reason that, whenever possible, early statisticians tried to simplify Pearson's general-purpose formula for *r*. For example, even though you can use Pearson's formula for the correlation coefficient to correlate nominal (i.e., categorical) as well as continuous variables, the most popular formulas for testing for associations between nominal variables are a lot simpler than the formula for Pearson's correlation coefficient. Here's Pearson's version:

$$r = \frac{N(\Sigma xy) - (\Sigma x)(\Sigma y)}{\sqrt{[N(\Sigma x^2) - (\Sigma x)^2][N(\Sigma y^2) - (\Sigma y)^2]}}.$$

Presumably, this is part of the reason why statisticians came up with the phi (φ) coefficient, which is *very much* like the correlation coefficient. The phi coefficient, for example, always ranges between 0 and 1. Furthermore, like an *r* of 0, a φ of 0 means two variables are unrelated, whereas a φ of 1.0 means that two variables are perfectly related. Thus, the only noteworthy difference between phi and *r* is simply that phi cannot be negative. This is because nominal variables don't signify direction or amount. To understand exactly what kind of an association a phi coefficient signifies, you just need to know *which* category of one categorical variable goes hand in hand with which category of the other categorical variable. For example, if male drinkers were twice as likely to prefer beer over wine but female drinkers were twice as likely to prefer wine over beer, there would be no way to place a meaningful sign on this association. This relation would yield a phi coefficient of .33, and you would need to describe the effect verbally (as we just did) to indicate which gender preferred which drink.

Early statisticians probably did not enjoy doing a lot of painstaking calculations. If they had, they probably would not have worked so hard to come up with a variety of computationally simple ways to describe the association between nominal variables. That is, they might not have invented some nonparametric statistics. In this chapter, we examine nonparametric statistics such as chi-square (χ^2), phi (φ), and the odds ratio (OR) as ways to express associations between nominal (i.e., categorical) variables. We begin with the chi-square test and then move on to phi coefficients and then odds ratios. As we do so, we point out how these nonparametric statistics compare with the Pearson correlation coefficient and with one another.

A Pilot Study of Name-Letter Preferences

A researcher interested in social cognition and self-esteem became interested in an automatic preference known as the **name-letter effect** (Nuttin, 1985, 1987). This effect refers to the finding that most people like the letters that happen to occur in their own names much more than they like other letters (and much more than other people like these same letters). For example, a hypothetical person named Brad Pittman should like the letters *B* and *P* more than a hypothetical person named Leonard DiCapri. This effect has been replicated in at least 20 cultures worldwide, and it occurs using a wide variety of measurement methods (e.g., forced-choice preferences for letter strings, Likert ratings of letters). The effect is most widely interpreted as a form of unconscious egotism or self-enhancement (e.g., see Pelham, Mirenberg, & Jones, 2002). That is, this effect seems to be grounded in the fact that most people possess positive, overlearned (perhaps classically conditioned) associations about themselves. In short, we seem to like ourselves a lot. If you have ever seen the episode of the TV show *Friends* in which Joey is delighted to have found his "hand twin" (a person who looks nothing like Joey but whose hands are identical to Joey's), you have a pretty good sense of what this effect is like (and how much higher order cognition it takes to produce it).

The researcher in question wanted to know if name-letter preferences might predict any real-world *behaviors* rather than merely predicting ratings of different letters. Accordingly, she conducted two preliminary studies that will be a part of your introduction to nominal data analysis. She began with a study of people's choice of a college. Specifically, she checked out an Internet college telephone directory and identified two small colleges whose names happen to correspond perfectly to common

Chapter 4 Nonparametric Statistics (Tests Involving Nominal Variables)

American surnames. The two schools were Brown University and Lewis University, respectively. Her hypothesis was that if people really do prefer the letters in their own names, then people should gravitate toward universities whose names happen to include their own surnames. Here are her data, based on the student directories at the two colleges (see Table 4.1). Incidentally, the researcher is fictitious, but the data are real; they were collected by the first author from Internet student directories at Brown and Lewis Universities circa 2000 or 2001 (*circa*, by the way, is Latin for "didn't keep very good research notes").[1]

Table 4.1 A Pilot Study of Surnames and Choice of a College

College	Student Surname	
	Brown	Lewis
Brown University	42	14
Lewis University	9	9

For our first test of the researcher's hypothesis, you will analyze some data from the SPSS data file [**brown-lewis college study.sav**]. Once you have examined the file carefully to see how the data are coded (e.g., by examining the **value labels**), you will want to conduct a χ^2 test to see if there is an association between people's surname and their choice of a college. Here's a peek at the most important portion of the SPSS data file you will use to run the χ^2 analysis.

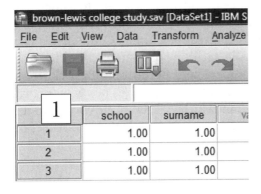 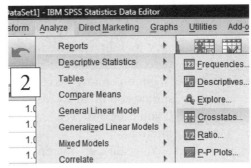

Image 1 merely gives you a glimpse of what your SPSS data file should look like. Image 2 shows that you begin a chi-square test of association by clicking the now familiar "Analyze" button and then going to "Descriptive Statistics." From there, click on "Crosstabs. . . ." Doing so will open the dialog box shown in Image 3. Image 3 shows that the first step to running a chi-square test of association is to choose your variables. To make your output resemble the data as we have summarized them for you here, make sure you send the *school* variable to your Row(s) and the *surname* variable to your "Column"(s) (as in Image 3). SPSS doesn't care where you put the two variables you choose to associate (as long as you don't try to put them in the *same* box), but where you put them dictates how they appear in your output. We tend to think of the column box as the place to put your *independent* (presumed driver) variables and the rows box as the place to put your *dependent* (i.e., presumed outcome) variables. If you now click on the "Cells..." button that is circled in Image 3, you will open the box shown in Image 4.

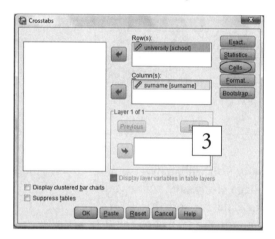 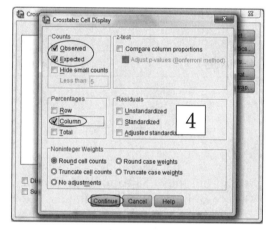

By default, the only thing that will be displayed in the cells of your Crosstab output is your *observed* frequencies. However, it is extremely useful to compare each observed cell frequency in a Crosstab output to the frequency you'd *expect* in that cell if the two variables are completely unrelated (as they should be if the null hypothesis is correct). Thus, you should make sure you check the "Expected" box (under Counts). It is also very useful to see the frequencies in each row or column converted to a percentage to make it easier to compare neighboring cells. By clicking on "Column" under Percentages, you will be asking to see percentages that should add up to 100 in each column, which will also make it easier to see if you have observed any kind of association between the two variables of interest.

You still are not done, though, because you need to request some specific statistical tests. After you click "Continue" from the box shown in Image 4, you'll be taken back to the main Crosstabs box. From here, click "Statistics..." (right above the "Cells..." button in Image 3). This will open the separate dialog box shown in Image 5. Here you should click on (and thus ask for) three statistics: "Chi-square," "Correlations," and "Phi and Cramer's V." If you click "Continue" again, you will return again to the main dialog box as shown in Image 6.

Chapter 4 Nonparametric Statistics (Tests Involving Nominal Variables) 105

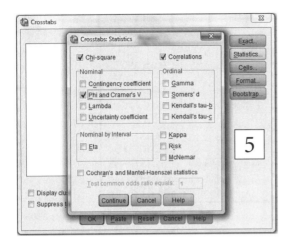

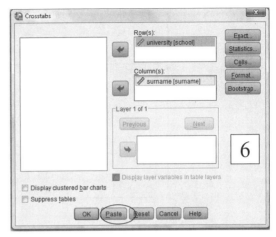

From here, click on "Paste" to send your analysis commands to a syntax file. You are now such a pro at running SPSS commands in syntax files that we are not going to insult you by reminding you that you should highlight the syntax command and then click on the run arrow. Instead, we'd like to inform you that you don't even need to highlight the section you wish to run (so long as you only want to run *one* section). Instead, merely *click your cursor anywhere in the syntax command.* If you look carefully at Image 7, you'll see that we clicked between "AVALUE" and "TABLES." After clicking your cursor anywhere in the "CROSSTABS" paragraph, just hit the run arrow (circled in Image 7). You will notice that the entire command becomes highlighted for a fraction of a second and then, if all goes well, SPSS will produce an output file that looks a lot like the one you see in Image 8.

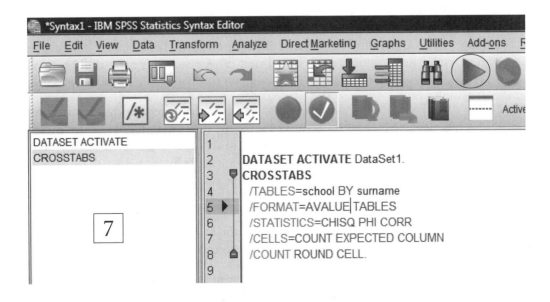

university * surname Crosstabulation

			surname		Total
			Brown University	Lewis University	
university	Brown	Count	42	14	56
		Expected Count	38.6	17.4	56.0
		% within surname	82.4%	60.9%	75.7%
	Lewis	Count	9	9	18
		Expected Count	12.4	5.6	18.0
		% within surname	17.6%	39.1%	24.3%
Total		Count	51	23	74
		Expected Count	51.0	23.0	74.0
		% within surname	100.0%	100.0%	100.0%

Chi-Square Tests

	Value	df	Asymp. Sig. (2-sided)	Exact Sig. (2-sided)	Exact Sig. (1-sided)
Pearson Chi-Square	3.974[a]	1	.046		
Continuity Correction[b]	2.893	1	.089		
Likelihood Ratio	3.788	1	.052		
Fisher's Exact Test				.077	.047
Linear-by-Linear Association	3.921	1	.048		
N of Valid Cases	74				

a. 0 cells (0.0%) have expected count less than 5. The minimum expected count is 5.59.
b. Computed only for a 2x2 table

We have marked four important things in the output screen capture that appears in Image 8. First, the circle shows you that whereas there were 9 people at Lewis University named Lewis (the **observed** frequency or "Count"), the **expected** frequency for this specific cell (based on the null hypothesis of no association between surname and school name) was 5.6 people. The very wide ellipse shows you that whereas 17.6% of the students at Brown University (9/51) were named Lewis, 39.1% of the students at Lewis University (9/23) were named Lewis. The triangle shows you that the chi-square value for this overall test of association is 3.97, and the small rectangle shows you that the p value for this chi-square statistic is $p = .046$.

QUESTION 4.1. Summarize the results of the college study. This includes creating a table of the results (just like we did earlier in this chapter, except that your table should include both observed and expected frequencies). Be sure to describe the results in plain, simple English. Also, make sure to discuss whether the results support the researcher's theory about implicit name-letter preferences. Point out at least one way in which this passive observational (i.e., nonexperimental) study is less susceptible than some other passive observational studies might be to concerns about reverse causality (e.g., Does x cause y, or does y cause x?).

QUESTION 4.2. Now provide a critique of these findings. That is, provide as many alternate explanations as you can for why you might have observed these findings—even if the researcher is wrong about the role of name-letter preferences in important judgments and decisions. Of course, the more obvious these alternative interpretations are, the better (remember you are playing the role of skeptical critic). Some clues: (a) What kind of things are correlated with people's last names or initials that might also predict where people go to college? (b) How do you suppose these two colleges got their names?

When you are done, by the way, you should first open a new (empty) SPSS data screen and *then* close the [**brown-lewis college data.sav**] data file. Because you may have never done this before, Image 9 gives you a quick clue about how to open a new (blank) data file—from any data file in which you happen to be working already. Just clicking on "Data" will do the trick. You shouldn't close the brown-lewis file until you've made this new data file because closing the only data file you have open will close SPSS completely, and if you did so, you'd have to start it up again to do anything new with a different data set. We say all this (open a blank file and then close your old data file) because unless you are an advanced user, keeping more than one SPSS data file open at the same time can create problems. It's easy to think, for example, that you're analyzing data from one SPSS data file when you're really asking SPSS to analyze data from another.

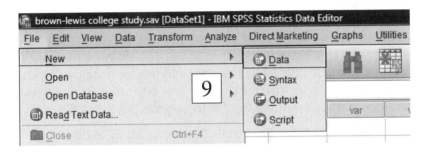

A Second Pilot Study of Name-Letter Preferences

Encouraged by her initial findings, the researcher decided to examine an even more important life decision—namely, whom people choose to marry. Of course, her hypothesis regarding choice of a marriage partner was very similar to her hypothesis regarding choice of a college. She expected that people would be disproportionately likely to marry other people whose surnames resembled their own. However, given the powerful cultural taboos against inbreeding, she didn't expect many people to marry people whose surnames were the same as their own. Instead, she hypothesized that people might be more likely to marry other people whose surnames merely began with the same letter as their own. To test her hypothesis, she found a genealogical website that provided exhaustive marriage records from the early 1900s for many of the counties in Florida. She then chose the county in which she grew up and created a "text-only" data file that contained a subset of this very large set of marriage records. To make her task manageable in this first study, she arbitrarily chose to focus on people (both brides and grooms) whose last names began with the letters *C* or *D*. She then created a data set that contained all of these names (it wasn't too hard to find all of the *C* and *D* names because the marriage records were organized alphabetically). Assume that all her collating and coding generated a text-only data file called [**SUBSET OF FL MARRIAGE FOR 721.txt**]. However, you will need to convert this text-only data file to an SPSS data set. Here are some instructions that will lead you through this process of converting text files to SPSS data files.

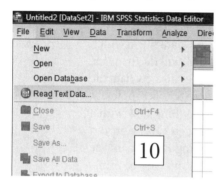
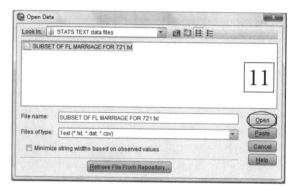

Notice, from Image 10, that instead of asking SPSS to open a regular data file using the "Open" button, you must ask SPSS to "Read Text Data...." Clicking the "Read Text Data..." button will take you to a dialog box where you can browse around until you find the text file you need (as shown in Image 11). When you open this "SUBSET OF FL MARRIAGE..." text file (by clicking the file name just once and then clicking the "Open" button that is circled in Image 11), SPSS will begin the "Text Import Wizard" to help you convert the text file to a regular SPSS data file.

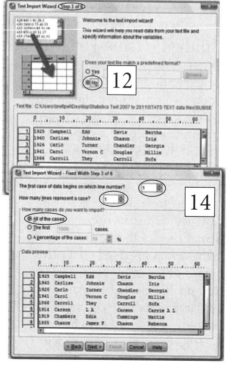
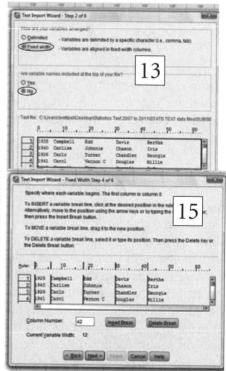

Chapter 4 Nonparametric Statistics (Tests Involving Nominal Variables)

The six steps in this text import process are numbered for you by SPSS (see the circled portion at the top of Image 12). In Step 1, you'll want to accept the circled default of "No" to indicate that your text file does *not* match a predefined format (one you've used and saved before). In Step 2 (Image 13), you need to tell SPSS that your data follow a "Fixed Width" format, where each variable starts in a certain column (you'll probably get a warning about the file containing tab characters; you may ignore it). Also, you need to click "No"—as you can see we did in Image 13—to indicate that variable names are *not* included at the top of your file.

In Step 3 (Image 14), you need to tell SPSS that (a) the first case in your file begins on line 1, (b) there is only one line to each case (in principle, there could be more), and (c) you want *all* of the cases. (This should be easy because these three options are all defaults.) If you click on "<u>N</u>ext >" at the bottom of the dialog box shown in Image 14, you will see the dialog box in Image 15, where you can adjust the **variable breaks** (the point where SPSS knows to start reading a new variable in a line of text). You can see the instructions for these adjustments in variable breaks toward the top of the box. In newer versions of SPSS, you may notice that a variable break line turns blue when you are adjusting it. *In Image 15, the adjustments have already been made so that there is a break immediately before each variable in the text file.* It is *extremely* important that you make these variable breaks exactly the way you see them in Image 15. If you were to leave a blank space before one of the key variables, for example, this could lead to a great deal of confusion later on. You will need to consult the Data preview (where you see the ruler) to make these same adjustments. Once you have done so, click on "<u>N</u>ext >" at the bottom of the dialog box shown in Image 15, and you will see the dialog box shown in Image 16.

Thankfully, the only thing you need to do in Step 5 is click "<u>N</u>ext >"—which will take you to dialog box 17. And the only thing you need to do in Step 6 is click on "Finish." If all goes well, doing this will generate an SPSS data file very much like the one you see in Image 18.

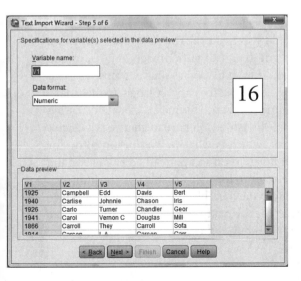

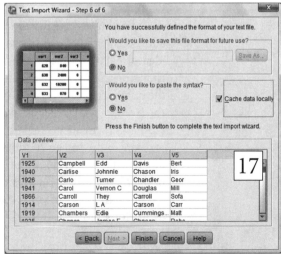

[Image 18 – SPSS Data Editor screenshot with columns V1–V5 showing years and names]

But these data are not ready for analysis. First, you need to *name* and then *label* all of your variables. In Image 19, the naming has already been done for you. Recall that the easiest way to rename variables is to get into the Variable View mode (using the toggle button near the lower left-hand corner of your SPSS data window). Once you are in Variable View mode, you can just type the variable names you want (e.g., "year") in place of the old (e.g., V1, V2) variable names.

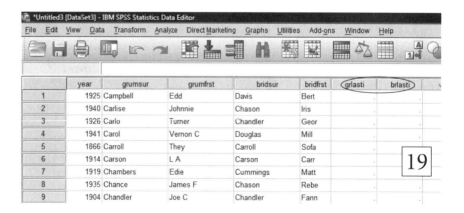

Once you've done that, you should also create the two new variables you see added (and circled) in Image 19: "grlasti and "brlasti." Image 20 gives you an idea of what your data screen should look like once you've almost finished changing all the variable names and adding all the variable labels. In these data, the only variable that should be Numeric is year. This means that after creating the two new variables ("grlasti" and "brlasti"), you have to change the default Numeric format into String format, which is what we had begun to do for "grlasti" in the Image 20 screen capture. To do so, we first had to click on the far right-hand section of the Type column for that variable (see the little oval in the highlighted cell). This called up the Variable Type box shown in Image 20. This box is where you'll convert the format of "grlasti" to a String variable (don't forget to click "OK")—and then repeat the same process for "brlasti."

Chapter 4 Nonparametric Statistics (Tests Involving Nominal Variables)   111

Once you have gone through all the trouble of creating this SPSS data file from your original text file, you still need to debug and clean the data. For gigantic files, this data-cleaning process would start by running **Frequency**

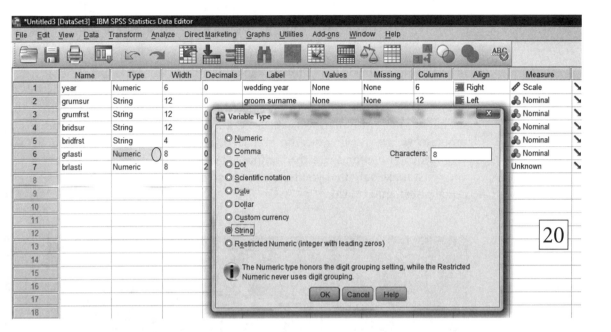

20

analyses on all of your variables. However, in a file this small, you can just visually inspect the entire file. When you discover the errors (they all occur for just one problematic case), you should correct the errors by simply typing the correct names over the erroneous names. By the way, because we aren't focusing at all on first names, you can live with the fact that the last few letters of many women's first names got lost. Needless to say, if we were doing a study of first names, we'd need to correct this serious problem. We hope it feels nice to be done with this painstaking process.

OK, now that you have savored this fleeting moment, let's get back to the painstakingness. If you're really going to analyze these names, we need to code them into simple, nominal categories (in this case, C or D) rather than complete surnames. You *could* do this by hand by just entering Cs and Ds for the two new variables (e.g., "grlasti"). But what if you had 100,000 cases instead of 50 or 60? This'd be a lot easier (and more accurate) if you could get SPSS to do the coding for you. But the coding here is a little tricky because the data you are dealing with are nominal (i.e., "String" or alphanumeric data).

To do your coding the right way, you need to create a syntax file. To create one from scratch, begin at "File," then click "New" and then "Syntax" (see Image 21). Once you've opened a new, empty syntax file, you'll be able to create surname initial scores of "C" or "D" for all the brides and grooms

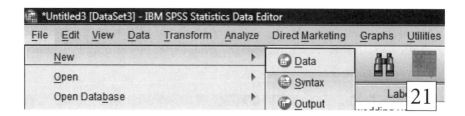

in the sample. Incidentally, for reasons SPSS software developers could probably explain, you can't make string variables from compute statements in syntax files without having first created the empty variables—while in the data mode. (Although you can also point and click your way out of the problem, we are showing you our preferred solution.) This just means that *before* you type in and run the compute statements you see in Image 22, you must make sure that you have already created the two string variables called grlasti and brlasti.

We don't want to belabor the wonders of the SUBSTR command *too* much, but the crucial logic of how it works is that it allows you to start and stop anywhere you want in a string variable and pull out the character (or set of adjacent characters) that you want. The two closely related commands you see highlighted in Image 22 tell SPSS to start in the very first position (1) of the original variables (grumsur and bridsur) and then to capture only one (1) character. If, for example, you replaced the "1,1" with "1,3," you'd select the first *three* characters of everyone's surname. Once you have carefully, nay, *very* carefully, copied the two COMPUTE commands shown in Image 22 into your own new syntax file (including the period at the end of each and every line), just highlight both commands, as illustrated in Image 22, and then click the now-familiar "Run" triangle. Doing this should populate both "grlasti" and "brlasti" with the single letters *C* and *D*.

Now, using what you learned earlier about running chi-square analyses in SPSS, you can analyze your data. In other words, if you go back to the instructions at the beginning of this chapter (for running a chi-square analysis in the Brown-Lewis study), you can run this chi-square analysis on these data, too. To make it easy on the person who has to check your work, please put the groom's last initial in the columns and the bride's last initial in the rows, so that the second line of your analysis command says "/TABLES=brlasti BY grlasti."

QUESTION 4.3. Provide the same kind of summary of the findings of the second (marriage) study that you provided for the college study. That is, describe these new results as if you were the researcher interested in name-letter preferences and discuss how, together with the first set of findings, they support your theory.

QUESTION 4.4. Now play the role of critic again. You may have some criticisms that overlap with your criticisms of the college study, but you should have some that are unique to this study, too. Don't limit yourself to possible confounds. For example, you can also criticize this study based on considerations of sampling or external validity.

The Chi-Square Statistic, Phi Coefficients, and Odds Ratios

Now that you've had some practice getting SPSS to calculate chi-square tests for you, we would like to move on to a discussion of two other statistics that are commonly used to analyze nominal data: phi coefficients and odds ratios. The first useful thing to know about phi coefficients is that they are easily derived from chi-square values. This includes the simple chi-square formula we reviewed in Chapter 1. To see how this works, let's examine an example for which the hand calculations are pretty easy. Table 4.2 contains some hypothetical data on the association between being a U.S. senator and being a millionaire. (These data probably aren't very far off the mark. U.S. senators don't have to disclose all of their assets to the public, but what they do have to disclose strongly suggests that most senators are millionaires.)

Table 4.2 A Hypothetical Study of Income and Occupation

	Nonmillionaires	Millionaires	Row Totals
Nonsenators	18	2	20
Senators	2	18	20
Column totals	20	20	40

Notice that whereas the overwhelming majority of senators are millionaires, the overwhelming majority of *non*senators are *non*millionaires. So there is a very strong association between this particular occupation and wealth. If there were no association at all between occupation and wealth, we'd expect nonsenators to be just as likely as senators to be millionaires. Because there are four cells in this design, because exactly half of those sampled are millionaires, and because exactly half are senators, the expected frequency of each of the four cells in this table under the null hypothesis is 10.0. You may also recall that multiplying the *column* total and the *row* total that corresponds to each cell (20 × 20, which is 400)—and then dividing by the total sample size (which is 40)—is a more formal way to get to these expected frequencies. So using the formula for the chi-square test that we reviewed in Chapter 1, let's see what we get. First, recall that

$$\chi^2 = \sum \frac{(f_o - f_e)^2}{f_e}.$$

This yields $\frac{(18-10)^2}{10} + \frac{(2-10)^2}{10} + \frac{(2-10)^2}{10} + \frac{(18-10)^2}{10},$

which is 64/10 + 64/10 + 64/10 + 64/10,

which is 6.4 + 6.4 + 6.4 + 6.4,

which is 25.6.

Here is the formula for converting a chi-square statistic to a phi coefficient:

$$\varphi = \sqrt{\chi^2_{obt} / N},$$

where χ^2_{obt} simply means the value of chi-square obtained in the χ^2 test for association. So in this case, φ equals the square root of 25.6/40, which is the square root of .64, which is exactly .80. So φ = .80. *In these hypothetical data,* there is a very strong association between being a U.S. senator and being a millionaire. If you were to convert these 40 frequencies into *x* (senator) and *y* (millionaire) scores dummy-coded as 0 and 1 (where, for example, 0 means is *not* a senator and 1 means *is* a senator), you could get SPSS to confirm that the correlation coefficient for these data is also exactly *r* = .80.

The phi coefficient is certainly easier to calculate than Pearson's *r*. As it turns out, however, there is another measure of association designed for use with nominal data that is even easier to calculate than the phi coefficient. This statistic is the **odds ratio.** An odds ratio is literally a **ratio** between the **odds** for (more loosely, the probability of) two potentially

Chapter 4   Nonparametric Statistics (Tests Involving Nominal Variables)

related but distinct events (e.g., the chances of dying if you take a drug compared with the chances of dying if you do not). One traditional way to represent this is to use the same kind of frequency table that you'd be likely to use for many χ^2 analyses. We'll use X and Y to label the columns and rows in such a table as a reminder that odds ratios are measures of association. We'll arbitrarily label the frequencies in the four cells a, b, c, and d.

	X_1	X_2
Y_1	a	b
Y_2	c	d

In each row of this 2 × 2 table, you could easily calculate the likelihood of a specific outcome. Thus, the odds of a (option x_1) at the value y_1 is simply a/b. This means, for example, that if outcome a occurred twice as often as outcome b, the odds of outcome a would be 2:1 (two to one). Along the very same lines, the odds of outcome c (option x_1) at the value y_2 is simply c/d. The ratio of these two odds is simply a/b divided by c/d. Thus,

$$OR = \frac{a/b}{c/d}$$

which can be simplified to

$$OR = \frac{ad}{bc}.$$

Notice that if the likelihood of event X *doesn't differ* at different levels of event Y, the OR = 1.0.

Consider eye color and gender:

	Brown Eyes	Blue Eyes	
Male	60	20	$ad = 60 \times 22 = 1{,}320$
Female	66	22	$bc = 20 \times 66 = 1{,}320$

Although brown eyes are three times more common than blue eyes, notice that the *odds of having brown eyes* are *identical* for women and for men (and notice that 60 × 22 = 20 × 66). Thus, the odds *ratio* for gender and eye color is exactly 1.0, which means that there is no association whatsoever between gender and eye color.[2] This means that if you were to compute a chi-square statistic rather than an odds ratio for these data, you would have to get a chi-square statistic of zero, which is the only way to get a phi coefficient of zero, indicating no association.

Although experimental psychologists do not make very frequent use of odds ratios, odds ratios are one of the favored statistics of medical researchers and health psychologists, who often care about associations involving nominal variables—such as whether a person has been diagnosed with (or died from) a particular disease. It is important to remember, however, that so long as you are dealing with associations between nominal variables, calculating odds ratios versus phi coefficients is largely a matter of taste.

Now that you have completed this brief review of phi coefficients and odds ratios, let's see if you can put your knowledge to good use to understand a very large archival study of marriage. In this case, we are going to forget about people's surnames and focus instead on more traditional predictors of marriage.

A Correlational Study of Interpersonal Attraction

One of the best documented findings in the literature on interpersonal attraction and close relationships is that people tend to be attracted to similar as opposed to dissimilar others. In other words, there is a great deal of support for the idea that "birds of a feather flock together" and very little support, by the way, for the idea that "opposites attract."

The next assignment in this chapter does not require you to use SPSS at all, but it does require you to handle and interpret a great deal of data. Table 4.3 summarizes these data, which are based on 10 years of Kentucky marriage records.[3] In a nutshell, the table shows you some real data that address whether people tend to marry others who are similar in (a) age and (b) ethnicity. Unfortunately, the search tool that came with these data did not allow us to download people's specific ages, and so we had to categorize people by "age decades" (specifically, we chose to study 20-somethings and 40-somethings). The data thus show you the degree to which people tend to marry others in their own versus the other artificially created 10-year age category. Focusing, for example, on White grooms who married White brides (the upper left-hand quadrant of Table 4.3—which is the group with the most marriages), you can see that people were *way* more likely to marry people who were in their same 10-year age group than they were to marry people who were much younger or older than they were. The odds ratio of about 240:1 is a combination of two different things. First, 20-something grooms were about 70 times as likely to marry brides in their 20s as to marry brides in their 40s. Second, 40-something grooms were more than 3 times as likely to marry brides in their 40s as brides in their 20s (the odds are 3.3:1 in the opposite direction). Putting this all together (*ad/bc*) gives you an odds ratio of about 240:1.

Chapter 4 Nonparametric Statistics (Tests Involving Nominal Variables)

Table 4.3 Matching Effects for Age and Ethnicity in Kentucky Marriages

	White Groom/White Bride Groom's Age		Black Groom/White Bride Groom's Age	
	20-something	40-something	20-something	40-something
Bride's Age				
20-something	136,565	6,356	1,997	147
40-something	1,907	21,441	48	241
	Odds Ratio: 241.6:1		Odds Ratio: 68.2:1	
	White Groom/Black Bride Groom's Age		Black Groom/Black Bride Groom's Age	
	20-something	40-something	20-something	40-something
Bride's Age				
20-something	483	28	7,092	322
40-something	10	43	100	1,496
	Odds Ratio: 74.2:1		Odds Ratio: 329.5:1	

As we already noted, you won't need to use SPSS to analyze these data. Instead, we'd like you to do some hand calculations that will require you to reorganize various aspects of the data to answer some specific questions about the similarity principle in interpersonal attraction. For example, if you wanted to use all of the data to determine whether people tend to marry others who are similar in age, you'd have to reorganize and simplify the data (which would require some addition) to produce the following table that ignores the ethnicity of both brides and grooms and focuses exclusively on age matching. Here are those reorganized data:

	Groom's Age	
	20-something	40-something
Bride's Age		
20-something	146,137	6,853
40-something	2,065	23,221

These figures for the total sample, organized by age matching, yield an odds ratio of 239.8:1, which indicates an extremely strong tendency for people to marry those who are similar in age. Given the sample size of $n = 178,276$ people, you can rest assured that this finding is statistically significant.

Now that you have seen how we can reorganize these data to ask different questions, we would like you to use the data to answer three additional questions.

QUESTION 4.5. Ignoring age, do people tend to marry others of the same ethnicity? Be sure to report the 2×2 frequency table and the accompanying odds ratio you used to answer this question. To give your readers an idea of how large this matching effect is, calculate a χ^2 test of association on these same data and convert it to a phi coefficient as well. Do the findings support the similarity hypothesis?

QUESTION 4.6. Were married couples who showed ethnic matching any more or less likely than usual to show age matching? Again, be specific about exactly which aspect of the data would allow you to test this idea, and report the appropriate odds ratio (without any conversion to χ^2 or to phi). Be sure to offer a brief psychological explanation for this finding.

QUESTION 4.7. Is the tendency for men in their 40s to prefer women in their 40s as strong as the tendency for men in their 20s to prefer women in their 20s? To answer this question, you can simply calculate and compare two *ratios* or two *percentages* rather than two odds ratios. To start with, what percentage of men in their 20s marry women in their 20s? Is this comparable to the percentage of men in their 40s who marry women in their 40s, or is there an asymmetry at work? If so, does this last finding qualify the general conclusion that people are attracted to similar others?

A Small Change of Pace: From Marriage to Mental Illness

In the following, you will find some hypothetical data that are loosely consistent with true prevalence rates for major depressive episodes (MDEs), substance abuse disorder rates, and BHSC rates for young to middle-aged adults (see Table 4.4).

QUESTION 4.8a. Using these data, calculate both (a) an odds ratio and (b) a phi coefficient for each 2×2 contingency table. Show your work for each hand calculation. For each 2×2 table, be sure to provide a brief summary and interpretation of your statistical analysis. You may need to consult tables of critical values for relevant statistics to determine which findings are significant. To get p values for phi coefficients, you may also enter the data in SPSS, run a chi-square "crosstabs" test (making sure to ask for a phi statistic), and consult the exact p values that SPSS gives you for each analysis.

Chapter 4 Nonparametric Statistics (Tests Involving Nominal Variables)

Table 4.4 Some Hypothetical but Realistic Associations Involving Gender

	Gender		
	Men (0)	Women (1)	Total
Current MDE?			
No (0)	355	368	723
Yes (1)	15	32	47
Totals	370	400	770
Lifetime MDE (Ever Diagnosed)?			
No (0)	309	308	617
Yes (1)	51	102	153
Totals	360	410	770
Lifetime Substance Abuse Disorder?			
No (0)	80	95	175
Yes (1)	13	7	20
Totals	93	102	195
Lifetime BHSC[a] (Ever Got BHSC)?			
No (0)	38	58	96
Yes (1)	57	42	99
Totals	95	100	195

Note: MDE = major depressive episode.

a. BHSC = Bad Haircut at SuperCuts (a likely addition to the *DSM-XV*).

QUESTION 4.8b. Now consider all of the results as a whole. When a dichotomous variable is "skewed" (i.e., when it takes on one of two values more often than the other), how does this seem to influence the value of phi? What about the value of an odds ratio?

QUESTION 4.9. Imagine that a researcher developed a new educational program to help recent PhDs pass a clinical licensing exam. (You may assume that everyone with a PhD from an APA-accredited university deserves to pass the exam—and that the exam exists merely to add a final insult to the painful experience of graduate school.) Here are the data from an experiment involving 24 people, 12 of whom took part in the program and 12 of whom did not. The data are coded so that 0 means "failed" and 1 means "passed" the clinical licensing exam.

Program: 1, 1, 1, 1, 1, 1, 1, 1, 1, 0, 1, 1
No program: 1, 1, 1, 0, 1, 0, 0, 1, 1, 1, 1, 1

Run a chi-square test on these data and summarize your conclusions. Be sure to discuss power. Assuming that the data from the control group are valid (assuming the normal passing rate is 75%), can you use SPSS to help you *prove* that the experiment had *virtually no power* to detect a real effect? (Doing so will require you to conduct an additional analysis in which you make a change in the data.)

How to Report the Results of a Chi-Square Analysis of Nominal Variables

Because knowing how to conduct and interpret a specific statistical analysis is not quite the same thing as knowing how to report one, we have included several appendixes in this text that illustrate how real researchers have reported the results of specific statistical analyses in published scientific journal articles. We begin that practice in this chapter's appendix by providing you with a couple of examples of how to report the results of a chi-square analysis.

Appendix 4.1: How to Report the Results of a Chi-Square Analysis

As you learned in this chapter, nonparametric tests such as the chi-square test are appropriate when a researcher studies outcomes that are nominal or categorical rather than continuous. Pollsters and political scientists, for example, often want to know which of several political candidates voters prefer. More interestingly, they might also want to know, for example, whether women prefer a specific candidate more than men do. Along similar lines, medical researchers and health psychologists often study categorical outcomes. You have either had cancer or you have not; you have a cold right now or you don't. As a third example, a great deal of research on judgment and decision making (including a great deal of marketing research) requires people to choose between two or more options or products. Choices and preferences (e.g., liking Coke versus Pepsi; reporting that you think suicide is less common than homicide) are usually categorical, and thus people who study choices often make use of chi-square or similar analyses.

Consider a classic example of judgment and decision making involving probability judgments.

To find out how people make certain kinds of probability judgments, Tversky and Kahneman (1972, p. 1125) asked people a question about two hypothetical hospitals:

> A certain town is served by two hospitals. In the larger hospital, about 45 babies are born each day, and in the smaller hospital, about 15 babies are born each day. As you know, about 50 percent of all babies are boys. However, the exact percentage varies from day to day. Sometimes it may be higher than 50 percent, sometimes lower.
>
> For a period of one year, each hospital recorded the number of days on which more than 60 percent of the babies born were boys. Which hospital do you think recorded more such days? (Circle one letter.)
>
> (a) the larger hospital
> (b) the smaller hospital
> (c) about the same (that is, within 5 percent of each other)

Before we discuss Tversky and Kahneman's (1972) results, consider your own answer to the hospital question. If you are like most people (which most people are, tautologically enough), you chose option c. Most people have a pretty strong intuition that 60% or more of a big number is the same as 60% or more of a small number. That intuition is quite correct, of course, but this *doesn't* mean that a random result that exceeds 60% is equally likely in a big versus a small sample. Results that deviate from an expected population value are much more likely in a small sample than in a large sample. Tversky and Kahneman argued that most people answer this problem incorrectly because they base their judgments on a useful but imperfect judgmental rule of thumb they dubbed the *representativeness* heuristic.

Without sweating the details of this specific heuristic, follow-up studies have shown that people do sometimes make careful use of logical decision rules about sample size. You just need to simplify the problem a little (or, arguably, you need to club people over the head a bit with information about sample size).

Consider the extreme example of counting the number of days in a year on which *all* (100% rather than 60%) of the babies born in a specific hospital are boys. And now consider a comparison between the same big hospital (where about 45 babies are born every day) and a really small hospital (where only about 2 babies are born every day). We hope you can see that the small hospital would have many, many more such days (about 91 per year) than the large hospital (about 0 per millennium). Pelham and Neter (1995) conducted an experiment in which they posed either (a) the original version of Kahneman and Tversky's hospital problem or (b) the simplified version involving extreme outcomes and a very small hospital to a couple of hundred UCLA undergraduates. Here is a simplified version of what they found (in their original study, there was an additional motivational manipulation that we're going to ignore):

> As expected, the analysis of participants' responses to the hospital problem revealed a large effect of the problem difficulty manipulation, χ^2 (1, $N = 280$) = 74.83, $p < .001$. As reflected in the error rates in Table 1, participants who received the original version of the hospital problem performed poorly. Only 31% answered the problem correctly. In contrast, 76% of those who received the transparent problem answered correctly. (Adapted from Pelham & Neter, 1995, p. 584)

Incidentally, because there are three rather than two possible answers to the hospital problem, Pelham and Neter (1995) also reported the results of additional statistical tests to see whether their experimental manipulations increased judgmental accuracy by decreasing the chances that people would pick the patently incorrect answer (answer a) versus helping people avoid the heuristic-consistent answer (answer c). In this and other studies, their problem difficulty manipulations did not usually help people avoid the heuristic-consistent answer. Instead, they made it a lot less likely that people would give the patently wrong answer. Check out Pelham and Neter (1995) for more details.

Apparently, then, most people are quite capable of thinking like statisticians. To paraphrase Pelham and Neter (1995), though, most people only seem to apply statistical decision rules to the hospital problem when they are presented with particularly hospitable versions of that classic

Chapter 4 Nonparametric Statistics (Tests Involving Nominal Variables)

problem. The first author's wildest dream is that this textbook will demystify statistics for you in much the same way that Pelham and Neter's problem difficulty manipulation demystified the hospital problem for half of their participants.

Notes

1. To be a little more serious, the first author did record the exact frequencies but failed to record the date. These pilot data eventually led to more serious studies that my students and I did a much better job of documenting (e.g., see Pelham, Carvallo, & Jones, 2005).

2. For a useful Web resource on odds ratios, see http://www.childrensmercy.org/stats/index.asp.

3. All marriages took place in the 1990s. Records were retrieved December 21, 2004, from www.vitalsearch.com.

Reliability (and a Little Bit of Factor Analysis) 5

Chapter Overview

In this chapter, we examine a couple of popular approaches to assessing the factor structure and/or reliability of psychological measures. We begin by examining a small hypothetical data file that should introduce you to some important concepts in factor analysis. Factor analysis is a statistical technique designed to uncover the underlying structure (i.e., the separable dimensions or basic constructs) in a set of many specific observations. Second, we look at a simple formula for computing one kind of reliability (split-half reliability, which is a form of internal consistency reliability). In the process of learning about this formula and this form of reliability, you will also learn a little about how to recode variables in SPSS. Doing these two activities in SPSS should prepare you for a third activity, which will require you to calculate both the internal consistency and the test-retest reliability of a 10-item self-esteem scale. Examining both internal consistency and temporal consistency (two distinct forms of reliability) will allow you to address how reliability is related to **validity.** Fourth, you will see how a researcher might use reliability analysis in a pilot study of scale development. Fifth, you'll work on a data set that focuses on interrater agreement. This activity should clarify how internal consistency and interrater agreement are similar as well as how they are different. Finally, you will examine some hypothetical data designed to illustrate some of the limitations of Cronbach's alpha (α) and give you a second chance to compare directly what you can learn from a reliability analysis and what you can learn from factor analysis.

Introduction: The Concept of Reliability

One of the most important issues in measurement theory and scale development is **reliability**, which has to do with the repeatability or consistency of a measure or observation. Ideally, all psychological measurements should be highly reliable. After all, if a measurement is *not* reliable, this means that there is disagreement or inconsistency with regard to the measurement. This could mean that different raters judge a behavior differently, that the same people score differently on the same personality measure at different times, or that people give very different answers to different questions that were designed to measure the same underlying construct. If Magnus Ver Magnusson can bench press 600 pounds and roll a car over on its side but fails the strength test for becoming a firefighter, it is very likely that the so-called strength test is not really measuring strength (e.g., perhaps it actually measures endurance or agility). In whatever form it comes, low reliability is usually a bad thing in measurement. Two key premises of good scientific thinking are empiricism (observability) and repeatability (consistency). You should get the same results in your lab experiment that I get in mine. If you don't, something is wrong (with you, with me, or with both of us). Similarly, what *you* call a major depressive episode, aggressiveness, or damage to the ventromedial hypothalamus should be very similar to what *I* call all of these same things. Another way of putting this is that scientific thinking requires agreement (a) about how things are being measured and (b) about what is being measured in the first place. Reliability is a way of formalizing exactly how much agreement there is in a set of independent measurements.

Reliability is also related to the scientific principle of parsimony (simplicity). Scientists are typically interested in making simple, concise, nonredundant (i.e., efficient, nonrepetitive, nonrepetitious, nonsuperfluous) statements about the relation between different variables. To be able to do so, scientists frequently try to develop a **single measure** of each variable they wish to study. Having a single measure of the variable you are studying makes your life much simpler. For example, it makes it much easier to summarize your research findings to another person. The logic of developing a single measure of what you are studying is similar to the logic of computing means from individual scores. In fact, it is no accident that the most common way of computing a single measure of a construct is to measure the construct in more than one specific way (or one more than one occasion) and to *average* the individual indicators of the measure together. However, it should go without saying that it is only appropriate to do this (to average items together) when the items are, in fact, measuring the *same* underlying construct (e.g., aggression, recession, depression, oppression). Conducting some kind of factor analysis or reliability analysis is the primary way researchers know whether their measures of a single construct really do tap into that single construct (rather than noise, or a similar but distinct construct).

Chapter 5 Reliability (and a Little Bit of Factor Analysis)

Because there are excellent conceptual discussions of reliability in many research methods textbooks (see Pelham & Blanton, 2013, as a self-serving example), we will forego a detailed discussion of different forms of reliability. However, if you are not familiar with the basic concepts of (a) interrater agreement (agreement between different raters), (b) internal consistency (agreement between different items that are added up to make a scale), and (c) test-retest reliability (agreement of a score with itself over time), you might want to read up on them a bit before you complete this chapter. For now, however, you should merely note that the logic of calculating and interpreting **interrater agreement** is essentially identical to the logic of calculating and interpreting **internal consistency. Test-retest reliability** (also known as temporal consistency) is a little different than internal consistency or interrater agreement. Although most of the activities in this chapter will focus primarily on internal consistency, we will also pay some attention to the other two forms of reliability. Before we do that, however, let's take a look at an analytic technique that often lays the foundation for a reliability analysis. This technique is factor analysis.

"Just the Factors, Ma'am"

In the classic TV show *Dragnet*, the show's main character, the sensible and plainspoken detective Joe Friday, became famous for his use of phrases such as "Just the facts, ma'am."[1] This phrase proved to be useful because many of Friday's witnesses couldn't resist throwing in a lot of redundant or irrelevant information along with the crucial information Friday needed to crack a case. Friday just wanted the facts—better yet, only the *highly relevant* facts. If Friday had been a statistician rather than a detective, he wouldn't have needed to change this famous phrase very much. "Just the factors" is the rough statistical equivalent of "just the facts." This is because researchers often wish to remove both (a) noise and (b) redundancy from a set of many different measures—to distill the basic essence of the measures. This is particularly true when researchers are trying to make people's answers to a very large number of questions more manageable. It is very hard to summarize people's responses to 100 questions, but if the 100 questions really boil down to only 4 questions (four basic concepts), the job gets 25 times easier.

As experts such as Dick Darlington[2] could tell you, the guy who invented factor analysis was the IQ expert, Charles Spearman. Spearman developed factor analysis about 100 years ago because he hoped to show that all the things people call intelligence (a good memory, processing speed, problem-solving ability, mathematical ability, reading comprehension, skill at *PRIME* the card game) really boil down to a single cognitive ability factor that Spearman called *g*. As it turns out, Spearman seems to have been profoundly wrong about *g*. There are *many* distinct varieties of intelligence. Furthermore, some kinds of intelligence are completely uncorrelated with others (e.g., see Sternberg's, 1988, discussion of practical intelligence). Ironically, however, Spearman's brilliant invention (factor analysis) was one of the very statistical tools that researchers used to invalidate Spearman's *g*-centric theory of intelligence. But what, exactly, does this invention called factor analysis do? Because there are so many different ways to do factor analysis, that's a tricky question. To simplify the answer, we are going to focus solely on one of the most basic forms of factor analysis—namely, **principal components analysis.**[3] A principal components analysis is simply a bottom-up (i.e., purely data-driven) effort to distill multiple observations down to the smallest meaningful number of dimensions (aka factors) that one can—while losing as little information as possible about the individual observations. It is important to remember that the *observations* of factor analysis (scores on specific variables in a data set) can be anything: attitudes, behaviors, emotions, memories, or personality judgments, to name only a few. Thus, almost any set of individual measurements can be factor analyzed (which *doesn't* mean almost any set *should* be, but more on that topic later).

One very useful way to think about factor analysis is that it is all about *eliminating redundancy*. To appreciate how beautifully a principal components analysis can sometimes do exactly this, let's examine an extremely unusual data set. Below is a screen capture of some hypothetical data that were collected by a highly intelligent but highly uninformed anthropologist from another planet. Imagine that the anthropologist got a grant to study earthlings' self-views. Because interstellar research is so expensive, let's assume that the anthropologist was only able to sample 14 representative earthlings and ask them six questions about how they viewed themselves. Image 1 is a screen capture of the anthropologist's SPSS raw data file (SPSS reputedly has an even larger market share in the Andromeda Galaxy than it does here). Image 2 is a matrix of the observed intercorrelations between the six measures. You can find these hypothetical data in the SPSS data file called **[factor analysis perfect 2 factor solution march 2011.sav]**.

The raw data are just for your benefit, by the way, because SPSS only needs a complete set of intercorrelations to perform a principal components analysis. Having said that, we hope you can see from both the raw data and the correlation matrix that once you know how smart all these earthlings think they are, you also know everything you could ever want to know about how intelligent, clever, and bright they all think they are—while learning nothing whatsoever about how honest and truthful any of them think they are. To put this concretely, *The first four columns of the raw data file are absolutely identical, and they are all completely uncorrelated with the last two columns, which are identical to one another.* Thus, if you look at row 14, you'll see that Jennifer (from Tampa) gave herself an 11 for smart and also gave herself 11s for intelligent, clever, and bright. Ditto for poor David (from Phoenix) whose self-rating of 1 for smart indicates that he thinks he is not at all smart. He also claims to be not at all intelligent, clever, or bright (see his ratings of 1, 1, 1, and 1). If you check out Image 2, you can also see that self-perceived intelligence is correlated zero (circled in the image) with self-perceived honesty. For instance, both Jennifer (the genius) and David[4] (the doofus) say they are 6s in honesty and truthfulness. To complete the picture, knowing how a person answered either one of the last two questions tells you exactly how the same person answered the other question. For instance, all four people who gave themselves 5s for honest *also* gave themselves 5s for truthful. To put this all a little differently, the anthropologist *thought* she was asking earthlings six different questions, but people's responses to the questions revealed that she was really just asking them two. One of the best ways to see how principal components analysis works is to see what a principle components analysis would tell us about these extremely unusual data.

Let's begin with the score that is probably most strongly associated with factor analysis—namely, **eigenvalues.** Eigenvalues are summaries of the degree to which any of the sets of items being factor analyzed tend to cluster together. A principal components analysis yields exactly as many distinct eigenvalues are there are specific items (e.g., survey questions) being factor analyzed. A principal components analysis of these hypothetical data will thus yield a total of six eigenvalues.

The basic output of a factor analysis is thus a set of n eigenvalues where n is the total number of items being analyzed. By convention, these eigenvalues are sorted from highest to lowest. Moreover, statisticians often like to plot the n eigenvalues in a **scree plot** like the one you see in Image 3. The component solutions are plotted in ranked order, from the one with the largest eigenvalue to the one with the smallest eigenvalue. Traditionally, researchers hoped to find an obvious break point in the scree plot that represented the point where the eigenvalues showed a dramatic drop (the point where the factors weren't very useful anymore). These days, sophisticated statistics are available for assessing when you have run out of

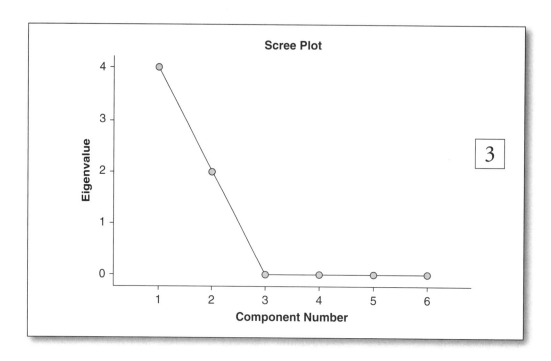

meaningful factors. However, in the case of our hypothetical data, we can live with the old school approach because it is patently obvious that there are two and only two components (aka, factors) in these data.

Another important thing to know about factor analysis is that, traditionally, eigenvalues greater than 1.0 tell you that you may have put your finger on a meaningful factor. Furthermore, all else being equal, an eigenvalue that is many times greater than 1.0 usually tells you that many different items load pretty heavily on (i.e., are pretty highly correlated with) the specific potential factor under consideration. Mathematically, each eigenvalue from a factor analysis is the sum of all the squared correlations between all of the individual items assessed and the presumed factor in question. Notice that, based on this definition, it's very easy to calculate the six eigenvalues in these data. There is one factor, which we'll call *self-perceived intelligence*, for which the relevant correlations are $1.0^2 + 1.0^2 + 1.0^2 + 1.0^2 + 0^2 + 0^2 = 4$. These six squared values reflect the fact that there is one construct in these data on which four questions load perfectly and on which the remaining two questions don't load at all. Moving on to the next construct, the eigenvalues for what we'll call *self-perceived honesty* are $0^2 + 0^2 + 0^2 + 0^2 + 1.0^2 + 1.0^2 = 2$. Because the sum of the six eigenvalues has to be six, this means that for the four remaining constructs (aka "components"), the eigenvalues are all zero (remember our first two eigenvalues were 4.0 and 2.0). This is as it should be because in these highly contrived data, two factors account perfectly for absolutely everything going on in the data.

If you were to conduct a principal components analysis of these data, you'd also be told the percentage of the variance in the data accounted for by each of the six components or factors. Notice that in these data, you can account perfectly for four of the six columns of data by knowing what is going on with the first factor. Fittingly enough, then, a principal components analysis would tell you that the first factor accounts for 66.7% (4/6) of the **total variance** in all of the scores. This means that the second factor accounts perfectly for the remaining 33.3% (2/6) of the variance in the data, and we are thus in the highly unusual situation of being able to say that two factors tell us absolutely everything there is to know about all of the variation in individual scores in these data.[5]

Caveats Regarding Real Data

The caveat we just offered also means that in real data sets, the factors with the largest eigenvalues will practically never account for *all* (100%) of the total variance in the data. This is because (a) none of the items in the real data would be *perfectly* correlated with any underlying factors and (b) quite a few of the specific items in the data might be completely unpredictable (or nearly so) from any of the other specific items. For example, accounting for 60% of the variance on a set of 20 individual scores with only two or three factors would often be a highly respectable outcome. At the other extreme, however, what if you could only account for half of the variance in a set of 20 measures by identifying seven factors? In this extremely wimpy state of affairs (i.e., with such weak intercorrelations among the variables), it simply might not be useful to do a factor analysis. Tabachnick and Fidell (2007) argue, for example, that if no correlation in your entire matrix of intercorrelations (**R**) exceeds $r = .30$, then the use of factor analysis "is questionable because there is probably nothing to factor analyze" (p. 614). They also warn that unless some of the correlations that go into a factor analysis are very large, factor analysis is highly inadvisable for small samples. Ideally, one would like to see a sample of $n = 1,000$ to feel extremely comfortable with the results of a factor analysis. It is common, however, to see researchers make do with much smaller samples (say, $n = 300$), especially when they can show that the same basic factor structure replicates well in different samples.

Perhaps one of the most common mistakes made by novice users of factor analysis is to interpret eigenvalues as relative indicators of the ultimate psychological importance of the multiple factors identified—as if the factor with the biggest eigenvalue is the most prominent or important thing out there in the real world. The most serious problem with this tendency is that eigenvalues are very strongly influenced by the number of conceptually related items that researchers happen to include in a particular study. For example, if the anthropologist from the Andromeda Galaxy had asked only two questions about self-perceived intelligence but six questions about self-perceived honesty, the eigenvalue for intelligence would shrink to a much smaller 2.0, and the eigenvalue for honesty would grow to a much larger 6.0. If you only ask your research participants questions about stones or turnips, you won't be seeing any eigenvalues for blood in your output.

There are other logical or conceptual caveats regarding factor analysis. As Kline (2005) astutely noted, we obviously have to give factor scores names. Otherwise, we won't be able to communicate our fascinating findings to others. However, the simple act of naming a factor may contribute to what Kline calls the **naming fallacy:** "just because a factor is named does not mean that the hypothetical construct is understood or even correctly labeled" (p. 176). Perhaps worse yet is what Kline calls the error of **reification,** "the belief that a hypothetical construct *must* correspond to a real thing." Kline specifically cites Spearman's *g* (a general cognitive ability factor) as an example. Assuming that *g* (or anything else that might seem to be suggested by the results of a factor analysis) *must* be real can often get a researcher into trouble.

Another caveat is that factor analysis (including principal components analysis) is extremely sensitive to the magnitude of correlation coefficients. Thus, it is particularly important to be sure that the correlations that serve as the basis of any factor analysis are free of error or bias. If your data include scores that deviate from a normal distribution (e.g., because of some extreme outliers), it is easy to observe spurious results that won't hold up well to replication. In Chapter 14, we address data-cleaning issues such as these in great detail. For now, let us merely note that a factor analysis of dirty variables is unlikely to yield clean (i.e., replicable) results.

Perhaps the biggest warning we should offer about factor analysis is reputational rather than computational. In our view, factor analysis has an undeservedly negative reputation. Perhaps because some of the early proponents of factor analysis promoted it as the holy grail for uncovering the nature of everything, some have focused on the fact that it does not live up to this reputation. For reasons that are less clear, some people also wrongly assume that if you to have to resort to an exploratory factor analysis, it means that you may have something up your sleeve (see Darlington's website, referenced in Note 2, for further details). Ironically, as Tabachnick and Fidell (2007) put it, "The very power of PCA [principal components analysis] and FA [factor analysis] to create apparent order from real chaos contributes to their somewhat tarnished reputations as scientific tools" (p. 609). Our view of factor analysis, like our view of every other statistical technique, is that there is a time and a place for it. Just as a hammer or screwdriver can be misused, statistical tools can be misused as well. The key to using any statistical technique effectively is (a) being careful about it in the first place and (b) then replicating any findings revealed by the technique to see if they really hold up.

Now that we've defined reliability and factor analysis, as well as warned you about some of the pitfalls of factor analysis, we'd like to spend the rest of this chapter exploring both reliability and factor analysis, with an occasional foray into what factor analysis and reliability analysis have in common. To do this, we will analyze the data from five different SPSS data files. A list of the five files follows.

1. [race and politics random subset for factor analysis.sav]
2. [Reliability Data (extraversion).sav]
3. [Reliability Data (GSE).sav]
4. [ADHD Camp Hypothetical QMCS-IV Data.sav]
5. [CRONBACHS ALPHA TWO FACTOR DEMO KEEPER.sav]

Principal Components Analysis With Real Data

Your first data analysis activity in this chapter involves a principal components analysis, which should allow you to uncover an underlying structure (i.e., a factor or factors) in a set of observations. To see how principal components analysis works, let's open the [**race and politics random subset for factor analysis.sav**] data file. This file is a random sample of records from a nationally representative telephone survey conducted at Berkeley in 1991. The data file should resemble what you see in Image 4.

Variables that begin with "va" are measures of people's **values.** For instance, as your SPSS variable labels will tell you, va3 is a measure of the degree to which participants reported that they valued "respect for authority." For the purpose of this first activity, we're going to assume that you are a researcher who eventually wants to know if people's values are related to (a) their age, (b) their political ideology (liberal vs. conservative), and (c) their attitudes toward abortion. The researcher is methodologically sophisticated (for a high school junior), but she doesn't know all that much about political attitudes or have a strong a priori hypothesis about how or whether the 11 different values that were measured in this study will form any kind of meaningful construct. She thus decides to perform a **factor analysis** on the 11 value items to see if she can discover any kind of underlying structure in the value data. That is, she simply wants to condense a large number of variables into a smaller set of variables, making her study a great candidate for factor analysis. In the images and instructions that follow, you can see how to go about conducting a **principal components analysis** (one particular kind of factor analysis) in SPSS.

First, go to "Analyze," "Dimension Reduction," and "Factor...." (as in Image 5). Then select all 11 of the "va" variables (values) and send them to the "Variables:" box (as in Image 6).

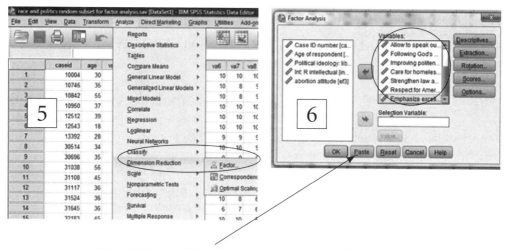

Now click on "Paste" to create a syntax file containing your factor analysis commands. By the way, we're not doing anything fancy, and so we didn't ask for any special "Descriptives..." or change the "Extraction..." method from the default of a *principal components* extraction method. This means that your syntax file should look very much like the one you see in Image 7.

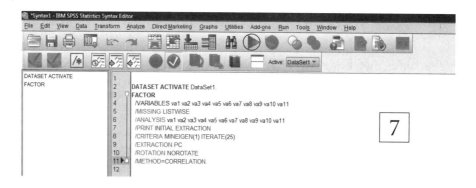

From the syntax file, you just run this "FACTOR" command. Remember, you can just click anywhere in the "FACTOR" statement set and then click the "Run" arrow (circled in Image 7). Doing this should generate an output file, the most important part of which you can see in Image 8.

Chapter 5 Reliability (and a Little Bit of Factor Analysis)   135

Checking Out the Eigenvalues

Your output file should contain three important things. Let's begin with the eigenvalues:

Total Variance Explained

Component	Initial Eigenvalues			Extraction Sums of Squared Loadings		
	Total	% of Variance	Cumulative %	Total	% of Variance	Cumulative %
1	3.757	34.157	34.157	3.757	34.157	34.157
2	1.532	13.926	48.083	1.532	13.926	48.083
3	1.055	9.587	57.670	1.055	9.587	57.670
4	.858	7.801	65.470			
5	.747	6.789	72.259			
6	.668	6.070	78.330			
7	.575	5.229	83.559			
8	.558	5.070	88.629			
9	.502	4.567	93.196			
10	.392	3.565	96.761			
11	.356	3.239	100.000			

Extraction Method: Principal Component Analysis.

[8]

The part of the output file depicted in Image 8 provides you with an eigenvalue (appearing under "Total") for each of the 11 factors (aka components) that were extracted from the data. As you know, an eigenvalue greater than 1.0 traditionally tells you that you may have found a factor—that is, a meaningful dimension that will allow you to lump together some items that are highly correlated with one another. Remembering some of the caveats we noted earlier, the size of an eigenvalue tells you something about how many of the specific items you analyzed are related to (i.e., correlated with) that particular factor. In the column labeled "% of Variance," you can see that each component also accounts for a certain percentage of the variance in the total set of scores. The bigger a factor's eigenvalue is, the greater the total percentage of variance accounted for by that factor.

You may also recall that the exact value of the eigenvalue for a given component is a direct reflection of all of the factor loadings for that specific component. To revisit exactly what this means, take a peek at Image 9, which shows you the section of your output labeled "Component Matrix[a]."

Component Matrix[a]

	Component		
	1	2	3
Tolerating diff beliefs 0-10	.005	.699	.293
Preserve tradition right and wrong 0-10	.683	-.042	.192
Respect for authority 0-10	.766	-.091	.146
Self-reliance 0-10	.452	.357	-.607
Allow to speak out 0-10	.074	.773	-.101
Following God's will 0-10	.682	-.301	.116
Improving politeness 0-10	.759	.019	.082
Care for homeless 0-10	.333	.346	.604
Strengthen law and order 0-10	.688	-.071	-.014
Respect for American power 0-10	.746	-.183	-.125
Emphasize excellence on job 0-10	.560	.245	-.362

Extraction Method: Principal Component Analysis.
a. 3 components extracted.

[9]

At the risk of being highly repetitive, let us remind you that the eigenvalue for each component in a factor analysis is the sum of all of the squared factor loadings on that component (added up across all items tested). Let's examine Component 1, whose 11 factor loadings are highlighted in Image 9. For Component 1, the 11 squared factor loadings are $.005^2 + .683^2 + .766^2 + .452^2 + .074^2 + .682^2 + .759^2 + .333^2 + .688^2 + .746^2 + .560^2$. If you were diligent enough to square and add up all 11 of these values, you'd get 3.759, which, with rounding error, equals the eigenvalue of 3.757 that you can see circled in the "Total Variance Explained" portion of Image 8. Recall that the sum of all of your eigenvalues will always be the number of items you entered into the factor analysis, regardless of how many meaningful factors there are in the data. If you had factor-analyzed eight items, the sum of your eight eigenvalues would have been 8.0. If you had factor-analyzed 100 items (using a very large sample size, we hope), the sum of your 100 eigenvalues would likewise have been 100.0. It is also worth noting that the percentage of variance accounted for by a specific factor is simply the eigenvalue for that specific component divided by the total number of components (which is the total number of items analyzed). Thus, as shown in Image 8, the first factor accounts for 34.157% of the total variance. This is simply 3.757 divided by 11 (expressed as a percentage).

You can also see in Image 8 that the eigenvalue for Component 1 (3.757) is quite a bit larger than any of the other eigenvalues. Thus, if you were to plot all the eigenvalues in a **scree plot,** you'd see a big drop in the size of the eigenvalues as we moved from Factor 1 (the "biggest" factor) to Factor 2 (the second biggest factor). Traditionally, this would mean that this is where we'd draw the line in our search for factors—and decide that there is probably only one meaningful factor in these data. As noted earlier, modern statisticians have developed much more formal ways to decide how many factors there are in a set of factor-analyzed responses. To learn more about this, you could read up on parallel analysis, which is beyond the scope of this intermediate text (e.g., see Lance, Butts, & Michels, 2006). If you want a more comprehensive overview of factor analysis, a good Web resource (in addition to Dick Darlington's site) is G. David Garson's: http://faculty.chass.ncsu.edu/garson/PA765/factor.htm#eigen. An excellent textbook on exploratory factor analysis is Fabrigar and Wegener (2011).

Because we have three components with eigenvalues bigger than 1.0, and because an eigenvalue bigger than 1.0 has been the traditional sign that a factor may account for some meaningful variance in a set of scores, we'll take at least a cursory peek at all three of the first three components. Specifically, we'll look at the **factor loadings** for each of the 11 value items on all three components to see if we can discover any meaning behind the three potentially informative factors.

Component Matrix[a]

	Component 1	Component 2	Component 3
Tolerating diff beliefs 0-10	.005	.699	.293
Preserve tradition right and wrong 0-10	.683	-.042	.192
Respect for authority 0-10	.766	-.091	.146
Self-reliance 0-10	.452	.357	-.607
Allow to speak out 0-10	.074	.773	-.101
Following God's will 0-10	.682	-.301	.116
Improving politeness 0-10	.759	.019	.082
Care for homeless 0-10	.333	.346	.604
Strengthen law and order 0-10	.688	-.071	-.014
Respect for American power 0-10	.746	-.183	-.125
Emphasize excellence on job 0-10	.560	.245	-.362

Extraction Method: Principal Component Analysis.

a. 3 components extracted.

Image 10

Each of the six value items highlighted in gray in Image 10 (which is just a shaded variation of Image 9) loads heavily on Component 1, *without* loading very heavily on either of the other two factors. We *might* be tempted to add the last item (about emphasizing excellence on the job), but in our opinion, this item correlates a little too strongly with Component 3.

It is also easy to see that Items 1 and 5 (see the dashed rectangles) load heavily on Component 2 and *do not* load heavily on Component 1 or 3. By the time we get to Component 3, though, it's not all that clear whether there's really a factor there at all.

From your second look at the Component Matrix output, then, we hope you can see that six items (whose variable labels and factor loadings we highlighted) load very well on the first factor, whatever it is. Having said this, we should add that it is often necessary to make judgment calls when using factor analysis. Luckily for us, the judgment calls in this particular case are not all that tough. Six of the 11 items load well on Factor 1.

But what exactly *is* Factor 1? For now, we're going to call it **"trism"**—so that we can create a new variable, name it "trism," and see if it correlates with the variables the researcher cares about—namely, age, political ideology, and abortion attitudes. Incidentally, we're ultimately going to ignore the other two factors. It looks like one of them (Component 2) is something like tolerance or support for freedom of expression. However, only two items load on this factor. Component 3 might also be of interest to those who study self-reliance, but if it indicates a factor at all, that factor seems to consist of only two items. Thus, we're going to ignore Factors 2 and 3 and focus on the first factor.

Reliability Analysis

Eventually, we are going to convert the six individual scores that all load on Component 1 into a single "trism" score. However, before we do that, let's take the opportunity to conduct an additional analysis, a **reliability** analysis. This analysis will tell us *exactly* how well the six items correlate with one another—that is, how reliable (i.e., *internally consistent*) our "trism" measure is. Of course, it *better* be pretty reliable, or else something must have been wrong with our factor analysis. But exactly how reliable is it? The most common way to answer this question about internal consistency is to compute Cronbach's (1951) alpha (α). Let's do that. Image 11 shows you how to get started, and Image 12 shows you how to select the variables of interest.

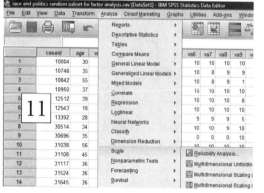
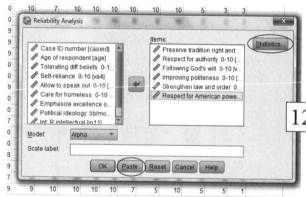

Notice in the "Items:" section of Image 12 that we chose *only* the six items that loaded well on the "trism" factor. We want to perform our reliability analysis *on these six items only*. This time, we want some specific statistics that don't come as a default, and so we need to click on "Statistics...." We're not going to show you the Statistics dialog box, but be sure to click on "Scale if item deleted" and "Means." Then click "Continue" in the same Statistics box, and then click "Paste" (circled in Image 12) after Continue takes you back to the main Reliability Analysis box. This should generate the following commands in your syntax file:

Chapter 5 Reliability (and a Little Bit of Factor Analysis)

RELIABILITY

/VARIABLES=va2 va3 va6 va7 va9 va10

/SCALE('ALL VARIABLES') ALL

/MODEL=ALPHA

/SUMMARY=TOTAL MEANS.

Once you run this reliability command, you should get an output file, part of which should resemble Image 13. The three most important things in this part of the output are marked up for you.

Reliability Statistics

Cronbach's Alpha	Cronbach's Alpha Based on Standardized Items	N of Items
.827	.836	6

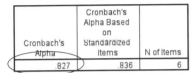

Summary Item Statistics

	Mean	Minimum	Maximum	Range	Maximum / Minimum	Variance	N of Items
Item Means	8.575	7.989	9.009	1.020	1.128	.166	6

Item-Total Statistics

	Scale Mean if Item Deleted	Scale Variance if Item Deleted	Corrected Item-Total Correlation	Squared Multiple Correlation	Cronbach's Alpha if Item Deleted
Preserve tradition right and wrong 0-10	42.44	62.070	.578	.387	.805
Respect for authority 0-10	42.75	58.522	.662	.482	.788
Following God's will 0-10	43.00	52.120	.598	.373	.804
Improving politeness 0-10	42.44	61.515	.620	.403	.798
Strengthen law and order 0-10	43.16	57.339	.555	.330	.808
Respect for American power 0-10	43.46	54.075	.630	.407	.792

The most important thing to note is that **coefficient alpha** (see the small oval) is very high for a six-item scale ($\alpha = .83$). It's also worth noting that there are no bad items in the scale. Deleting any single item from the scale would *reduce* the overall reliability of the scale (see the dashed rectangle and compare each of these alpha values with $\alpha = .83$). Each of these things is a reflection of the fact that the **corrected item-total correlation** for each item is very high (see the big oval). Incidentally, if you are wondering why item-total correlations need to be *corrected*, rest assured that it has nothing to do with them behaving badly. The correction has to do with the fact that if you were to add up all six of the items and correlate each individual item with that total, one sixth of that total score would always include whichever item you were correlating with the total score. This would obviously inflate the individual correlations. Moreover, the fewer items you have in a scale, the worse this inflation would be. Imagine correlating (x) how much you like ice cream on a 7-point scale with (y) the *average* of how much you like ice cream and how much you like the movie *Casablanca* (both on the same 7-point scale). You'd expect a pretty positive correlation even if the correlation between the two individual items is zero. Correcting for this problem is simple. You simply remove each item from the total scale when correlating that specific item with the total scale. Thus, in these data, the $r = .578$ corrected item-total correlation for VA2 ("right and wrong") is the simple correlation between VA2 and the sum of the other five VA items, with VA2 removed. Of course, this would be a pain to have to create 6 (or 40) different corrected versions of the total scale score, one for each corrected item-total correlation, but SPSS is quite happy to do this for you.

Putting all this together, this reliability analysis just tells us that principal components analysis is not a sham or a disgrace. It allowed us to identify six items that go together well.

Adding Items Together to Make a Scale

So let's average the six items together to make a "trism" scale. Just type the commands directly into the syntax file. We recommend putting them at the top of the same syntax file you already created.

COMPUTE TRISM = (VA2+VA3+VA6+VA7+VA9+VA10)/6.

EXECUTE.

Make sure you highlight both the COMPUTE statement *and* the EXECUTE statement before you click the Run button. Otherwise, if you only highlight the compute statement, you may compute the new "trism" variable without actually adding it to your data file. If you're successful, every participant who answered the six value items will now have a trism score (and it'll have become part of your data file).

Now you are ready to see if trism is associated with the three key variables that the researcher was studying: age, political ideology, and abortion attitudes. As a quick reminder of how to conduct a correlation, start at Analyze, and then go to Correlate and Bivariate.... As shown in Image 14, this will lead you to a screen that lets you choose your four variables and *then* pick your Options....(to ask for means and standard deviations). (Don't forget "Continue" and then "Paste.")

Chapter 5 Reliability (and a Little Bit of Factor Analysis) 141

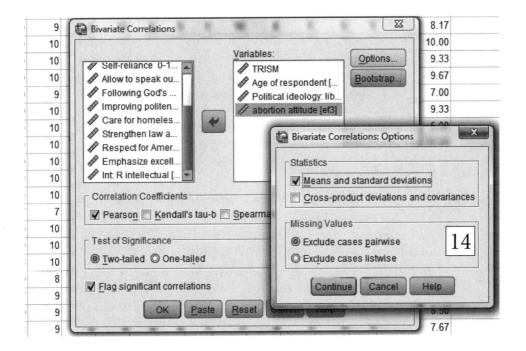

So assuming you used Image 14 as a guide and selected your four variables from the list in the order in which we selected them (e.g., trism first), your pasted correlation statement should read:

CORRELATIONS

/VARIABLES=TRISM age ideo ef3

/PRINT=TWOTAIL NOSIG

/STATISTICS DESCRIPTIVES

/MISSING=PAIRWISE.

Notice that everything except EF3 (abortion attitudes) has a pretty intuitive label. Be sure to check out the variable labels for both ideo and EF3, though, to see how they are coded (e.g., do higher numbers for EF3 correspond to greater *approval* or greater *dis*approval of abortion?). You are now ready to run this set of correlations in SPSS and summarize what you've learned.

QUESTION 5.1. Come up with a meaningful label for the trism variable. That is, based on your reading of the six items in the scale, what do these six value items seem to have in common?

QUESTION 5.2. Summarize *briefly* how you extracted this variable in the principal components analysis, and then report the reliability (Cronbach's α) of the trism scale as you would in a journal article.

QUESTION 5.3. Report and interpret any significant correlations between the trism variable and the three other variables. Make sure you spend some time and effort here, but do not worry too much about how the three variables are related to one another. Do the three correlations make intuitive or theoretical sense? Do they tell us anything about whether there is a general, meaningful "trism" factor in human values? Do they help tell you what trism probably is? (If so, feel free to modify your answer to Question 5.1.)

A Comparison of Cronbach's Alpha and Split-Half Reliability

Now for an easy activity. After you save both your edited data file and your syntax file, open the second data file [**Reliability Data (extraversion).sav**]. This little file contains hypothetical data from a small project with only 20 participants. The researcher who collected these data simply wanted to develop a reliable two-item measure of extraversion. As an aside, he was also curious to see if the formula he learned for calculating the reliability of a measure based on a *split-half correlation* would yield the same reliability score as a test for Cronbach's alpha (based, of course, on the same two items). This simple split-half formula is as follows. And yes, to answer your question, the same Charles Spearman who helped invent factor analysis also helped to figure out the formula for split-half reliability:

$$\text{Spearman-Brown reliability} = 2r_{yy'}/1 + r_{yy'},$$

where $r_{yy'}$ refers to the correlation between one half of a test (e.g., one test form) and the other half of a test, assuming that the halves are similar or parallel. Notice that in the case of a test consisting of only two individual items, split-half reliability and alpha become conceptually identical—a measure of reliability based on the degree to which two specific items are correlated with each other. For instance, if the two items in a two-item measure were correlated .50, our estimate of the reliability of the **total measure** would be $2 \times .50/(1 + .50)$, which is 1/1.5, which is .67. If you are wondering why this value is greater than .50, remember that you are *adding together two items*, each of which (alone) has a reliability estimate of $r = .50$.

You may also be wondering if it is OK to use this formula when two conceptually related items, such as outgoing and shy, are *negatively* correlated. Reliabilities cannot normally be negative. To deal with this issue, researchers *reverse code* items that they expect to be negatively correlated with the other items in a scale. In the case of a two-item scale with items that are opposite in meaning (e.g., "tall" and "short"), this means reverse coding whichever item you wish. Because this researcher labeled his scale "extraversion," it made the most sense to reverse code the "shy" item. But how do you this in SPSS? There are two ways. We'll learn an easy but inflexible way now and a complex but more flexible way in a later chapter. The easy way to do this is to use the same kind of COMPUTE command you use to sum or average variables. Because the researcher measured his variables using a 7-point scale that ranged from 1 to 7, he needs to convert 7s into 1s, 6s into 2s, and so forth. The easiest way to do this is simply to subtract the original "shy" scores from 8, like this:

COMPUTE RESHY = 8 − SHY.

EXECUTE.

Doing so will create a new variable called *reshy*. If you are wondering why we subtract from 8 rather than 7, do the simple subtraction and you'll see why. If you're wondering why we gave the recoded variable a new name, rather than just replacing the old *shy* scores with new shy scores, the answer is that it's not necessary to use a new variable name, but it greatly reduces the likelihood that we'll ever get mixed up about whether the variable is recoded or whether it appears the way it did in our survey.

Once you have your new, recoded variable, you are ready to run a reliability analysis, exactly as you did with the "race and politics" data. Remember to ask for "Scale if item deleted" and "Means." Notice that the item-total correlation is now the same for each of the two items in the scale.

QUESTION 5.4. First, briefly explain why this is the case (why the item-total correlation is the same for the two different items in the scale).

QUESTION 5.5a. Second, compute the overall reliability of the extraversion scale, using the Spearman-Brown formula.

QUESTION 5.5b. Now compare this split-half reliability estimate value with the value of alpha (α) that you obtained in the reliability analysis, and explain why the two values compare the way they do.

Finally, before you move on to the next part of this chapter, make sure to save both your edited data file and your syntax file for the extraversion study.

Applying What You Learned to a Hypothetical Study of Self-Esteem

Now let's apply what you've learned so far to a data set involving a popular 10-item measure of self-esteem. This data set [**Reliability Data (GSE).sav**] will require you to do several different things.

First, let's take a look at the specific questions that are a part of the Rosenberg Self-Esteem Scale:

```
    1        2        3         4        5        6        7
Disagree              Neither agree                      Agree
very much             nor disagree                       very much
```

___1. I feel that I am a person of worth, at least on an equal basis with others.
___2. I feel that I have a number of good qualities.
___3. All in all, I am inclined to feel that I am a failure.
___4. I am able to do things as well as most other people.
___5. I feel I do not have much to be proud of.
___6. I take a positive attitude toward myself.
___7. On the whole, I am satisfied with myself.
___8. I wish I could have more respect for myself.
___9. At times, I feel that I am useless.
___10. At times I think I am no good at all.

You will notice that half of these self-esteem items (e.g., Item 3) are negatively worded. Create a syntax file and recode all five of the negatively worded items so that higher scores on all 10 items will correspond to *higher* self-esteem. We recommend giving the five recoded items names such as "re_rse3."

QUESTION 5.6. Correlate each of these 10 individual items (five original and five recoded) separately with "outspokn" (a course instructor's blind rating of how outspoken people were during class discussions—see the variable labels). Comment briefly on whether the 10 correlations (rse1 with outspoken, rse2 with outspoken, etc.) make sense as a whole. Then compute (by hand) and report the average of these 10 individual correlations along with the highest and lowest correlation (don't bother to report all 10 individually).

QUESTION 5.7. Next, determine whether the 10 individual items form an internally consistent self-esteem scale by calculating Cronbach's alpha. If there are any bad items in the scale, identify them, delete them from the scale, and explain how you knew the items were bad. Report the alpha level for your final scale and comment briefly on the acceptability of this alpha level. (For a 10-item scale, a Cronbach's alpha level greater than .70 is generally considered acceptable.[6])

QUESTION 5.8a. Now calculate an overall self-esteem score (the average of the items you decided to keep in your self-esteem scale) and correlate this single score with the "outspokn" variable. Is this correlation any higher or lower than you might have expected based on the average correlation you reported in Question 5.6? Explain this, with an emphasis about what this tells you about reliability. What do we gain or lose by lumping items

Chapter 5 Reliability (and a Little Bit of Factor Analysis)

together into a single scale? Does reliability seem to contribute in any way to validity? If you're not sure, and if you are not embarrassed to be spoon fed (we're certainly not), you can always consult Appendix 5.1 at the end of this chapter for some very big clues.

QUESTION 5.8b. Now that you have a single Time 1 self-esteem score, correlate this score with the Time 2 self-esteem score (which is already calculated for you). Report this correlation as you might in a research report, and give a one-sentence summary of what it tells you about the temporal reliability (i.e., test-retest reliability) of your self-esteem scale.

QUESTION 5.8c. Now create a composite measure of chronic self-esteem by averaging the Time 1 and Time 2 self-esteem measures. Correlate this new self-esteem measure with the "outspokn" variable. Discuss how this correlation compares with the average of the correlation between (a) the Time 1 self-esteem measure and "outspokn" and (b) the Time 2 self-esteem measure and "outspokn." Again, does reliability seem to contribute in any way to validity? If you're not sure, and if you are not embarrassed to be spoon-fed again, you can always consult Appendix 5.1 at the end of this chapter.

A Return to Extraversion: Reliability Analysis as a Tool for Item Development

In the early phases of many research projects, researchers often use reliability analyses to refine and improve multiple-item measures of (a) some variable they have never assessed before or (b) some variable they wish to reassess in a novel way. For example, suppose you were an experimenter who studies heat and aggression. If you had a trained confederate insult your participants in a hot versus a cool room, you'd probably want to assess people's hostile feelings toward the confederate using something more reliable than a single hostility question. Imagine, though, that you had some serious conceptual concerns about the existing (highly reliable) four-item measure of hostility previously used by many of your colleagues in their experiments. Thus, you might want to run an experiment to show that your superior, equally reliable, four-item measure (a) was only modestly correlated with the previously used measure of hostility and (b) yielded much stronger results (i.e., was more sensitive to the heat manipulation and was a better predictor of how much shock participants administered to the confederate). Admittedly, this isn't the most groundbreaking idea ever, but after all, you've only been a postdoc working on this research topic for 9 years. If you needed to develop your own reliable three- or four-item measure of hostility, you might begin by conducting a pilot study in which you asked your pilot participants 20 or 30 questions that you felt would assess hostility. That is, you'd begin by assuming that only a few of your carefully chosen, clearly worded, related but distinct hostility questions would prove to hang together well (and capture your novel conception of hostility). After completing the pilot study, you could run a reliability analysis to decide which three to four questions produced the most reliable scale. This pilot work would all be necessary before you ran a full-fledged experiment.

So one common reason why researchers run reliability analyses is to develop or refine new measures. With this in mind, we'd like you to take a look at some quasi-real data from a very informal study a researcher might have run to begin to develop a short measure of extraversion. (Yes, when real research participants aren't readily available, researchers sometimes *pilot* their *pilot* studies on people, such as family members, who happen to be handy.) In this case, the first author piloted the pilot measure on some of his immediate family members. Incidentally, these data are only quasi-real because the first author had to be a proxy for a couple of people he could not reach (by answering for them). Finally, we should note that we did not create an SPSS data file that corresponds to these data. To keep you on your toes, we'd like you to create this tiny data file yourself. The response scale follows:

```
    1       2       3       4       5       6       7       8       9
Not at all like me          Somewhat like me                Exactly like me
```

Here are the responses:

	Jason	Stacy	Melanie	Rhonda	Barry	Bill	Dottie
Outgoing	7	9	4	6	3	6	2
Pleasant	8	9	9	9	8	9	9
Gregarious	5	4	5	8	5	4	5
Sociable	9	6	3	6	6	7	3
Talkative	6	8	3	4	2	8	2
Convivial	4	6	7	5	3	8	5
Uninhibited	7	8	4	5	7	8	1

QUESTION 5.9. Run a reliability analysis on these data, decide which items to retain, report your final reliability, and explain why any bad items were likely to have been bad. Briefly discuss any concerns you may have about the susceptibility of this short measure to a positive ("yes") response bias. How could one minimize this concern? (A clue: Did you have to recode any questions?)

Chapter 5 Reliability (and a Little Bit of Factor Analysis)

QUESTION 5.10. Imagine that you run a lab in which your grad students make observations of children at a summer camp for kids with ADHD (Absurdly Delirious Hebephrenic Disorder), otherwise known as extreme silliness. One of the important things your graduate students do is to rate each student daily for how he or she performed during a delay of gratification task known as EQM (Extreme Quiet Mouse). All of your students have been carefully trained, and so you know that they are up to speed on coding with the highly sophisticated QMCS-IVR (Quiet Mouse Coding System–4, Revised). Nevertheless, you have become concerned that, after winning $1 million on the reality TV show *American Idolatry,* Jessica, whom you previously considered a top student, has become disinterested in her job. Below are the EQM ratings that your six graduate students made for all 12 children over a 1-week period in this program. In this data analysis activity, you will use the tools in your reliability analysis toolbox to see if Jessica has really been falling down on the job. Incidentally, higher scores indicate greater ability to delay gratification.

EQM Raters	AJ	BP	CH	DB	ET	FF	GS	HB	IB	JH	KM	MC	Rater Mean
Jada	2.2	1.9	2.5	3.5	3.8	3.8	5.4	2.4	5.4	3.2	4.7	3.0	3.48
Janelle	1.2	2.7	3.0	3.6	5.6	5.8	5.4	4.7	6.0	0.8	3.3	1.5	3.63
Jessica	3.0	2.2	4.5	5.6	5.6	6.3	5.6	4.4	5.7	3.5	5.8	2.5	4.56
Jocelyn	2.4	1.8	3.5	3.1	4.1	4.0	5.2	3.2	5.3	2.2	4.5	3.0	3.52
Jolie	3.0	2.7	4.9	3.0	3.5	4.0	4.9	4.0	2.1	3.7	2.7	3.3	3.48
Josefina	2.6	1.5	4.3	2.8	3.8	4.8	3.8	4.4	4.5	2.8	2.7	1.8	3.32
Child mean	2.4	2.1	3.8	3.6	4.4	4.8	5.1	3.9	4.8	2.7	4.0	2.5	

Before you perform any reliability analyses in SPSS, take a careful look at these mean ratings for each rater and give some thought to whether it is worrisome that there is a range of almost 1.25 points in the mean ratings (with a high of 4.56 and a low of 3.32—for the same set of behaviors).

Now you are ready to perform a reliability analysis on these data. The data file you'll need is titled [**ADHD Camp Hypothetical QMCS-IV Data. sav**]. Make sure that when you get to the "Reliability Analysis" window, you click on "Statistics...." When the statistics dialog window opens, select the same four options you see selected in Image 15.

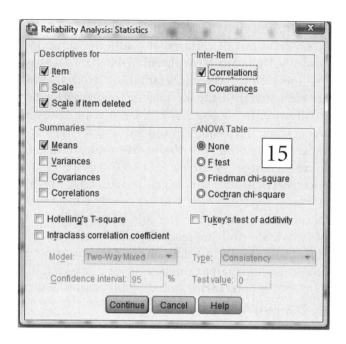

If you click "Continue" and then "Paste", and then run the analysis, you should have a clear answer to the question of whether Jessica was doing her job during the week of your reliability check. Comment briefly on three issues:

QUESTION 5.11a. Take a close look at all of your raters. Was anyone doing a poor job? For example, if you wished to maximize the reliability of your EQM scores, would you be better off deleting Jessica's ratings? If you're having trouble deciding if there were any bad raters, remember that in this design, individual *raters* are very much like individual *survey items* in a questionnaire.

QUESTION 5.11b. Even when a rater's scores are highly reliable, it is still possible that the rater may need some retraining. For example, even if one rater's judgments correlate very highly with those of other well-trained raters, there could be problems with *depression* or *elevation* in that rater's scores. These biases occur when, relative to other raters, a specific rater tends to give very low (depression) or very high (elevation) ratings *across the board* (see Cronbach, 1955). In cases in which researchers only care about relative scores, this is not usually much of a problem. However, in cases in which specific numerical cutoffs lead to some kind of categorization or clinical diagnosis, depression or elevation can be huge problems. Should you have any concern about depression or elevation in the ratings of any of these raters? (Incidentally, there are specific statistics you could calculate to say this definitively; we're just asking you to do this logically and/or intuitively.)

Chapter 5 Reliability (and a Little Bit of Factor Analysis)

QUESTION 5.11c. Based on everything you learned in this activity, which child should be awarded the highly coveted Quiet Mouse trophy during the week in which you conducted your reliability check? You may wish to use an SPSS "COMPUTE" command to help you decide. Briefly justify your answer.

Limitations of Cronbach's Alpha

In the many years since Lee Cronbach championed alpha (α) as the most sophisticated way to assess the internal consistency of a multiple-item measure, this useful and intuitive reliability statistic has become *incredibly* popular. Virtually any time you read about a multiple-item scale in a research report, especially a *new* scale, you can expect to see the researchers report the internal consistency of the scale by reporting Cronbach's alpha. It is much less common to see reports of the test-retest correlation of a measure. This is too bad, because as McCrae, Kurtz, Yamagata, and Terracciano (2011) have argued, the **test-retest** reliability of a scale may be a much more important predictor of the validity of the scale than is **internal** consistency. For example, in cultures all over the world, McCrae et al. focused on some common measures of personality and assessed judgmental congruence—the degree to which people's own reports of their personality traits agree with what other people say about them. The retest reliability of a personality measure was usually a very good predictor of the strength of judgmental congruence on that measure. The internal consistency of the same personality measures usually was not.

As McCrae et al. (2011) astutely note, one obvious reason why researchers may overrely on internal consistency (e.g., alpha) as an index of reliability is the simple ease of computing it. If you can get a group of third graders to fill out your six-item measure of how much they fear their teacher, you can compute alpha. If you can get a group of surfers, PSY 101 students, illegal immigrants, or overworked CEOs to fill out your five-item measure of how much they love playing the card game *Cliffhanger,* you can compute alpha. Getting most of these participants *back* to your lab in 5 years (or even 5 weeks) so that you can compute test-retest reliability is much more challenging. Along similar lines, it is sufficiently difficult to get measures of interrater reliability for many behaviors that researchers often calculate interrater agreement only for a subset of their participants and then assume that the estimate for the subsample gives a good indication for the total sample.

Furthermore, even if we accept the idea that it is very useful to know about the internal consistency of a measure, it is not at all clear that maximizing alpha is always ideal. In many cases, it might be better to assess a construct with as many unique questions as possible (to do justice to the breadth and richness of the construct) rather than merely trying to maximize alpha (see Vandello & Cohen, 1999, for an excellent example of how diversity or "bandwidth" in measurement can be more important than specificity). Conversely, inexperienced or unscrupulous researchers sometimes create scales that are extremely high in internal consistency by simply making use of items that are not-so-subtle paraphrases of one another. Consider the three-item scale "I like to eat pizza," "I enjoy eating pizza," and "Pizza is something I enjoy eating." A reliability analysis of this scale would likely yield an *extremely* high Cronbach's alpha! However, we hope it is clear that this scale is probably *not* a good measure of either addiction to junk food or sexual attraction to Italian grandmothers.

As it turns out, another serious limitation of Cronbach's alpha is its extreme sensitivity to the number of items in a scale. All else being equal, the more items you have in a scale, the more reliable you'd expect it to be. So far, so good. However, alpha turns out to be extremely sensitive to the number of items in a scale. This means, for example, that even if a scale has as few as 20 or 30 items, a high alpha may be no guarantee that all of the items in the scale are really assessing the same thing. With this in mind, the final assignment in this chapter is a cautionary tale about the importance of going beyond Cronbach's alpha to assess the psychometric nature of a multiple-item measure. You can find the hypothetical data for this activity in [**CRONBACHS ALPHA TWO FACTOR DEMO KEEPER.sav**].

This data file is based loosely on Christopher Peterson's research on well-being (e.g., see Peterson, Park, & Seligman, 2005). We should note, by the way, that Peterson has argued cogently that well-being is a multifaceted construct. However, we are going to assume that the researcher who collected the data you're about to analyze was not aware of Peterson's research. Instead, he simply conducted a focus group (not a bad idea, of course), took careful notes about the ways in which different people defined "life satisfaction," and then generated a list of 88 survey questions on life satisfaction. Luckily for you, he conducted two pilot studies in which he pared down the original 88 questions to the 27 questions you see spelled out in the "Labels" in this SPSS data file. By the way, you should assume that this considerate researcher already went to the trouble to reverse code all of the negatively worded life satisfaction items so that you don't have to do all this recoding yourself.

We'd also like to save you a little trouble by telling you when you conduct a basic reliability analysis on these data, you are going to find that a few of the items in this 27-item scale have pretty low corrected item-total correlations. Furthermore, we should add that if you were to run this analysis iteratively—deleting the three or four most offensive items that have the lowest item-total correlations and then rerunning the reliability analysis to see if any other items now had low item-total correlations—you could end up (after three or four cycles of this iterative process) getting rid of quite a few of these 27 items. However, we *don't* want you to do this (though you may wish to do it for your own edification) because it would defeat the purpose of this assignment. That purpose is to see whether a respectable alpha always guarantees that a scale really contains a set of items that all load on the same single underlying factor.

QUESTION 5.12. With this goal in mind, conduct a reliability analysis on the 27 life satisfaction items and report the overall Cronbach's alpha.

QUESTION 5.13. Now conduct a principal components factor analysis on the same 27 items to see if the items all load on a single "life satisfaction" factor. To make this task a bit less ambiguous than it otherwise might be, please restrict your interpretation to whether the results are more consistent with a one-factor or a two-factor solution. Be sure to report and then interpret the two highest eigenvalues. Furthermore, make sure to include a conceptual label for the factor or factors that you come up with based on your careful analysis of all the item loadings.

QUESTION 5.14. Regardless of how you label them, create two different scores in these data: (a) a 15-item score based on the items (ls1–ls15) and (b) a 12-item score based on the items (ls101–ls112). How well do these two scores map onto the factor(s) that you identified in your factor analysis? What is the simple correlation between the two composite scores? Given your original reliability score (for all 27 items), is this correlation consistent with what the researcher probably would have expected? What does this same correlation tell you about the limits of alpha? Finally, what does this correlation tell you about the nature of life satisfaction? Is life satisfaction one thing?

Appendix 5.1: Why Psychological Scales Are More Reliable Than the Average of Their Imperfectly Reliable Components

One of the truisms of scale development is that when you sum or average all of the items in a multiple-item scale (or average across two or more measurement periods when repeatedly using the same scale), you increase the reliability of the scale. In some ways, this principle is counterintuitive. Why, exactly, should a bunch of individual items that were not-so-perfectly measured produce a highly reliable scale when you sum them all together? That is, why should the reliability of a total scale score be higher than the *average* reliability of the individual items that go into the scale? As one past student put it, this means that the whole is somehow greater than the sum of its parts. There are a lot of ways to answer this question, but one of the simplest answers is that when you pool different items together into an overall score, you minimize the sources of noise or *error* that go into each individual's overall score. Different sources of noise that are associated with different individual items cancel each other out when you lump all the items together. Let's look at a couple of concrete examples to see how this process might work.

Imagine that you were developing a new pencil-and-paper scale to assess self-reports of depression. Ideally, you would want this scale to tap a single, coherent dimension. If you included multiple items in this scale, you would expect the measure to be more reliable than if you asked people only a single question. To keep it simple, let's imagine that you asked people only the five questions that appear below. To keep it even simpler, imagine that you asked people to simply endorse an item or not (rather than using a 5-point or 7-point scale).

In the past four weeks I have . . .

1. ____ lost weight
2. ____ had thoughts of harming myself
3. ____ had difficulty concentrating
4. ____ blamed myself for things that go wrong
5. ____ had difficulty sleeping

Here are the responses of two different hypothetical people to these five questions (0 means the person did *not* endorse the item, and 1 means the person did endorse it). To the right of these responses, you can see each person's total (composite) score for the five-item measure, followed by ratings an objective observer made about whether the person seemed depressed. The observer might be a psychologist who did a structured clinical interview for depression or simply a friend who was in a good position to know whether the person seemed depressed. In any event, you should assume that the observer made valid ratings. Sadsack is clinically depressed; Pollyanna is not.

	Weight	Self-Harm	Focus	Blame	Sleep	Total	Observer Rating
Sadsack	1	1	1	1	0	4.0	1
Pollyanna	1	0	0	0	1	2.0	0
$r_{observer}$	0.00	1.00	1.00	1.00	–1.00	1.00	NA

The row beneath the two hypothetical participants contains the correlation between each person's score on that item (or set of items) and the overall depression rating made by the objective observer. For instance, in the case of the weight question, both people reported that they had lost weight recently, and thus there was no correlation at all between whether people said they had lost weight and whether the observers said the two people seemed depressed (remember that the observers made only one rating of whether the person seemed depressed—not individual ratings of specific symptoms). In contrast, the items involving self-harm, focus, and blame all correlated perfectly with the observer rating! Next, the sleep item actually correlated negatively with the observer rating because Sadsack (the depressed guy) said he hadn't experienced trouble sleeping, and Pollyanna (despite being a veritable cornucopia of happiness) said that she *had* experienced trouble sleeping. Finally, the composite indicator of depression (the sum of the five items) correlated perfectly with the observer rating, and this perfect correlation is noticeably higher than the average correlation of .40 observed for the five individual depression items $(0 + 1 + 1 + 1 - 1)/5 = .40$.

Is there some trickery at work in this example? No, there is not. Although the data are bogus, and the example is highly simplified, the data show that composite scores can correlate better with a meaningful outcome than does the average of the individual items in the score. Of course, the fact that composite scores *can* correlate better than you would expect with a meaningful outcome doesn't guarantee that composite scores usually *do*, and it doesn't really explain *why* this might generally be the case. To get back to our main point, then, let's examine why Sadsack, Pollyanna, and a couple of other hypothetical participants might have responded as they did to the five individual items in the depression scale. In so doing, we will be able to see that each person's response to each item in the scale consists of some "true score" for that item and some idiosyncratic source of *error* for that specific item.

To begin with, if Pollyanna isn't depressed, why did she say she had lost weight? Perhaps it was because she had. For instance, maybe she was getting ready to go to her 10-year high school reunion, and she was *trying* to lose weight. Now for the items involving self-harm, focus, and self-blame, we can assume that Sadsack's and Pollyanna's responses did not contain any idiosyncratic sources of error—that their actual scores were, in fact, their true scores. However, it's important to keep in mind that if we examined the responses of a large number of participants, we would certainly come across an occasional nondepressed person who endorsed some of these three items and some depressed people who failed to endorse some of the items. For instance, your extremely depressed friend Jean Paul might fail to report that he has had difficulty keeping his focus lately (even though he has) because he is a philosophy major who prides himself in his amazing powers of concentration. This is an idiosyncratic source of noise that applied to people like Jean Paul for this item but did not apply to most other people for this item. As a final example, why did both Sadsack and Pollyanna give seemingly incorrect answers to the question about having difficulty sleeping? Maybe Pollyanna just bought a new mattress that proved to be too firm. Maybe Sadsack was so sleepy that he dozed off a little while reading this question—and checked the wrong response by mistake! We could list a lot of maybes like this. As long as these maybes are idiosyncratic to only a few people, and as long as *most* truly depressed people respond in a predictable way to each of the individual items in the scale, each individual item will have a good item-total correlation. And here is the main point—these idiosyncratic sources of noise will tend to *cancel each other out* when we average across the total set of items. For example, a hypothetical person who is experiencing a major depressive episode might fail to endorse at least one item he probably *should* have endorsed, but he might also endorse at least one other item he probably *shouldn't* have endorsed. These unique sources of noise will typically reduce the individual correlations of these specific items with the observer ratings quite a bit, but they will often cancel each other out when you lump them all together into the composite depression score.

As a more familiar example of this whole process, consider what happens when students who vary widely in their knowledge of course material take a multiple-choice exam. There are two and only two reasons why an individual student could answer a specific question correctly. One reason is that the student simply knew the answer to the question. That is, the student's correct response might reflect her **true score** (her true level of knowledge). However, the correct response might also be nothing more than a lucky guess (i.e., the correct response reflects noise or **error**). Along similar lines, an incorrect response to a question might reflect either a low true score or a bit of bad luck on a particular question (e.g., this is one of the few things the student didn't know very well or, worse yet, the student *did* know this material well but talked himself or herself out of the correct answer by overanalyzing this specific question). As was the case for the hypothetical measure of depression, the sources of error that play a role in the individual questions on the exam will frequently cancel each other out when you lump all the questions together. The lumped score will thus be more reliable than the individual scores that went into it.

In short, there is a simple, logical reason why adding items to a scale typically increases the reliability of the scale, especially in comparison with the reliability of any single item that happens to be part of the scale. To provide you with a final way to think about this, let's visualize it. The figure below represents factor loadings or item-total correlations for the five depression items that we examined previously. Each item is labeled, and you can see that each item consists of both a person's **true score** for depression and that item's own idiosyncratic source of noise or **error.** Researchers who create psychological measures are typically ignorant of these sources of noise, but they are labeled here to help you see that the composite depression score will consist mainly of (a) a bunch of true depression scores added together and (b) a bunch of more or less random sources of noise that are not likely to be correlated with one another and thus will not work together to systematically inflate or deflate a person's overall depression score.

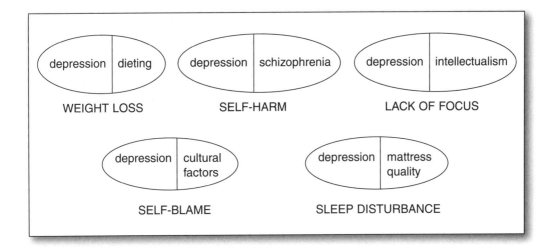

When you add all of these items together to form a composite, you can see that any one source of noise will be only a small portion of the composite score, whereas it may have been a substantial portion of the score for the individual item that it affected. Moreover, the net effect of all sources of noise will, on average, be very small. This means that for most people, most of the time, these sources of noise will tend to cancel one another out. Of course, this canceling-out process will often be imperfect, but all else being equal, the more items there are in a scale, the more likely it is that the various sources of noise that are unique to each item will cancel one another out. In the example you see here, imagine that a person's true depression score is 3.0. The fact that the person failed to report having difficulty sleeping (because of having just gotten a new mattress), for instance, is balanced out by the fact that the same person said she had lost weight recently (but said so because she was dieting rather than because she had lost her appetite). As it turns out, the whole scale usually *is* more reliable than the sum of its parts.

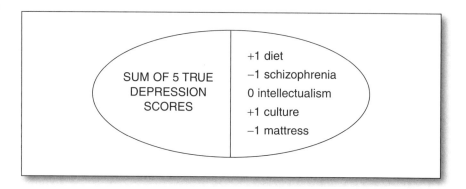

Appendix 5.2: Reporting the Results of a Factor Analysis and a Reliability Analysis

As we noted earlier in this chapter, one of the most popular indicators of the internal consistency of a multiple-item measure is Cronbach's alpha (α). Remember that if you transform your raw data properly, you can calculate an alpha for any set of multiple indicators of a construct (e.g., multiple items in a personality scale, ratings of children's politeness that come from different independent raters, or even multiple z-transformed social or economic indicators that come from different nations). The key idea is simply whether all of the individual indicators of interest are tapping into the same underlying construct. You could say the same thing about factor analysis. Although researchers *usually* use factor analysis to identify specific survey or personality items that tend to hang together, you could use factor analysis, for example, to decide which exact subset of 20 national-level economic indicators hang together best. The result of this unusual kind of factor analysis might be an index that proved to be a better way to compare nations on economic productivity, consumer confidence, or economic equity.

In this appendix, we provide you with the results from two high-profile papers in social psychology that both report the results of a reliability analysis. The first set of results is based on a traditional psychometric analysis that helped the author develop a new measure of individual differences. The second is a less traditional reliability analysis of how different social and cultural indicators can be combined across different U.S. *states* to create a state-level measure of how individualistic or collectivistic a specific U.S. state is. Let's begin with the more traditional (individual differences) example.

In the late 1990s, Elizabeth Pinel became interested in a classic question that is at the heart of a great deal of research on stereotyping and prejudice. Two different people often observe the same behavior and come to very different conclusions about it. For example, John might see Jennifer floundering at a stereotypically male task and conclude that Jennifer is simply incompetent. In contrast, Brenda might observe the same behavior and conclude that the social situation in which Jennifer finds herself (e.g., a critical supervisor who administered the task, coworkers who looked on disapprovingly) was the real cause of the problem. Pinel's insight was that it is important not only to know what John and Brenda think but to also know what *Jennifer* thinks. Pinel's basic idea was that there are large individual differences in "stigma consciousness," which is the chronic tendency to believe that one is likely to be stigmatized (viewed negatively on the basis of one's social category) by other people.

Before we share the results of Pinel's (1999) stigma consciousness scale development studies, let us inform you that Pinel presumably did several pilot studies that preceded her first full-blown study. The pilot studies presumably allowed her to begin with 16 candidate self-report items that she hoped would tap into a unitary measure of stigma consciousness. Furthermore, she was careful to balance these 16 questions evenly between those that reflected a high level of "stigma consciousness" and those that reflected a low level of stigma consciousness. Thus, 8 of her original 16 items were intended to be reverse coded. A couple of typical items in the *gender* version of the Stigma Consciousness Questionnaire (SCQ) are "Most men have a problem viewing women as equals," and "I almost never think about the fact that I am female when I interact with men." This second item, of course, was reverse coded prior to analysis, so that higher scores always mean greater stigma consciousness. Here are Pinel's primary results, which, luckily for you, include both a factor analysis and a simple reliability analysis.

> I began by factor analyzing the 16 SCQ for women items. Specifically, I conducted a principal-axis factor analysis with varimax rotation. Only one factor with an eigenvalue of greater than one (actual eigenvalue = 2.92) emerged, accounting for 83% of the common variance and 11% of the total variance. I retained 10 items that loaded .33 or higher on the single factor. These items and their factor loadings are presented in Table 1.
>
> Having honed the original questionnaire down to 10 items, I followed two data-analytic strategies. First, I conducted another principal-axis factor analysis, this time on the 10-item scale. Again, one factor emerged, this time accounting for 96.5% of the common variance and 24% of the total variance. The factor pattern revealed that all 10 items loaded .32 or higher on the single factor, with the average loading being .48. I supplemented this factor analysis by examining the internal consistency of the 10-item scale using the technique proposed by Cronbach (1951). The results indicated that no deletion of an item would increase alpha (which was .74). (p. 116)

It is worth noting that, after trimming the original measure down to only 10 items, Pinel wanted to be sure that she had not merely capitalized on chance in so doing. If you go to her original paper, you'll see that her unitary factor structure and her overall reliability results replicated extremely well in a completely different sample. For example, the reliability of the 10-item SCQ in a completely different sample ($n = 302$) was $\alpha = .72$.

To see a very different approach to scale creation, consider the research problem faced by Vandello and Cohen (1999). They wanted to develop a reliable measure that would let them identify an important dimension of culture in the 50 U.S. states. This dimension is *collectivism*, which has to do with the degree to which a culture promotes a focus on groups (aka "collectives" such as families, religious groups, and ethnic or national groups) rather than individual people. Although the United States is a highly individualistic country relative to the rest of the world, Vandello and Cohen argued that some U.S. states are probably much less individualistic (i.e., more collectivistic) than others. In an ideal world, Vandello and Cohen might have located 30 or 40 different state-level measures and then factor

Chapter 5 Reliability (and a Little Bit of Factor Analysis)

analyzed these state-level indicators (with state as the unit of analysis) to see which 5, 8, or 12 measures loaded onto a general collectivism construct.

The problem with this approach is that it's really difficult to get reliable state-level measures of most things, at least for those things one might expect to be markers of collectivism. So Vandello and Cohen (1999) identified some specific measures they expected would be relevant to collectivism and focused on these specific measures. They submitted the (standardized) measures to a traditional reliability analysis to see whether they should retain all the measures they chose and to see how reliable (i.e., internally consistent) their final set of measures might be. For example, one state-level indicator was the percentage of households in a specific state with grandchildren living in them (a measure of how many people in a given state live in an "extended family"). Another indicator was the percentage of people in a given state who reported having no religious affiliation. Because high degrees of religiosity are thought to be associated with a high level of collectivism, this item was reverse scored (fewer nonreligious people means more overall religiosity, without regard for which particular religion we're talking about).

Here are Vandello and Cohen's (1999) basic results for the internal consistency of their eight-item (state-level) index of individualism, preceded by a brief justification of how they chose the eight items.

> *Summary statistics for the index.* We chose items that were moderately positively correlated (i.e., not too low as to be unrelated but not too high as to be redundant). We also chose items such that the alpha coefficient of reliability would increase with the inclusion of the item in the index (see Baron & Straus, 1989). In addition, we sought indicators that would reflect a broad scope of social domains (family, job, politics, religion, etc.). Thus, although we wanted reasonably high reliability, we were willing to sacrifice some internal consistency to increase the breadth of the measure.
>
> As can be seen in Table 2, the overall standardized alpha for our eight-item index was .71. Also, the inclusion of each item increased the reliability of the measure (and the exclusion of any item resulted in a lower reliability). Table 3 shows the correlation matrix of the eight indicators. As desired, the indicators were, in general, moderately positively correlated, with all of the corrected item-total correlations being positive. Given the diversity of indicators, these correlations are actually quite respectable. The general pattern of low to moderate correlations suggests that although the individual items are not measuring the same thing, they are tapping into a similar construct. Overall, the eight-item index adequately met our statistical criteria of acceptability. (p. 283)

One nice thing about this particular example of a reliability analysis is that the researchers explicitly discussed some of the criteria they used for item selection in the first place. This kind of discussion and analysis is probably particularly important when researchers are relying on data collected by others—and thus do have the luxury of designing the specific items or individual indicators that go into a scale or index.

Notes

1. According to Wikipedia, there are no known episodes of *Dragnet* in which Friday actually uttered the phrase for which he is best known. The closest thing was apparently "All we want are the facts, ma'am." See http://en.wikipedia.org/wiki/Dragnet_(series).

2. See http://www.psych.cornell.edu/darlington/factor.htm.

3. There is some disagreement among experts about whether principal components analysis is truly a form of factor analysis. Dick Darlington, for example, seems to consider principal components analysis a specific form of factor analysis. In contrast, Field (2005) argues that principal components analysis is not technically factor analysis. However, Field also notes that "the two procedures often yield similar results."

4. In the interest of truth in advertising, there is a real David (who lives in Phoenix), and in addition to being extremely honest, he is a genius rather than a doofus. If you don't believe us, check out Boninger, Gleicher, and Strathman (1994).

5. We should note that with real data, it would be unwise to factor analyze any individual items that were so highly correlated with one another—especially if we hoped to *separate* any such highly correlated items into different factors. In almost any kind of multivariate statistical analysis (e.g., factor analysis, multiple regression analysis), it is difficult, if not impossible, to get a stable answer to your statistical question when you are trying to parse apart two or more very highly correlated variables. Statisticians refer to this worrisome situation as **multicollinearity** (usually defined as correlations between variables whose absolute values exceed $r = |.85|$ or so). With real rather than hypothetical data and large rather small sample sizes, however, it is highly unlikely that you will ever have to worry about such extreme multicollinearity. We address multicollinearity in later chapters. In this case, however, we gave ourselves permission to have factor analyzed these highly multicollinear responses because of the instructive value of this exercise—and because we did not wish to pull apart the perfectly correlated (aka **singular**) variables.

6. We say generally because researchers often set the bar for reliability (and validity) much higher if they are planning to use a score to help make an important real-world decision (as opposed to testing a theory). For example, if you were planning to make a hiring decision based on people's scores on a multiple-item personality measure, you'd probably want to see that your measure of this personality trait had a much higher reliability than $\alpha = .70$. Furthermore, even if you had good reason not to be too worried about internal consistency, you'd almost certainly be very worried about test-retest (temporal) reliability, because you'd be hoping to predict worker behavior over a long period.

Single-Sample and Two-Sample *t* Tests

6

Introduction

The simple *t* test has a long and venerable history in psychological research. Whenever researchers want to answer simple questions about one or two **means** for **normally distributed variables** (e.g., neuroticism, daily caloric intake, height, rainfall), a *t* test will often provide the answer to such questions. In the simplest possible case, a researcher might want to determine whether a *single* mean is different from some known standard (e.g., an established population mean). For example, a researcher interested in the academic qualifications of Dartmouth students might want to see if the SAT scores of Dartmouth students are higher than the national average. If the researcher had access to a small but representative sample of Dartmouth students, she might obtain their SAT scores (let's say the mean was 1190) and conduct a statistical test to see if this mean is significantly higher than the national average (it fluctuates around 1000; let's say it's 1023). As you may recall, the formula for the single-sample *t* test is

$$t = \frac{M - \mu}{S_M},$$

where *M* is the sample mean about which you would like to draw an inference (in this case, 1190), μ is the hypothetical population mean (1023) against which you would like to compare your observed sample mean, and S_M is the **standard error of the mean.** The standard error of the mean is closely related to the estimate of the population **standard deviation,** but it also takes the size of your sample into consideration. More specifically, it's

$$S_M = \frac{S_x}{\sqrt{N}}.$$

As your sample size gets *larger,* your estimate of the standard error of the mean gets *smaller,* and this is why single-sample t tests (like almost all other statistics) become more powerful (i.e., more likely to detect a true effect, if there is one) as sample size increases. As a concrete example, suppose that you only had access to a measly sample of four Dartmouth students, whose mean SAT score was 1190 (as noted previously). Furthermore, suppose you didn't know that the population standard deviation for the SAT is 100 points—and thus you had to estimate the standard error of the mean for the SAT by first estimating the population standard deviation for the SAT (based on your sample of four randomly sampled Dartmouth students). Let's say this estimate of the standard deviation turned out to be 114 points. According to the formula for a single-sample t test, the t value corresponding to your mean of 1190 would be

$$\frac{1190-1023}{57},$$

which turns out to be 167/57, which is 2.93.

The critical value of t with 3 df ($\alpha = .05$) is 3.18. Thus, based on this single-sample t test, we would *not* be able to conclude that the mean SAT score of Dartmouth students is significantly greater than the national average. However, if we obtained the same mean and the same estimate of the standard deviation in a random sample of nine rather than four Dartmouth students, our new estimate for the standard error of the mean would be 38 rather than 57 (you should be able to confirm why this is so with some very quick calculations). The t value corresponding to this new test would be 4.39, and the new critical value for t with 8 df is 2.31. Even if we set alpha at the more stringent .01 level, the new critical value would be 3.36, and we'd be able to conclude (quite correctly, of course) that Dartmouth students have above-average SAT scores. Finally, if we had a random sample of 100 Dartmouth students at our disposal, and we observed the same values for the mean and the standard deviation, the resulting value for a single-sample t test would be 14.65, a very large value indeed. We would have virtually no doubt that Dartmouth students have above-average SAT scores.

So the single-sample t test can answer some important questions when you have sampled a single mean. To be blunt, however, many questions of this type are pretty darn boring. Most interesting psychological questions usually have to do with at least *two* sample means. In an experiment, these two means might be an experimental group and a control group. In a survey study of gender differences, these two means might represent men's and women's problem-solving strategies or the strength of their preferences for the initials in their first names (by the way, women like their first initials even more than men do).

Back in the good old days, a *lot* of psychological studies could be analyzed using a two-sample t test. That is, many studies focused on (a) a single

Chapter 6 Single-Sample and Two-Sample *t* Tests

independent variable and (b) a normally distributed, or quasi-normally distributed, dependent variable. For example, a classic social psychological study by Hastorf and Cantril (1954) examined people's reactions to a bitterly contested football game between Princeton and Dartmouth (back in the days when these schools had highly respectable teams). The only independent variable in this classic study was whether the observers of the game were enrolled at Princeton or at Dartmouth. One of the main dependent variables was students' estimates of the total number of penalties Dartmouth committed during the game (the researchers also looked at estimates of the number of penalties Princeton committed, but we'll focus on Dartmouth because these results were the most dramatic). Because you live in the era of controversial murder trials, controversial NFL playoff games, and controversial presidential election counts, you probably won't be too surprised to learn that the Princeton students and Dartmouth students strongly disagreed about just how many penalties Dartmouth had committed—*even after reviewing a film of the game.* However, back in the 1950s, many people subscribed to the view that there is a single reality out there that most honest, self-respecting people can easily detect. How hard is it, really, to keep track of blatant infractions *on film* during a football game? Hastorf and Cantril documented that it could be quite hard indeed. Whereas the average Dartmouth student estimated that Dartmouth committed 4.3 infractions during a sample film clip, the average Princeton student put the total number of infractions at 9.8 after observing the same film clip. A two-sample *t* test confirmed that these different estimates were, in fact, different. This *t* test thus indicated that, even after viewing the same segments of game film, the student bodies of Princeton and Dartmouth strongly disagreed about the number of infractions committed by Dartmouth.

In this chapter, we are going to examine the *t* test and explore some of its most common uses. We begin by examining a simple data set that requires us to conduct a one-sample *t* test. Because the single-sample *t* test is conceptually similar to certain kinds of χ^2 tests, we will compare and contrast the single-sample *t* test with a χ^2 analysis based on the same simple study. Next, you will examine some simulated data from a classic field experiment using both an independent samples *t* test and a two-groups χ^2 test. After warming up on these two activities, you will use *t* tests to examine three other data sets, including a name-letter liking study, an archival study of heat and violence, and a blind cola taste test. Let's begin with the data involving inferences about a single score or mean.

Bending the Rules About Happiness

A cognitive psychologist was interested in the function rule that relates happiness to the objective favorability of life outcomes. Because of the preliminary nature of her research, she wanted to begin by testing her most basic assumption about the perception of happiness. In particular, the researcher assumed that as the *objective* favorability of a positive outcome increases in a linear fashion, people's *subjective*, psychological responses to the outcome will *not* increase in a linear fashion. Instead, she felt that subjective happiness will follow what might be termed a *law of diminishing returns*. As events become objectively more favorable in a linear fashion, people will only experience them as more favorable in a diminishing, curvilinear fashion. For example, based on past research, this researcher knew that if we wish to double the *perceived* intensity of a light presented in an otherwise dark room, we will have to increase its physical (i.e., objective) intensity by a factor of about 8. To be perceived as twice as bright as Light A, Light B must emit about eight times the amount of light energy as Light A (e.g., see Billock & Tsou, 2011; Stevens, 1961). Does this simple rule of diminishing returns apply to emotional judgments such as happiness? As a first test of this idea, the researcher asked people to think about how happy they would be if they were given 10 dollars. She then asked them to report exactly how much money it would take to make them exactly twice this happy (see Thaler, 1985). Her reasoning was simple. If the law of diminishing returns applies to happiness, it will take *more* than $20 to make the average person twice as happy as he or she would be if he or she were to receive $10. Image 1 depicts an SPSS data screen containing data from 20 hypothetical participants who took part in this simple study. *You should convert the screen capture in Image 1 to an SPSS data file of your own.*

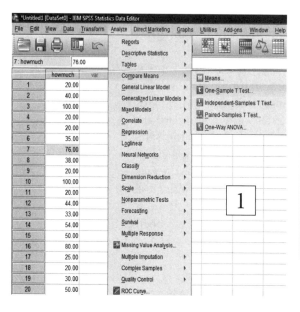

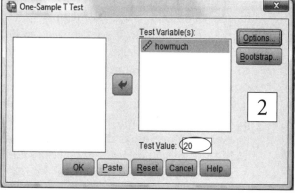

After making sure the scores in your data file look exactly like those in Image 1, click "A̲nalyze" and then go to "Compare M̲eans" and "One-Sample T-Test. . . ." Clicking on this analysis button will open a dialog box like the one you see in Image 2. Send your only variable ("howmuch") to

Chapter 6   Single-Sample and Two-Sample *t* Tests

the Test Variable(s) list, and enter 20 as your Test Value (circled in Image 2). Now click the "Paste" button to send this one-sample *t* test command to a newly created syntax file. Your syntax file (especially the text) should look very much like the one you see in Image 3.

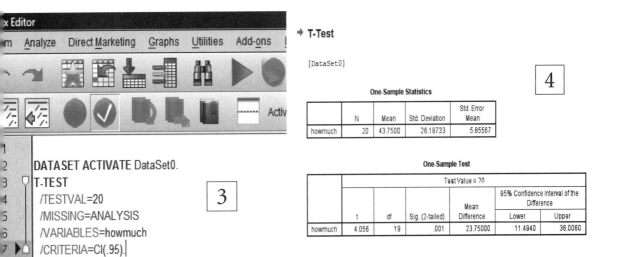

Running this one-sample *t* test command (by clicking the "Run" button) should produce an output file that looks a lot like the one you see in Image 4. We will leave it up to you to decide exactly what each piece of information in this output file tells you about the results of the researcher's study.

QUESTION 6.1a. Summarize the results of this study as if you had conducted the study yourself. Make sure to include everything you would want readers to know about your results. First, briefly explain why you set 20 (vs. 0, 100, etc.) as your comparison standard. Defend this choice against the (incorrect) criticism that one-sample means are supposed to be compared with some kind of *population value* (e.g., population SAT scores). Second, be sure to describe your findings relative to this comparison standard (e.g., "Dartmouth students exceeded the population mean on the SAT by more than 30%."). In other words, what, exactly, did you find? Third, was this finding statistically significant? Report your observed *t* value the way you would report it in a research paper. In case you are not familiar with the format for reporting the results of a one-sample *t* test, some examples appear in Appendix 6.1.

QUESTION 6.1b. Just to keep you on your toes, work from the information given in your printout and check to be sure that SPSS (a) correctly converted the standard deviation to the standard error of the mean and (b) correctly calculated the value of the *t* test itself. Make sure to show your work for these hand calculations. (Do not start from scratch. Instead, merely use the values for the sample mean and standard deviation that you are given in your SPSS output file.)

Simplifying the Outcome

Imagine that you showed your data to a collaborator who insisted that the most important question in your study is not the exact value of people's monetary self-reports but whether these reports *exceed* the hypothetical standard of $20. In short, the collaborator wants to know what happens if you simplify people's responses by coding them for whether they exceed the $20 standard. The syntax file depicted in Image 5 provides an example of how you might perform this recode. Be sure to highlight the entire set of new syntax commands (including "execute," as we did in Image 5) before you click the "Run" arrow. After you have run this "compute" command, your revised SPSS data file should look something like the one depicted in Image 6.

Now you want to analyze this new "exceeds" variable. Because this new variable is dichotomous (0 or 1), Karl Pearson would be quick to remind you that you can run a simple χ^2 test to see if more than half of your participants exceeded this critical value. After all, if people's scores just represent random noise around a true mean of 20, then half of the scores should exceed 20—just as about half of all coin tosses land tails up and about half of all z scores exceed the hypothetical mean of zero. In Images 7 through 10, you'll get a peek at how to do this. By the way, the way you select and modify nonparametric statistical tests changed pretty radically in Version 20 of SPSS, so if you have an older version of SPSS, you may wish to consult Appendix 6.2 for a couple of screenshots based on an earlier version of SPSS. However you do it, though, you'll want to conduct a simple chi-square test in which you compare your observed frequency for "exceeds" with the frequency you'd expect if "exceeds" had a probability of .50 in the population.

As you can see in Image 7, you begin a one-sample chi-square test by selecting "Analyze," "Nonparametric Tests," and then "One Sample. . . ." As

Chapter 6 Single-Sample and Two-Sample *t* Tests

shown in Image 8, after you click "One Sample..." SPSS will open a dialog box that has an Objective tab, a Fields tab, and a Settings tab (circled near the top of Image 8). You should also notice that the default objective (also circled in Image 8) is to "automatically compare observed data to hypothesized." This is great because this default is exactly what you want. From here, you *could* select run, but SPSS would assume that you want to analyze every single variable in your data file! This is rarely what people want to do. So let's *not* do that. Instead, let's focus on the newly created "exceeds" variable.

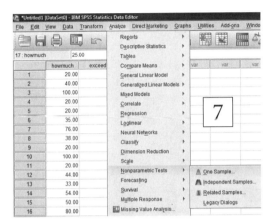
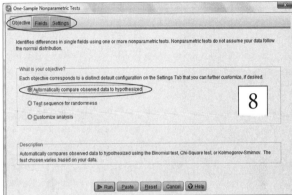

To do this, click the Fields button (the second of the three circled buttons at the top of Image 8), and this will take you to the screen you see in Images 9 and 10. Notice in Image 9 that the default is to include both of your variables in the "Test Fields." Of course, you don't want to analyze the original "howmuch" variable, and if you click on this variable as we did in Image 9, you can use the arrow button (circled for you in Image 9) to send it back from "Test Fields:" to "Fields:" (see Image 10)—where SPSS will now politely ignore it.

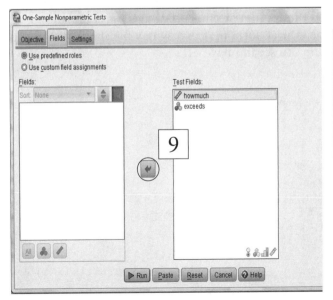
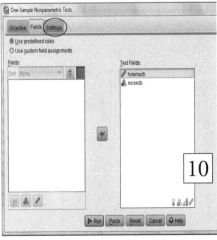

Unfortunately, you still are not quite ready to run your analysis. You'll now want to click the "Settings" button (circled in Image 10) to open the "One-Sample Nonparametric Tests" dialog window, which you can see in Image 11.

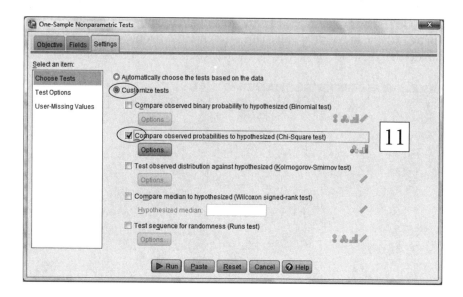

Instead of accepting the really interesting default, which is to let SPSS choose what test to run based on the properties of the "exceeds" variable that you chose in the last ("Fields") dialog box, click "Customize tests" (circled near the top of Image 11). This will allow you to decide, for example, whether to run a binomial test or a chi-square test. To keep things simple, let's only ask for the chi-square test, which you can see we requested by checking the appropriate box in Image 11. If you now click "Paste," you should send the appropriate chi-square analysis command to your syntax file, and the crucial portion of that syntax file should look a lot like the screen capture you see in Image 12.

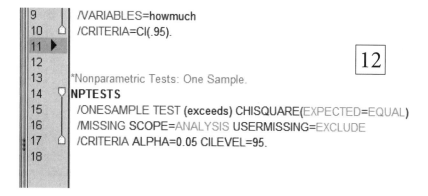

Chapter 6 Single-Sample and Two-Sample *t* Tests

Running the appropriate portion of this syntax command should produce the portion of the SPSS output file you see in Image 13.

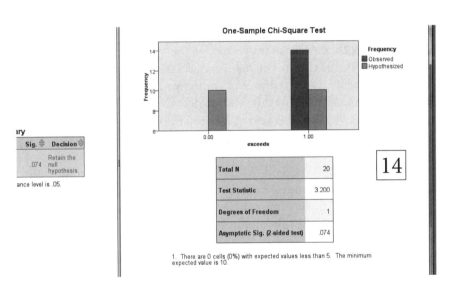

From here you can probably already see that the results of this test are *not* statistically significant ($p = .074$). However, it would also be nice to know the exact value of the chi-square statistic (rather than merely knowing its *p* value). To see the actual chi-square value associated with this *p* value, just double click anywhere inside the "Hypothesis Test Summary" box you see in Image 13. This will open up the more detailed part of your output file you see selectively captured in Image 14.

You now have all the information you need to make a complete report of your findings using this alternate measure.

QUESTION 6.2a. Compare the results of your new χ^2 analysis with your original findings based on the one-sample *t* test. We'll leave it up to you to figure out the output file and to draw your own conclusions about the pros and cons of this alternate analysis.

But we can't resists making one helpful suggestion. To figure out if any differences between your new (χ^2) results and your old (*t* test) results are due to the nature of the *statistical tests* themselves (versus being due to the way you *recoded the data*), we suggest that you copy the entire *t* test syntax command that you initially conducted and paste it below the χ^2 command. Now edit the copied *t* test command by hand in two ways. First, replace "howmuch" with "exceeds." Then replace "20" in the TESTVAL = line with "0.50." This will mean that the new *t* test analyzes the "exceeds" variable rather than "howmuch" and compares it with the standard of 0.50 rather than the old standard of 20. Here is how that new command should read

DATASET ACTIVATE DataSet0.

T-TEST

/TESTVAL=0.50

/MISSING=ANALYSIS

/VARIABLES=exceeds

/CRITERIA=CI(.95).

Running the edited *t* test command will generate a new section in your SPSS output file.

QUESTION 6.2b. Compare the results of the χ^2 test and the *t* test when you perform the two tests on *the same scores*. Based on this additional comparison, what is the best way of coding your data in this study (continuously or dichotomously)? Be sure to defend your answer.

Incidentally, you may be having a kneejerk reaction that says it's a big mistake to analyze dichotomous data using a *t* test. Like most other jerks, this particular jerk is wrong. Remember that Question 6.2b is more of a diagnostic exercise in seeing how statistics work than it is a recommended way of analyzing dichotomous data. If that isn't good enough for you, we'd like to reassure you that many parametric tests (e.g., *t*, *z*, *F*) are more robust than most people think to the assumption that dependent variables are normally distributed, especially when the nonnormal distribution is platykurtic (i.e., flat). At any rate, it certainly is *not* the case, for example, that merely switching from a parametric statistic to a nonparametric statistic when your data are not normally distributed will solve all of your distributional dilemmas (e.g., see Harwell, Rubinstein, Hayes, & Corley, 1992; Zimmerman, 1987; Zumbo & Zimmerman, 1993). Nonetheless, if you wish to avoid painful and unnecessary debates with reviewers and editors when trying to publish your own research with nominal variables, we recommend—for purely pragmatic reasons—that you lean toward the default of using nonparametric tests such as χ^2.

Before you move on to the next data set, make sure that you save both your data file and your syntax file using the same descriptive name stems. We recommend "perceived happiness study (one-sample t-test)." If you haven't done so already, you should also give your variables very clear labels in your data file before you save everything one more time.

The Independent Samples (Two-Samples) *t* Test

One of the most widely cited studies in social and educational psychology is Rosenthal and Jacobson's (1966) classic study of how teachers' expectations can influence children's academic performance. Like a lot of other classic studies, this study involved a very simple manipulation (with only two levels), and thus, it qualifies as a good illustration of the two-sample *t* test. Rosenthal and Jacobson went into an elementary school and received permission to give all of the children in the school a new intelligence test. The test was a recently developed test of nonverbal intelligence that none of the students or teachers in the school would have been likely to have ever seen before. Thus, Rosenthal and Jacobson were able to create teacher expectancies regarding the test without worrying much about any naturally existing expectancies the teachers may have already had. In particular, they informed the teachers (erroneously, of course) that the new test was a "test for intellectual blooming." They further explained to the teachers that children who perform well on this test typically experience what might be thought of as an academic growth spurt. That is, the researchers led teachers to believe that students who did well on this test possessed a special kind of intelligence—one that might normally go undetected in the classroom. And then they informed teachers of exactly which students had achieved exceptional scores on the test. Of course, these students were *not* identified on the basis of their actual performance on the test. Instead, Rosenthal and Jacobson *randomly* chose about a dozen students from each of six large classrooms (composed of about 50 students each). In other words, they conducted a **field experiment** on teacher expectancies.

Eight months later, the researchers returned to administer the same test again in each of the classrooms. Back in those days, people typically analyzed change by simply examining difference scores. In this case, the researchers simply developed a single score for IQ change by subtracting children's (true) initial IQ scores from their IQ scores on the same test administered 8 months later. Not surprisingly, most children showed increases in performance on the nonverbal IQ test (after all, they had completed a year of formal schooling). The question, of course, was whether the children who were randomly dubbed intellectual bloomers would show greater increases in IQ than the remaining children (who had not been labeled as bloomers). Your data for this part of the assignment are a *very close approximation* of the Rosenthal and Jacobson (1966) data for first-grade children only (if you want to see their findings for children in other grades, see the original article). The SPSS data file is called [**simulated teacher expectancy data (2-sample t-test).sav**]. A small portion of the data file appears in Image 15.

Before you go any further, you should check out the variable labels for these data. Doing this will show you, for example, that for the independent variable "bloomer," scores of 0 refer to kids in the control group and scores of 1 refer to kids randomly labeled "bloomers." You can also see a reminder that "iqchange" is the difference between children's nonverbal IQ scores at Time 2 and these same IQ scores at Time 1 (it's Time 2 minus Time 1 IQ, so more positive scores indicate greater increases in IQ).

Because you are probably becoming quite adept now at selecting statistical analyses, we offer you only a couple of clues in Images 16 through 18 about how to select and run an independent samples (two-samples) *t* test. However, one novel aspect of the two-sample *t* test window is that SPSS requires you to tell it exactly how you defined (coded) the two levels of your independent variable. In this case, you probably recall that the codes for the control and "bloomer" groups, respectively, were 0 and 1. So after defining bloomer as the "Grouping Variable" and "iqchange" as the "Test Variable," as we did in Image 17, click on "Define Groups..." from the box you see in Image 17, and then enter these values of 0 and 1 in the "Define Groups" box you see in Image 18.

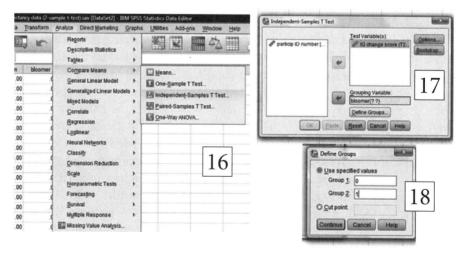

The "Continue" button will now take you back to the main analysis box (Image 17), and if you click "Paste" from this box, you should generate the specific syntax command that appears below:

DATASET ACTIVATE DataSet2.

T-TEST GROUPS = bloomer(0 1)

/MISSING = ANALYSIS

/VARIABLES = iqchange

/CRITERIA = CI(.95).

Running this command in the usual way will produce the results of this independent samples t test.

Results of the Teacher Expectancy Study

The part of your output file that matters most should strongly resemble what you see in Image 19.

This output file will tell you everything you need to know about one set of findings from this classic study.

Group Statistics

	was child labeled bloomer?	N	Mean	Std. Deviation	Std. Error Mean
IQ change score (T2 - T1)	no	19	12.0000	16.39444	3.76114
	yes	11	27.3636	12.57197	3.79059

Independent Samples Test

		Levene's Test for Equality of Variances		t-test for Equality of Means					95% Confidence Difference
		F	Sig.	t	df	Sig. (2-tailed)	Mean Difference	Std. Error Difference	Lower
IQ change score (T2 - T1)	Equal variances assumed	.744	.396	-2.678	28	.012	-15.36364	5.73623	-27.11378
	Equal variances not assumed			-2.877	25.599	.008	-15.36364	5.33992	-26.34839

19

QUESTION 6.3. Summarize the results of this study. Include a table of your findings and summarize your results in the same way you might in a very short journal article or research talk. Make sure that you include a brief discussion of the meaning and significance (social, educational, etc.) of these findings.

More Simplification

The same colleague with whom you collaborated on the happiness study argues that, to simplify this study, you should simply code for whether the students showed an increase (of any size) in IQ over the course of the academic year. Based on what you recently learned about the syntax commands for computing and recoding data (in the happiness activity), write some coding statements that will create a new variable called "gain." Code gain as 1 if students showed any increase at all in IQ, no matter how small. Code it as 0 if a student failed to show such an increase (including a loss or no change at all). Now using the skills you have developed in this text, conduct both a χ^2 test and a new independent samples t test (on the same simplified "gain" variable) to see if your results are any different when you use this much simpler way of coding and analyzing the data. You might be tempted to conduct a χ^2 test using the same set of commands that you just used for the one-sample t test. You cannot easily do this, however, because this specific test is designed for the χ^2 equivalent of one-sample t tests. The χ^2 test you want to run now is the kind of χ^2 contingency test you conducted a couple of times in Chapter 4 (the tests having to do with name-letter preferences in relation to choices about colleges and marriage). As you may recall, the easiest way to run this χ^2 test is to start at "Analyze," then to go to "Descriptive Frequencies" and then "Crosstabs." You can always return to Chapter 4 if you want more detailed instructions.

QUESTION 6.4. To get back to the point of this exercise, summarize the results of both the χ^2 test and the t test for the "gain" variable. Be sure to include a brief statement about which coding scheme (original or simplified) you think is better for this study.

If you haven't already done so, remember to save your syntax file, ideally giving it the same prefix that your data file has (your data file should be OK the way it was when you opened it; so there is no need to save it). There is really no need to save your output file, by the way, unless your instructor asks you to submit it as proof that you ran the analyses that will support your answers.

Yet Another Name-Letter Preference Study

The famous Chilean social psychologist Mauricio Carvallo (he says he's *particularly* famous in Santiago but refuses to say why) once convinced the first author of this text to collaborate on a name-letter study (i.e., a study of implicit egotism) in which we sampled every person named Cal or Tex who lived in either California or Texas. (His most persuasive argument for doing the study, by the way, is that he would do all the hard work.) Mauricio thus sampled about 900 men from an Internet email directory, asking them to report how much they liked each of the 26 letters of the alphabet on a 9-point scale (where higher numbers indicate higher levels of letter liking). In addition to replicating the finding that people were disproportionately likely to live in states that resembled their names, we wanted to see whether men who lived in states that resembled their names would like their initials more than men who lived in states that did *not* resemble their names. The idea would be to gather some direct evidence that strength of name-letter preferences plays a role in the strength of behavioral preferences based on name letters.

Here are some hypothetical data that very strongly resemble our actual data. The SPSS file containing these data is called [**Cal Tex name letter liking study.sav**].

Analyze these data to see if we were correct that relative to people living in states that do not resemble their first names, people who live in states that do resemble their first names have stronger preferences for their first initials. Here's the tricky part: Find a way to do this that allows you to analyze a *single score* for each participant, and code the score so that higher numbers reflect higher levels of name-letter liking. Although there are more complex ways to analyze the data, you can and should be able to conduct a simple, two-groups *t* test when you complete all the recoding for this activity. Table 6.1 gives you some ideas about how to think about and analyze these data. As shown in Table 6.1, each participant (all of them named Cal or Tex) has two important scores: liking for the letter *C* followed by liking for the letter *T*.

Table 6.1 Liking for the Letters *C* and *T* Among People Named Cal Versus Tex Who Live in California Versus Texas

	FIRST NAME	
STATE	Cal	Tex
California	8 4, 9 5, 9 8, 3 4, 9 9, 3 4, 8 5, 7 8, 7 4, 9 8, 4 4, 6 5, 8 5, 6 5	5 6, 7 9, 9 9, 7 4, 8 9, 9 9
Texas	7 8, 8 8, 3 4, 9 7	5 7, 7 7, 9 9, 3 7, 5 8, 8 7, 3 4, 9 9, 5 8, 8 9, 3 4, 5 6, 7 7, 5 5, 6 8, 6 7

Here are some further clues: First, you'll need to create a new variable that reflects whether men do or do not live in a state that resembles their own first name. For this purpose, a Cal living in California is no different from a Tex living in Texas. But they are both different from Cals living in Texas and Texes living in California (who belong in the same group). Second, you will also need to create a new variable that reflects the strength of people's liking for their first initials. To do so, you need to take a difference score (e.g., liking for *C* minus liking for *T*), but this difference score will be calculated differently for people with different names. Because we have not asked you to do a lot of independent syntax programming in this text, we include a brief primer on SPSS COMPUTE statements and IF statements in Appendix 6.2. This will probably come in very handy for this data coding and data analysis activity.

QUESTION 6.5. Report what you did and what you found. Be sure to include (a) exactly how you coded your data to allow for a test of our hypothesis. This should include a copy of the part of your syntax file that makes the crucial calculations (i.e., that creates the two new variables). In addition, (b) discuss the *results* of your *t* test, including a discussion of whether these findings support the basic assumptions behind the idea of implicit egotism. Finally, (c) be sure to discuss any critiques or limitations of your findings. If it seems relevant, discuss the issue of one-tailed versus two-tailed hypothesis testing.

An Archival Study of Heat and Aggression

A researcher interested in aggression identified the 10 hottest and coldest U.S. cities and assessed murder rates in each city. His hypothesis was that because heat facilitates aggression, the hottest cities would have higher murder rates than the coldest cities. Each city name below is followed by the number of murders per 100,000 people committed in that city in 2006. Conduct a *t* test with city as the unit of analysis ($n = 20$) to see if there is support for the researcher's hypothesis. (You'll need to create your own SPSS data file for these real data.)

Hot Cities	Murder Rate	Cold Cities	Murder Rate
1. Key West, Florida	4.1	1. International Falls, Minnesota	0.0
2. Miami, Florida	19.6	2. Duluth, Minnesota	1.2
3. W. Palm Beach, Florida	17.1	3. Caribou, Maine	0.0
4. Ft. Myers, Florida	25.2	4. Marquette, Michigan	0.0
5. Yuma, Arizona	3.4	5. Sault Ste Marie, Michigan	0.0
6. Brownsville, Texas	2.9	6. Fargo, North Dakota	2.2
7. Orlando, Florida	22.6	7. Williston, North Dakota	0.0
8. Vero Beach, Florida	0.0	8. Alamosa, Colorado	0.0
9. Corpus Christi, Texas	7.2	9. Bismarck, North Dakota	1.7
10. Tampa, Florida	7.5	10. St. Cloud, Minnesota	0.0

Note: Murder rate data were taken from http://fargond.areaconnect.com/crime/compare.htm. The hottest and coldest U.S. city list can be found at http://web2.airmail.net/danb1/usrecords.htm.

Chapter 6 Single-Sample and Two-Sample *t* Tests

QUESTION 6.6. Summarize the results of this study. Did it support the researcher's hypothesis? Be sure to point out any methodological concerns you have about the study. A good answer will use the word *confound*. A good source that may address some of the confounds that are relevant to this research is Anderson (2001).

A Blind Cola Taste Test

During about 20 years of teaching research methods, the first author of this text has frequently conducted blind taste tests in which he served people different popular colas and allowed the taste testers to see whether they can tell the different colas apart. People in such taste tests are always told, truthfully, that they will get two or three different colas (e.g., Coke, Pepsi, and a generic cola) without being told which cola is which. I always use sequentially numbered cups, and I serve the colas to participants in different random (counterbalanced) orders. I've now done this blind taste test so many times that I don't usually bother to save the data anymore, and so I am going to give you some hypothetical data that reflect the typical findings when I serve three colas (more often I serve only two). Fortunately for you, I entered these data into an Excel file rather than an SPSS data file, and so you get to learn a new skill (reading an Excel file into SPSS). A screen capture from the Excel file appears in Image 20.

idnum	gender	predcoke	predpepsi	predgeneric	gotcoke	gotpepsi	gotgeneric
1	1	0.75	0.75	0.75	0	1	0
2	2	0.75	0.75	0.75	1	1	1
3	1	0.8	0.9	0.6	1	1	1
4	1	0.55	0.33	0.22	1	0	0
5	2	0.35	0.33	0.22	1	0	0
6	1	0.33	0.48	0.4	1	0	0
7	1	0.99	0.99	1	1	0	0
8	1	1	1	0.8	0	0	0
9	2	0.6	0.7	0.6	1	1	1
10	2	0.58	0.58	0.44	1	1	1
11	2	0.66	0.88	0.5	0	0	1
12	1	0.33	0.33	0.33	0	0	0
13	1	0.33	0.33	0.34	0	0	1
14	2	0.75	0.9	0.8	1	1	1
15	2	0.33	0.33	0.33	0	0	0
16	1	0.68	0.5	0.5	0	1	0
17	2	0.33	0.33	0.34	0	0	0

Fortunately, SPSS reads Excel data files almost as fluently as it reads SPSS data files. Thus, to convert an Excel data file to SPSS, you do much the same thing you would do to open an SPSS data file. The screen captures for this simple series of steps follow:

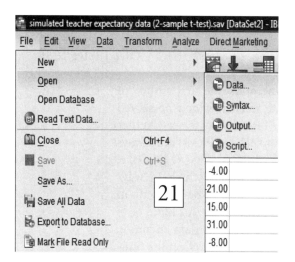
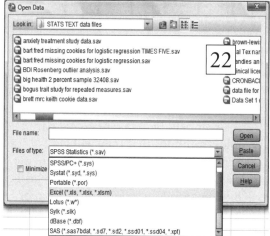

Once you've clicked on the "D<u>a</u>ta. . ." button shown in Image 21, you'll see a dialog window that resembles the one in Image 22. From this window, you'll need to let SPSS know, using the "Files of type:" box, that you want to open an Excel file rather than the default SPSS data (.sav) file. Once you've done this, just click "<u>O</u>pen" and then accept the default (shown in Image 23), which is to allow SPSS to read the variable names from the first row of data in an Excel file. Of course, if your Excel file had no such variable names in the first row, you'd have another job to do, but luckily for you, the Excel file you are reading was created by a caring nurturer who wanted your data analysis life to be as wonderful as possible.

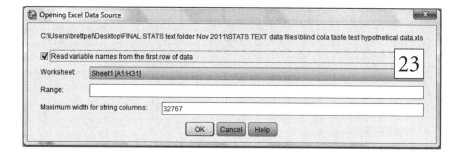

The Excel file for the blind cola taste test data is called [**blind cola taste test hypothetical data.xls**]. So you merely need to go to the folder where this Excel file is located and then read the data directly into SPSS. Incidentally, because this Excel file came from a caring nurturer who is *not very organized*, there are no variable labels in this file. Thus, we should tell you that gender is coded 1 = M, 2 = F. Also, the three variables that begin "pred" are people's *predicted* accuracy scores (assessed before people got feedback) for Coke, Pepsi, and a generic cola. Thus, the answer .75 for "predcoke" for the first participant means that he said the chances that he had correctly labeled what the experimenter knew to be Coke was 75%. Finally, the last three variables (e.g., "gotcoke") reflect whether a person actually did label a given brand of cola correctly. Thus, the first participant labeled Pepsi correctly but mixed up the Coke and the generic cola (thus mislabeling them both). Now that you understand what all the scores indicate, you should enter all of your variable labels and save the SPSS data file you created from the original Excel file. Then you'll be ready to analyze the data so that you can answer the questions that are soon to follow. Before you do so, however, you will need to use your developing variable creation (COMPUTE) skills to create two new variables. One is the average *predicted accuracy* score for each participant. The other is the average accuracy score for each participant. If you do these calculations correctly, the first participant, who we are pretty sure was the well-known psychologist Keith Maddox, will have an average predicted accuracy score of .75 (the average of .75, .75, and .75) but will have a true accuracy score of only .33 (the average of 0, 1, and 0).

QUESTION 6.7a. Write up these results for labeling accuracy as if your data were real. Specifically, can people identify the three colas at an above chance level in a blind taste test? A tricky issue is how to define the chance standard used for purposes of statistical comparison in your one-sample *t* test. To decide on this, assume that everyone always got all three brands of cola (they always do in our three-cola tests) and that the experimenter presented each of the six possible orders of the three colas (counterbalanced across all of the participants). Now assume that a typical participant was *arbitrarily guessing* the identities of the colas and label her guesses 1, 2, and 3 (or C, G, and P, for the three brands). If you compare this random answer pattern with the list of every possible order of delivery, you should be able to see how many correct answers the average guesser would get across all of the trials. In case you are not sure what we mean by this, the six possible orders of cola delivery are indicated for you in Table 6.2. All you need to do to figure out the average accuracy score produced by chance guessing is to always assume that any one constant set of guesses (e.g., P, G, C) is used in all six scenarios (for all six orders). Just make sure to use the *same* constant guess pattern for all six orders. Incidentally, to prove that the arbitrary guess order you choose doesn't matter, try the exercise with more than one order (e.g., first with C, G, P and then with P, G, C). Once you have converted this average to a proportion, you will know what value to enter as your comparison standard in the one-sample *t* test.

Table 6.2 Six Possible Orders of Cola Delivery in a Fully Counterbalanced Blind Cola Taste Test

Actual Cola Delivery Order	Number of Correct Answers
C, G, P	_____
C, P, G	_____
G, C, P	_____
G, P, C	_____
P, C, G	_____
P, G, C	_____
Average:	_____

QUESTION 6.7b. Do people have an accurate appreciation of their ability to do this task? That is, is the degree of *confidence* the average person reports for the cola-labeling task appropriate for the average level of *accuracy* people show during the blind taste test? A clue: Check out research on "overconfidence" by Fischhoff, Slovic, and Lichtenstein (1977). Another clue: To really answer this question thoroughly, you should compare the average confidence score not only with chance performance (in one analysis) but also with actual accuracy (in a different analysis). Your standards of comparison then (against which you should compare confidence rates) in these one-sample *t* tests are (a) chance performance in one case and (b) observed accuracy in the other case (whatever it proves to be).

QUESTION 6.7c. Are men and women equally accurate? Are they equally confident? Answering this will require two separate *t* tests.

Appendix 6.1: Reporting the Results of One-Sample and Two-Sample *t* Tests

This appendix contains the results of some studies, both hypothetical and real, that make use of *t* tests.

> The results of this marketing study showed that consumer liking for the card game *PRIME* ($M = 6.3$ on the 7-point Coopersmith liking scale) significantly exceeded the established population estimate for *UNO* ($M = 5.1$), $t(438) = 138.43, p < .001$.

The value in parentheses (438) right after the *t* is the degrees of freedom (*df*) for the *t* test. For a one-sample *t* test, this value is always $n - 1$. So the above analysis tells you that these marketing researchers sampled 439 consumers. It is traditional to report actual *t* values to two decimals and to report *p* values to three decimals. So even though a *t* value this large would certainly have a *p* value with many leading zeros after the decimal point, it would be sufficient just to say $p < .001$.

Here are two examples of how you might report a *nonsignificant* result:

> The results of a one-sample *t* test did *not* support our prediction that sexual attraction to the middle-aged experimental social psychologists would exceed the negative scale endpoint of 1.0. Recall that on this 9-point scale, a score of 1 corresponds to *I'm about as attracted to the target as I am to a pile of smelly gym socks*. The sample mean was only $M = 1.2$, which did not significantly exceed 1.0, $t(19) = 1.29, p = .214$. In fact, only 2 of the 20 supermodels studied failed to choose 1 as their answer, and all 20 of the attraction scores were below the scale midpoint of 5 (*I'm about as attracted to the target as I am to my kind but unattractive cousin*).

> There were only six U.S. presidents for whom we were able to obtain WAIS-R intelligence test scores. The average IQ score for these six presidents ($M = 116.3$) was descriptively higher than the U.S. population mean of 100. Perhaps due to the very small sample size, however, this value did not reach conventional levels of significance, $t(5) = 2.16, p = .083$.

Here is an example of how to report a result for a **two-sample (i.e., independent samples) *t* test**. Unlike the hypothetical examples above, this real example comes from an early draft of a research paper by Shimizu, Pelham, and Sperry (2012). The manipulation in the study was whether participants who tried to take a nap were subliminally primed with sleep-related or neutral words immediately prior to the nap session. That is, prior to trying to nap, some participants were *unknowingly* exposed (for just a few milliseconds per word) to words such as *relaxed, restful,* and *slumber.* In contrast, others were subliminally exposed to neutral words such as *bird, chair,* and *paper.* (The means, incidentally, represent minutes of self-reported sleep.)

> For self-reported sleep duration, the effect of prime condition was significant, $t(71) = 2.17$, $p = .03$. Participants primed with sleep-related words reported having slept about 47% longer ($M = 9.11$, $SD = 6.38$) than did those primed with neutral words ($M = 6.18$, $SD = 5.14$).

Because this was a simple two-groups experiment, analyzed by means of a two-sample *t* test, the *df* value of 71 you see in parentheses above tells you that there were 73 participants. In the case of a two-samples *t* test, $df = n - 2$ rather than $n - 1$.

Appendix 6.2: Some Useful SPSS Syntax Statements and Logical Operands

There are many times in data analysis when researchers need to modify a variable or variables from an SPSS data file. For example, in Chapter 5, you computed a single self-esteem score from the 10 separate self-esteem questions that make up the Rosenberg Self-Esteem Scale. At other times, a researcher might want to break a continuous variable (such as age) into meaningful categories rather than dealing with 80 or 90 specific scores. The two most useful SPSS syntax commands for this purpose are the COMPUTE statement and the IF (...) statement, each of which we briefly review here, along with some very useful logical or mathematical operators.

COMPUTE: The COMPUTE command creates a new variable in the active SPSS data file according to whatever logical or arithmetic rules follow the command. The new variable name always follows the COMPUTE command.

COMPUTE quiz_1 = q1+q2+q3+q4+q5.

EXE.

For every case with complete data on q1 to q5 (items on a quiz), this syntax command will produce a new variable called quiz_1 and add it to the SPSS data file.

COMPUTE quizp_1 = (q1+q2+q3+q4+q5)/5.

EXE.

This researcher decided to divide by 5 to calculate the **proportion** of correct quiz responses rather than the number of correct responses. Thus, you can see that mathematical commands can be used with COMPUTE.

IF (...): The IF (...) command creates a new variable based on whether some already existing variable (or variables) satisfies some condition or conditions. For example, you can use the IF statement to create age groups from a continuous "AGE" score. The syntax statement series that follows would create a new variable consisting of four age groups. If you highlighted and ran this entire statement set, the new variable would become the last variable in your SPSS data file.

IF (AGE LE 34) age_group = 1.

IF (AGE GE 35 and AGE LE 49) age_group = 2.

IF (AGE GE 50 and AGE LE 64) age_group = 3.

IF (AGE GE 65) age_group = 4.

EXE.

Notice that this IF set uses "LE" and "GE" ("le" and "ge" would also be acceptable) to indicate "is less than or equal to" and "is greater than or equal to," respectively. SPSS also recognizes other logical operands. Here is a list of some very useful ones along with their meanings:

"**LT**"   "is less than"

"**GT**"   "is greater than"

"**EQ**"   "is equal to"

"**NE**"   "is not equal to"

"**AND**" "and"

"**OR**"   "or"

Thus, the following syntax command

IF (GENDER EQ 2 AND NUM_KIDS GE 1) MOTHER = 1.

IF (GENDER EQ 2 AND NUM_KIDS EQ 0) MOTHER = 0.

EXE.

would only create "MOTHER" scores for women (assuming 2 means women for the gender variable). Furthermore, women with kids (regardless of how many) would be coded as mothers (1) and those without kids would be coded as nonmothers (0). Men would not receive a score at all for this new variable.

As a very different example, imagine that a researcher wanted to identify people who have been victims of any of three common crimes in the past 6 months. She might use the following IF statements:

IF (assaulted EQ 1 OR theft_victim EQ 1 or vandal_victim EQ 1) crime_victim = 1.

IF (assaulted EQ 0 AND theft_victim EQ 0 AND vandal_victim EQ 0) crime_victim = 0.

EXE.

This would create two mutually exclusive groups: (a) recent crime victims (regardless of how many times or in how many ways they were victimized) and (b) recent nonvictims.

Here is an example that combines both the IF (...) and the COMPUTE commands.

```
COMPUTE TALL = 0.
IF (SEX EQ 1 AND HEIGHT GE 74) TALL = 1.
IF (SEX EQ 2 AND HEIGHT GE 69) TALL = 1.
EXE.
```

This combination of COMPUTE and IF commands creates a new variable called TALL that has a default value of zero—and then replaces the zero with a value of 1 for all men who are 6′2″ (74″) or taller and all women who are 5′9″ (69″) or taller. A danger of this otherwise elegant approach is that if a person failed to report his or her gender and/or height, the variable TALL will still be set to zero for cases for which there are no data for height and/or gender. To avoid this problem, the researcher could have revised the command set as follows:

```
IF (SEX EQ 1 AND HEIGHT LT 74) TALL = 0.
IF (SEX EQ 1 AND HEIGHT GE 74) TALL = 1.
IF (SEX EQ 2 AND HEIGHT LT 69) TALL = 0.
IF (SEX EQ 2 AND HEIGHT GE 69) TALL = 1.
EXE.
```

Only people who report both their gender and their height will now receive a score on the dichotomous variable TALL.

RECODE: Sometimes (as was the case in Chapter 5 for some of the self-esteem items) you want to recode an item (e.g., to reverse score it). The RECODE statement is a very useful way to do this.

```
RECODE sex (0=2) (1=1) INTO gender.
EXECUTE.
```

In this example, the researcher may have wished to recode the variable "sex" from someone else's data file to use the familiar variable name and value labels that she prefers. As you will see later, it would also be easy to add a VALUE LABELS command that would specify exactly which gender is 1 and which is 2.

Here is an IF (...) command that uses the EQ command and the OR operand with alphanumeric (i.e., "String" or word) variables. Notice that you must use quotations marks with the EQ operand to specify alphanumeric variables.

```
IF (country EQ "Japan" OR country EQ "China" OR country EQ "India") East_West = 1.
```

IF (country EQ "Italy" OR country EQ "Germany" OR country EQ "Spain") East_West = 2.

EXE.

Although the IF (...) and COMPUTE commands are very flexible, they cannot do everything. Often, researchers need to perform mathematical, rather than logical, calculations on variables. When this is the case, you'll usually want to combine the COMPUTE command—and occasionally the IF (...) command—with some kind of mathematical operand. By far, the most common computations researchers use are simple addition (+), subtraction (−), division (/), and multiplication (*). However, many other mathematical calculations are possible, including, for example, log transformation, squaring or cubing variables, and taking the absolute value of a number. Here are a few quick examples of some useful mathematical operands along with brief explanations of what they do.

MEAN: The MEAN command is extremely useful because you cannot always get a mean of a set of scores by simply adding up the scores and dividing by the number of scores. For example, suppose you asked people whether they had been victims of three different crimes over the past year. What if a substantial number of respondents left one of the three crime questions blank? What if some respondents were only asked two of the three questions in the first place? For people missing only *one* crime question response, it would probably be reasonable to use the data they provided you for the two questions they did answer. The easiest way to do this would be to use the SPSS MEAN command, as shown below:

COMPUTE crime_index = MEAN.2 (assault, theft_vic, vandal_vic).

EXE.

This version of the mean.x command would give you a "crime_index" score for everyone who answered at least two (any two) of the crime questions—because of the ".2" suffix. So if Bubba *failed to answer* the vandalism question and said that he had *not* been assaulted but said that he *had* been a theft victim, his crime index score would be 0.50. Importantly, if you just use the MEAN command without a ".x" suffix, SPSS will calculate a mean score even if a respondent only has one score on the specific scale items listed in parentheses. So be careful what you specify and be especially careful not to omit the suffix (unless you're willing to live with a single score to represent a composite or index).

ABS: The ABS mathematical operand can also be very useful. Here is an example:

COMPUTE age_diff = ABS (wife_age − husb_age).

EXE.

This researcher seems to be studying married couples. The command that creates the new variable "age_diff" will take the **absolute value** of the difference between a husband's age and his wife's age, *without regard for which member of the couple is older*.

LG10: The base-10 logarithm command returns the logarithm of whatever value or variable is placed in parenthesis after the LG10 command (the power to which you need to raise the number 10 to yield the number in question). This can be useful for reducing skew so that extreme outliers are not weighted disproportionately in an analysis. The base-10 log of 100 is 2. The base-10 log of 1,000 is 3.

COMPUTE log10_height = LG10 (height).

EXE.

Note that it is the LG10 command (*after* the equal sign) that is recognized by SPSS.

SD or SD.X: The SD.X command calculates the standard deviation of whatever scores or variables are listed in parentheses. A researcher who collected negative mood scores on eight different occasions might be interested in the variability of people's negative moods. Imagine that she only wanted mood standard deviation scores for participants who had mood data on at least six occasions. Note that if you use SD (rather than the SD.X), SPSS will make $n = 2$ scores per participant the default requirement to produce a score.

COMPUTE st_dev_mood=SD.6 (mood1, mood2, mood3, mood4, mood5, mood6, mood7, mood8).

EXE.

SQRT: The square root command returns the square root of whatever value or variable is placed in parenthesis after SQRT. This can be useful for reducing the skew of a variable so that extreme outliers are not weighted disproportionately in an analysis.

COMPUTE sqrt_height = SQRT (height).

EXE.

VALUE LABELS: When you create new variables, it's often useful to remind users of the variables exactly what different scores on the variables indicate. The VALUE LABELS command allows you to do this (although you can also do it manually when in the SPSS Variable View mode). An example for body mass index (BMI) follows.

COMPUTE BMI = (weight*703)/ (height*height).

EXE.

IF (BMI LT 21) BMIcat4 =1.

IF (BMI GE 21 and BMI LT 25) BMIcat4 = 2.

IF (BMI GE 25 and BMI LT 30) BMIcat4 = 3.

IF (BMI GE 30) BMIcat4 = 4.

EXE.

VALUE LABELS BMIcat4 1 'underweight' 2 'normal' 3 'overweight' 4 'obese'.

EXE.

Note that the VALUE LABELS command stands on its own. It does not have to follow any IF (...) statements, and it could be used, for example, to label the unlabeled values of a variable that someone else previously created.

Needless to say there are many other things you can do to perform calculations or logical operations in SPSS syntax files. Recall, however, that whatever you plan to do with these statements, the statements will only create new variables in your SPSS data file once you highlight them and then run them (including the EXE commands, complete will all of your periods). Finally, we should add that anything you can do by typing in the syntax commands we reviewed here can *also* be done from SPSS dropdown menus. For example, you could do a COMPUTE command by clicking "Transform," then "Compute Variable...." Assuming your instructor is a seasoned SPSS user, he or she could lead you through this alternate approach to variable manipulation and variable creation. For the time being, though, we hope this primer will get you started.

Appendix 6.3: Running a One-Sample Chi-Square Test in Older Versions of SPSS (SPSS 19 or Earlier)

After you select "Chi-square..." as shown in Image 24, you'll open the dialog box you see in Image 25, where you'll need to (a) send "exceeds" to the "Test Variable List," (b) click the "Options" button that you can see in the lower right-hand corner of Image 25, (c) click "Descriptives" from the Options box depicted in Image 26, (d) click "Continue" from that same Options box, and (e) click "Paste" when you get back to the "Chi-Square Test" box shown in Image 25.

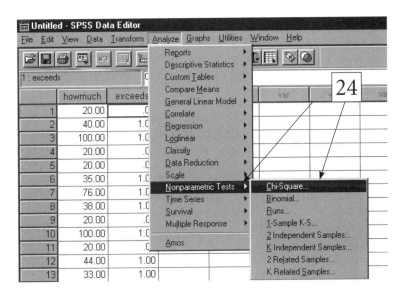

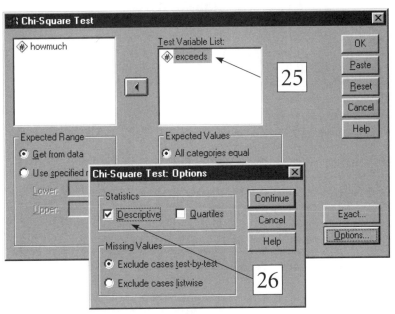

The changes this will create in your syntax file are shown in Image 4. As you know by now, you can run this new analysis by highlighting *only* the section that begins with "NPAR TEST" and then clicking the "Run" arrow (circled in Image 27).

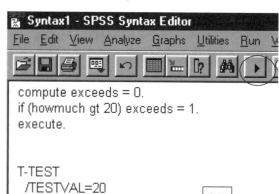

```
compute exceeds = 0.
if (howmuch gt 20) exceeds = 1.
execute.


T-TEST
 /TESTVAL=20
 /MISSING=ANALYSIS
 /VARIABLES=howmuch
 /CRITERIA=CIN (.95) .

NPAR TEST
 /CHISQUARE=exceeds
 /EXPECTED=EQUAL
 /STATISTICS DESCRIPTIVES
 /MISSING ANALYSIS.
```

7

One-Way and Factorial Analysis of Variance (ANOVA)

Introduction: The Trouble With Levels

In Chapter 6, we learned that a lot of simple research hypotheses can be tested by performing *t* tests. In particular, any time a researcher wants to (a) compare a single sample mean with a hypothetical standard (e.g., a population mean) or (b) compare two different sample means with one another, he or she can perform a one-sample or two-sample *t* test, respectively. If the world were very, very simple, we might need nothing more than *t* tests to test most research hypotheses involving means for normally distributed variables. However, the world is *not* very, very simple, and most contemporary research designs reflect this fact. Even studies that only focus on a single independent variable often require researchers to expose participants to *more than two levels* of this variable. For example, consider a researcher who studies the effects of alcohol consumption on social judgment. The researcher might want to know whether being intoxicated causes people to underestimate risks such as driving under the influence or having unprotected sex. However, upon close inspection, a seemingly simple variable such as "intoxication" turns out to be more complex than you might think. As an obvious example, it is possible, in principle, that mild and moderate levels of intoxication have dramatically different effects on judgment. Alternately, the effects of intoxication on judgment might prove to be linear (as you get more intoxicated, bit by bit, your judgment becomes more impaired, bit by bit).

In either case, an experimenter who wanted to make a sophisticated statement about the precise relation between intoxication and judgment would need to study people who varied step by step in their level of intoxication, and this would require the experimenter to create *more than two* experimental groups. For instance, an experimenter might decide to study four levels of intoxication in the same study (e.g., by studying the judgments of people whose blood alcohol levels were 0.00, 0.04, 0.08, and 0.12). To be more specific, suppose the researcher expected higher levels of intoxication to be associated with greater judgmental impairment but had no firm ideas about exactly which groups in her study would differ from one another. Or suppose she *did* have such an idea. Suppose she expected that each additional step toward intoxication would lead to a significant reduction in the quality of people's judgment. Couldn't she just conduct a series of t tests in which she compared each specific group with each other specific group?

Well, she *could*, but one problem with this approach is that she would run the risk of observing a seemingly significant effect here and there just by chance. Think about the extreme example of a researcher who created 10 groups of participants who differed in their precise level of intoxication. Making *every possible* two-groups comparison between the 10 specific groups would mean conducting 45 different t tests (Group 1 vs. Group 2, Group 1 vs. Group 3, etc.)! In addition to being extremely tedious, conducting 45 separate t tests could easily yield some very misleading findings. Specifically, the more comparisons you make between many different sets of means, the more likely it is that you will observe a "significant" difference or two because of chance alone. Remember that if you conducted 100 different t tests between 100 different pairs of means—with all pairs drawn at random from the *same* population—you should expect about 5 of these 100 tests to be "significant" due to nothing but dumb luck (also known as sampling error, also known the Vegas principle: "bad luck happens").

The basis of this problem is **experiment-wise error:** The more statistical tests a person runs in an experiment, the greater the likelihood that the person will obtain some seemingly significant effects due to chance alone. To step outside experiments for a second, run a hundred thousand t tests (or correlations, or χ^2 tests) involving randomly generated numbers and, by definition, about 5,000 (5%) of these tests will be "significant." Of course, if you only run 3 or 4 such tests, the problem won't be nearly this serious, but it'll be a problem nonetheless. How can you avoid this problem? By taking it into account that you are comparing multiple means whenever you do so—that is, by computing a more sophisticated cousin of the t test. The eminent statistician R. A. Fisher was one of the first people to recognize (and solve) this problem involving experiment-wise error rates, and the statistical test that he developed to fix the problem was appropriately named the F test (after R. A. Fisher; remember that r and R were already taken by Karl Pearson). The general analytic approach that statisticians developed to deal with experimental designs more complex than the simple two-groups design is called the analysis of variance or **ANOVA,** for short. ANOVAs thus generate "*F*" statistics using variations on Fisher's *F* test. The first half of this chapter examines one-way ANOVA, which is the analysis you conduct when (a) your dependent measure is continuous and (b) you have a single independent variable that has more than two levels. Let's look at a couple of relevant research examples.

Understanding One-Way ANOVAs by Experimenting With Alcohol

Let's begin with an example that might be relevant to the alcohol researcher described above. Suppose she conducted an experiment in which she administered three different doses of alcohol. A control group got a placebo drink that only looked and tasted like it contained alcohol. One group got enough alcohol to bring their blood alcohol levels to .06% alcohol, and a second group got enough alcohol to bring their blood alcohol levels to .12%. After administering this manipulation, the researcher gave participants (a) a series of thought problems that required abstract logical thinking (e.g., "Brothers and sisters have I none, but this man's father is my father's son." From the perspective of the quoted speaker, who is "this man"?) and (b) a measure of creativity that required people to list as many uses as they could for a brick (e.g., use it as a paperweight, use it as a weapon against a statistician). On the basis of past research, the researcher expected that increasing levels of blood alcohol would be associated with increasingly poor judgments on the logical thinking task (in a linear fashion). However, in the case of the creativity task, the researcher expected that the lower of the two doses of alcohol would actually improve performance (by disinhibiting people), whereas the higher dose would disrupt performance (because people would be muddleheaded, wouldn't care so much about what they were doing, or both). The SPSS data file for this analysis is called [**alcohol experiment for one-way ANOVA.sav**]. The data for this hypothetical study appear in Image 1, and Image 2 begins to show you how to conduct a one-way ANOVA to see if there are any effects of the alcohol manipulation. By the way, an overall *F* test such as the one you'll be conducting here is an **omnibus test,** meaning that it addresses whether *anything significant* is happening in a data set (i.e., if any means are different from one another). Thus, if an omnibus test such as this one is significant, it just means that there is a significant effect to be found *somewhere* among the various means. It does not tell you exactly where. Deciding exactly where often requires at least one follow-up test.

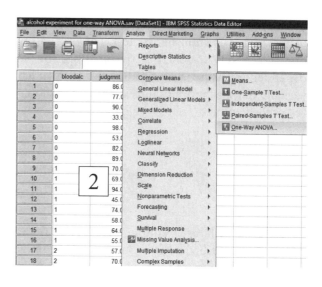

To conduct a one-way ANOVA, go to your SPSS data window and, as usual, begin at the "Analyze" button as depicted in Image 2. Next, go to "Compare Means," and then click "One-Way ANOVA...." This will open the window you see in Image 3. Following the example provided, send both of the dependent variables—"creative" and "judgmnt"—to the "Dependent List" box. Then send "bloodalc" (your independent variable) to the "Factor" box (see Image 3). For now, we're going to run the simplest possible kind of one-way ANOVA, and so the only remaining thing to do is to click on the "Options" button (circled in Image 3) and then click "Descriptive" under "Statistics" in the Options box that you can see in Image 4. As usual, clicking "Continue" will take you back to the main dialog box, where you can click "Paste" to send your one-way ANOVA command to a syntax file.

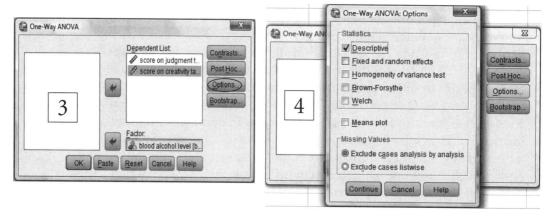

The crucial part of the syntax file appears next (although it will be much more colorful in your actual syntax file).

Chapter 7 One-Way and Factorial Analysis of Variance (ANOVA)

DATASET ACTIVATE DataSet1.

ONEWAY judgmnt creative BY bloodalc

/STATISTICS DESCRIPTIVES

/MISSING ANALYSIS.

If you run this ONEWAY command from the syntax file, you should get an output file that is a more complete version of what you see in Image 5.

Descriptives

		N	Mean	Std. Deviation	Std. Error	95% Confidence Interval for Mean Lower Bound	95% Confidence Interval for Mean Upper Bound	Minimum	Maximum
score on judgment task	0% blood alcohol	8	76.0000	21.92194	7.75058	57.6728	94.3272	33.00	98.00
	.06% blood alcohol	8	66.1250	14.65252	5.18045	53.8752	78.3748	45.00	94.00
	.12% blood alcohol	8	56.0000	15.55635	5.50000	42.9946	69.0054	32.00	78.00
	Total	24	66.0417	18.83678	3.84504	58.0876	73.9957	32.00	98.00
score on creativity task	0% blood alcohol	8	.7500	.70711	.25000	.1588	1.3412	.00	2.00
	.06% blood alcohol	8	1.2500	.70711	.25000	.6588	1.8412	.00	2.00
	.12% blood alcohol	8	.5000	.53452	.18898	.0531	.9469	.00	1.00
	Total	24	.8333	.70196	.14329	.5369	1.1297	.00	2.00

ANOVA

		Sum of Squares	df	Mean Square	F	Sig.
score on judgment task	Between Groups	1600.083	2	800.042	2.561	.101
	Within Groups	6560.875	21	312.423		
	Total	8160.958	23			
score on creativity task	Between Groups	2.333	2	1.167	2.722	.089
	Within Groups	9.000	21	.429		
	Total	11.333	23			

5

QUESTION 7.1. Summarize the results of this study. Create a single table that includes the results (means and standard deviations) for each of the two dependent variables. To what extent did the findings support the researcher's predictions? Check out the lower portion of your output file. This is where you will find the results of your two significance tests. Be sure to address the important issue of statistical power.

Finding Meaning in Means: Using Contrasts

Recall that in its simplest form, a one-way ANOVA is an omnibus test that lets you know whether *any* of your means differed from one another. In this particular case, the omnibus F test controlled for the fact that there were three different means in the study and thus three different pairs of mean comparisons one could make (.00 vs. .06, .00 vs. 12, and .06 vs. .12). However, in this researcher's defense, she made a very clear a priori prediction about exactly *how* her means would be different. That is, she wasn't simply going on a fishing expedition and wondering whether any of her means would differ. Instead, she expected them to differ in a very precise way. Recall that she expected (a) a linear reduction in the quality of logical judgments as blood alcohol levels increased and

(b) an inverted V-shaped trend in creativity as blood alcohol level increased. Shouldn't the researcher get some credit for making this precise prediction? (People who bet correctly on horse races sometimes do.) Doesn't this negate the problem of experiment-wise error that caused statisticians to develop the *F* test? Most statisticians think it does, and they have developed more focused tests—called **contrasts**—that test for specific patterns of mean differences, rather than any possible pattern of mean differences. In other words, testing for a highly specific pattern of mean differences, as opposed to *any* pattern of mean differences, shouldn't yield significant results due to chance alone. With this in mind, let's run these one-way ANOVAs again and add the specific *contrasts* that are appropriate for each of these two specific hypotheses regarding blood alcohol levels and performance.

To do this, you should repeat the analysis you just conducted, except that you should also click the button labeled "Co*n*trasts" (see Image 6). This will open the dialog box you see in Image 7, where you'll want to do

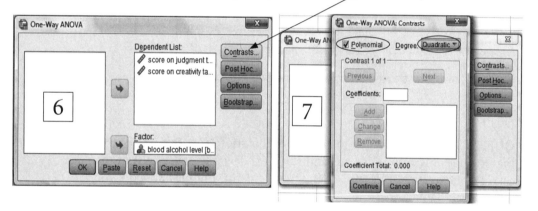

two important things. As shown in Image 7, first check the "*P*olynomial" box. Second, make sure to choose "Quadratic" from the "*D*egree" box. Normally, we'd instruct you to run *separate contrast analyses* for "creative" and "judgment," focusing solely on the specific contrast that is appropriate for each variable (linear for "judgmnt" and quadratic—aka curvilinear—for "creative"). However, it should be instructive to see both a linear and a quadratic contrast for both dependent measures—to get a good feel for the data patterns that linear versus quadratic contrasts are really designed to detect.

If you're curious to know exactly what a **quadratic** contrast is, remember that a quadratic function is simply a squared function. If you square a set of *x* scores that (a) range from negative to positive and (b) are plotted on a Cartesian (*x-y*) grid, you will notice that the lowest scores become big (e.g., $-3^2 = 9$, $-2^2 = 4$), scores of zero do not change ($0^2 = 0$), and the highest scores also become big ($2^2 = 4$, $4^2 = 16$). If you plotted this function rule on graph paper, it would produce a parabola—roughly

Chapter 7 One-Way and Factorial Analysis of Variance (ANOVA)

speaking, a *U* or a *V*. (Incidentally, this contrast analysis doesn't care whether the *U* or *V* is right side up or upside down.) We hope you can see that an inverted *U* or *V* describes perfectly the curvilinear pattern of data the researcher was expecting to see for the "creative" variable. By the way, whenever you ask for a quadratic contrast, you'll also get a linear contrast as well because you can't make a strong statistical statement about the quadratic effect of a variable without first taking into account whether there is a linear effect of the same variable. So in your case, asking for a quadratic contrast for both dependent variables covers all of the bases (and then some).

Returning to more pragmatic issues, after you click "Continue" in Image 7, click the "Paste" button you see in Image 6 to send this new contrast command to your syntax file. Now your syntax file should contain your original ONEWAY command, plus the new syntax command you see below. Make sure to highlight and run only that new contrast command (so as not to duplicate your original analysis).

ONEWAY judgmnt creative BY bloodalc

/POLYNOMIAL=2

/STATISTICS DESCRIPTIVES

/MISSING ANALYSIS.

When you run this new command, you should get an output file that looks very much the way it would have if you had run a regular one-way ANOVA—without any special contrasts. However, the box containing your significance tests for "judgmnt" should now include not only an omnibus *F* value but also the *F* test that corresponds to your linear contrast for this variable—as well as the unnecessary test for the quadratic contrast for "judgment." In the case of the "creative" variable, you'll have the same two contrasts, complete with significance tests, but now the quadratic contrast is the one you care about.

The part of your output file you should care about most appears in Image 8.

ANOVA

				Sum of Squares	df	Mean Square	F	Sig.
score on judgment task	Between Groups	(Combined)		1600.083	2	800.042	2.561	.101
		Linear Term	Contrast	1600.000	1	1600.000	5.121	.034
			Deviation	.083	1	.083	.000	.987
		Quadratic Term	Contrast	.083	1	.083	.000	.987
	Within Groups			6560.875	21	312.423		
	Total			8160.958	23			
score on creativity task	Between Groups	(Combined)		2.333	2	1.167	2.722	.089
		Linear Term	Contrast	.250	1	.250	.583	.454
			Deviation	2.083	1	2.083	4.861	.039
		Quadratic Term	Contrast	2.083	1	2.083	4.861	.039
	Within Groups			9.000	21	.429		
	Total			11.333	23			

QUESTION 7.2. Summarize the results of the study based on these more sophisticated contrasts. Was the researcher correct in her predictions? Pretend you had never done the first set of analyses, and describe the results from scratch. Finally, assume that the researcher had decided to conduct *correlations* rather than one-way ANOVAs on these data (by simply correlating each of her dependent variables with her independent variable). How would the results for these two correlations compare with the results from the two contrasts? (To find out for sure, you'll want to conduct the appropriate correlations in SPSS.) Based on your understanding of the correlation coefficient, do your results for the two different correlations make sense?

QUESTION 7.3. Some students are surprised to see that results that are not significant when analyzed by means of a one-way ANOVA may be significant when analyzed using planned comparisons. The key to using planned comparisons fairly is taking seriously the word *planned*. A good metaphor for understanding the logic of planned comparisons comes from betting on horse races. A person can win a little money in a horse race by correctly betting on a specific horse to win, place, or show (come in first, second, or third), but a person can win a *huge* amount of money by correctly betting a "trifecta" or "superfecta." *Use this fact about gambling payoffs to justify the logic behind giving researchers extra credit for the appropriate use of planned comparisons.* If you are not a fan of horse racing, check out www.wisegeek.com/in-horse-racing-what-is-a-trifecta-bet.htm.

Looking at More Than One Independent Variable: Factorial ANOVAs

In addition to conducting studies examining one independent variable that can take on multiple values, researchers often conduct studies that examine two or more independent variables *at the same time*. These days, that is, many experiments and quasi-experiments make use of **factorial designs.** In a factorial design, two or more independent variables are completely **crossed.** This means that each level of each independent variable occurs along with each level of the *other* independent variable or variables. As you may already know, this process of creating every possible combination of every level of each independent variable in a study makes it possible to see if two independent variables work *together* to produce certain outcomes. Let's first consider this idea in the abstract and then move on to consider a specific hypothetical study that makes use of a factorial design.

Does your dog bite? Will this pain reliever make me feel better? Are people attracted to those who say good things about them? When people ask important questions such as these, they usually want simple, straight-forward answers. That is why people typically become frustrated when they ask psychologists these kinds of questions. That's because psychologists often give the same answer to every question: "It depends." "Spot usually bites men, but he almost never bites women." "Sodium salycarbanol is a great pain reliever, but it'll kill you if you take it with antihistamines." "Most people like to be flattered, but people who are very low in self-esteem prefer to be viewed negatively rather than positively."

Experimenters (especially experimental psychologists) are so intrigued with the "it depends" notion that they have developed a special statistical analysis to tell us *when* it depends. They have developed techniques, that is, for detecting **interactions.** Technically, an interaction means that a particular independent variable has different effects on a dependent variable at different levels of some *other* independent variable. It means, in short, that we cannot predict the effects of A on B without knowing something about variable C. The key is that the two independent variables work together to determine behavior. Information about one variable by itself may tell you very little, but information about both variables together tells you a lot.

To test for an interaction between two variables, you have to manipulate both of the variables independently in a **factorial design.** In its simplest form, a factorial design is a research design that includes every possible combination of two independent variables. It is easiest to illustrate both an interaction and a factorial design with an example. Medical doctors and pharmacologists often use the phrase "drug interaction precaution." Such medical experts are using the term *interaction* exactly the way a statistician would. What they usually mean is that Drug A by itself is good, Drug B by itself is good, but Drugs A and B together are bad, perhaps very bad. In Table 7.1, you will find the results of a hypothetical study that illustrates this point. The study makes use of a 2 (Drug A: placebo vs. treatment) × 2 (Drug B: placebo vs. treatment) factorial design (the "×" is pronounced *by*), and the dependent variable is the amount of pain that migraine sufferers reported 30 minutes after receiving their specific combination of treatments (higher numbers mean more self-reported pain—on a 7-point scale).

Table 7.1 A Statistical Interaction in a Hypothetical 2 × 2 Factorial Study of Two Pain Relievers

	Level of Drug A	
Level of Drug B	**Placebo**	**Treatment**
Placebo	5.8 (ouch!)	1.2 (aaah . . .)
Treatment	1.5 (aaah . . .)	6.3 (ouch!)

Notice, first of all, that the design includes *every possible* treatment combination of the two drugs. Based on their random assignment to one of four experimental conditions, (a) some patients got Drug A only (along with a placebo that they thought was Drug B), (b) some got Drug B only (along with a placebo that they thought was Drug A), (c) some got neither drug (they got two placebos), and (d) some got both drugs. You might also notice that all of the patients *thought* that they got both drugs; the manipulation had to do with what drugs the patients actually got. If we had a large sample in this hypothetical study, a statistical test would definitely indicate that there was an interaction between the two treatments. When given alone, each drug was very effective, but when given together, the two drugs canceled one another out.

Of course, interactions do not always take the form described previously. For instance, some drugs (including some common pain relievers) probably work better in combination than they do alone. An interaction means that the effects of variable A are *different* at different levels of variable B. The effects of A could be strengthened, weakened, or totally reversed at a particular level of variable B, and you would still say that there was an interaction between the two variables. If all of these different patterns qualify as interactions, does this mean that factorial designs always reveal some kind of interaction? Definitely not. Some drugs are absolutely unaffected by the presence of other drugs, just as some dogs are equally likely to bite people of either sex. According to self-enhancement theory, it should also be the case that everybody (whether high or low in self-esteem) prefers positive feedback over negative feedback. In statistical terms, straightforward ideas and theories such as this produce *main effects* rather than interactions. The presence of a main effect of variable A in a factorial design (without an interaction) means that you don't need to know anything about variable B to describe fully the effect of A on your dependent variable. If the drug experiment described above had yielded only a main effect of Drug A (no main effect of B and no interaction), the results might look something like those in Table 7.2.

Table 7.2 A Single Main Effect (Only) in a Hypothetical 2 × 2 Factorial Study of Two Pain Relievers

Level of Drug B	Level of Drug A	
	Placebo	Treatment
Placebo	6.2 (ouch!)	1.4 (aah . . .)
Treatment	6.3 (ouch!)	1.5 (aah . . .)

Notice that the effects of Drug A had nothing to do with Drug B. People who got Drug A felt better. Period.

A Hypothetical Example of When and How "It Depends"

To give you a better feel for how interactions work and to see how you might conduct a test for an interaction in SPSS, we'd like you to consider some data from a hypothetical study involving feedback and self-esteem. In this quasi-experiment, 10 high self-esteem participants and 10 low self-esteem participants were randomly assigned to receive either positive or negative feedback from a confederate (e.g., "You don't seem to be very good at this. It seems like something you're just not cut out for" vs. "Wow! You did a really great job on this task. You really seem to have what it takes!"). The dependent variable was participants' liking for the confederate measured on a 9-point scale (on which higher scores indicate greater liking). Thus, the design was a 2 (self-esteem: high vs. low) × 2 (feedback: positive vs. negative) factorial. The raw data from this hypothetical study appear in Table 7.3.

Table 7.3 Hypothetical Data From a 2 × 2 Factorial Study of Self-Esteem and Interpersonal Feedback

	Self-Esteem	
Feedback	Low	High
Negative	7, 8, 7, 9, 9	3, 2, 3, 4, 3
Positive	2, 3, 4, 5, 1	6, 9, 9, 7, 9

Before you run any SPSS analyses, compute the mean liking score in each cell of this design by hand and compare these four means. If you are person of worth, you will also want to calculate both the *row* and *column* means as well (to help you make decisions about main effects). For example, a comparison of the 10 scores in the left-hand column of Table 7.3 and the 10 scores in the right-hand column of Table 7.3 will tell you whether there is a main effect of self-esteem (i.e., it will tell you whether, on average, people who differed in self-esteem differed in their liking for the confederate, ignoring the feedback manipulation). Based on these hand calculations, decide whether these results suggest that you observed (a) a main effect of self-esteem, (b) a main effect of feedback, and (c) an Esteem × Feedback interaction. In case you are not absolutely sure, let's examine these same data by conducting a two-way ANOVA on them. In Image 9, you'll see a screen capture from an SPSS data file that includes all the data from Table 7.3 [**self-esteem study for interactions n=20.sav**]. In Image 10, you'll see how to get started conducting a 2 (self-esteem: low vs. high) × 2 (feedback: negative vs. positive) ANOVA on these data.

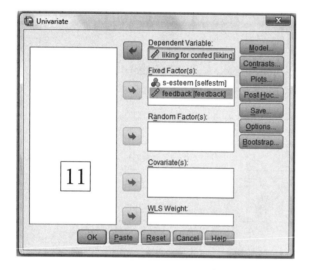

Clicking "Univariate…" will open up the window you see in Image 11.

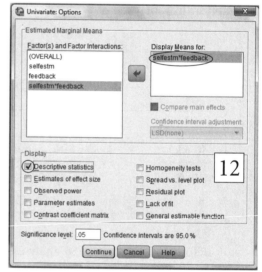

After sending your two independent variables to the "Fixed Factors:" box and sending your only dependent variable to the "Dependent Variable:" box, click the "Options" button. This will open the window you see in Image 12. Notice that three different terms are listed under "Factor(s) and Factor Interactions:," meaning that you could ask SPSS to give you descriptive results (means) that would showcase the means that are relevant to (a) the self-esteem main effect, (b) the feedback main effect, and/or (c) the Self-esteem × Feedback interaction. Because we are emphasizing interactions in this activity, you only need to send the "selfestm*feedback" interaction term to the "Display Means for:" box.

However, before you click "Continue" to get back to dialog box 11, make sure you also ask for descriptive statistics by checking the appropriate box you can see circled in Image 12. Then click "Paste" from the main analysis window. Now you can run the two-way ANOVA from your syntax file.

Chapter 7 One-Way and Factorial Analysis of Variance (ANOVA)

QUESTION 7.4a. Let's return to the question at hand. Did you observe any main effects? Comment on why you think the *F* values for each of your main effects took on the exact values that they did. (A clue: Examine whichever set of means corresponds to each main effect.) What about an interaction? Did you observe one?

QUESTION 7.4b. Assuming that you did observe an interaction, you'll need to conduct a set of **simple effects tests** (also known as **simple main effects tests**). These are follow-up tests conducted to see exactly what *kind* of interaction you have observed. Statistical purists would dictate that you only conduct these tests if your test for an interaction is, in fact, significant. In a 2 × 2 ANOVA such as this one, simple effects test almost always consist of two separate *t* tests. In this study, we recommend (a) conducting a *t* test for people low in self-esteem and then (b) conducting the same *t* test for people high in self-esteem. This will require you to split your data file into two separate files. The following image should help you do this. You'll start in the data file, which is, after all, the file you wish to temporarily split in two. If it seems a little odd to split your data file in two, remember that the meaning of an interaction is that a variable such as feedback has different effects for different people (or in different experimental conditions). Splitting the file up based on that other variable (in this case self-esteem) lets you examine this idea directly by doing two separate statistical tests. The SPSS "Split file…" command is, by far, the easiest way to do these two separate simple effects tests.

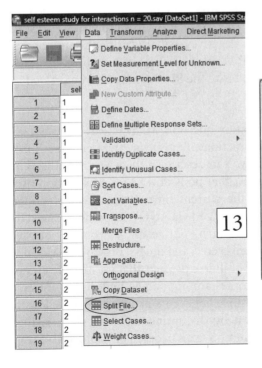
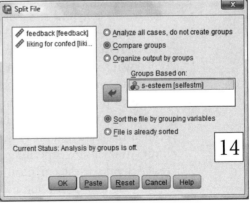

After you click on "Data" and then "Split File..." (see Image 13), you'll see a dialog box like the one in Image 14. Before you do anything else, click on "Compare groups." Because you wish to create two different *self-esteem* groups, you should now click on the self-esteem variable and send it to the "Groups Based on:" box (as we have already done in Image 14). In so doing, you'll be telling SPSS to create a separate data subfile for each level of the self-esteem variable (thus, if you had 10 levels of self-esteem, you'd temporarily create 10 different data subfiles). Now, *instead of clicking on "Paste,"* we recommend simply clicking "OK" to run the split file command. You won't see much happening, but if you did this correctly, two different SPSS data files will now secretly await your analysis. This means that if you now run just one *t* test (or one correlation, or one anything), this *t* test will be conducted separately for the low self-esteem participants and for the high self-esteem participants.

To get back to the question, report and describe the results of the two independent samples *t* tests that make up your simple effects tests. You may need to refer to Chapter 6 to recall how to conduct an independent samples *t* test. *Remember, though, that because you have created two different self-esteem groups, the only independent variable in your* t *test will be "feedback."*

QUESTION 7.4c. Consider your results in their entirety. Report the separate results of these two different *t* tests. What do these results tell you about the nature of self-esteem (e.g., is there evidence for self-enhancement theory?)?

QUESTION 7.4d. Consider your results further. If you had failed to include self-esteem as a factor in this study and had simply run a *t* test to see if people preferred positive or negative feedback (using the entire sample), what would you have found? Would these results have been misleading? Reminder: Did you name and save the SPSS syntax file you created to analyze these hypothetical data?

QUESTION 7.5. Research application question. Many research hypotheses boil down to questions about interactions. To give you more practice thinking about interactions, we'd like you to describe a theory or hypothesis in your own personal area of interest that boils down to the prediction of a statistical interaction. Although the current introduction to interactions emphasizes data that are typically analyzed via ANOVA, you do *not* need to limit yourself to experimental research. For example, consider survey research on self-esteem, positive life events, and illness. Brown and McGill's (1989) identity disruption model of illness predicts that positive life events only lead to positive health outcomes among people high in self-esteem. Their theory states that when positive life events happen to people low in self-esteem, this creates confusion by disrupting these people's existing sense of self. Apparently, this disruption taxes people's physiological as well as cognitive resources. For example, this could occur (a) by stressing people out and directly taxing people's immune systems or (b) by causing people who are low in self-esteem to engage in maladaptive, unhealthy behaviors (e.g., smoking, drinking) that people hope will reduce their anxiety. Regardless of one's preferred spin on exactly *why* people low in self-esteem should suffer in the wake of positive life events, the model predicts that a higher frequency of positive life events (e.g., getting a good grade, getting a raise at work) will be associated with *lower* levels of illness among people high in self-esteem but will be associated with *higher* levels of illness among people low in self-esteem. This is clearly an example of an interaction. When you write up your answer, provide about the same level of detail that we provided in this research example. If you are unsure of whether your example constitutes an interaction, make a table or figure of your results. If you have created a proper figure using lines (as would be appropriate in the study we just described), a sign that you have graphed an interaction is that the two lines you've drawn will not be parallel. Along similar lines, if you've created a 2 × 2 bar graph, the difference between one set of two bars should be bigger than (or even different in direction than) the difference between another possible set of two bars.

More Practice Understanding Main Effects and Interactions

To give you more practice thinking about main effects and interactions, we would like you to analyze the following two data sets and decide whether each set of findings reveals any main effects and/or an interaction. Recall that if there is a main effect of variable *A* on your dependent measure, it must be the case that, averaging across levels of the other independent variable (e.g., variable *B*), there is an effect of variable *A*. For example, if, averaging across a measure of verbal aggression and a measure of physical aggression, boys behaved more aggressively than girls did, this would mean that there was a main effect of gender on aggression, with boys generally behaving more aggressively than girls. Of course, it would be possible, in principle, to observe just the opposite kind of main effect. Girls or women might generally behave more aggressively than do boys or men. How might you need to qualify any statements about main effects, though, if you observed a Gender × Type of Aggression interaction?

Practice Study 1: A Lab Study of Aggression Among Kids

Imagine that 3- to 4-year old boys and girls were observed in a structured situation in which they interacted with an obnoxious puppet that bossed them around and insulted them while playing *Candy Land* (a kids' board game). The dependent measure is kids' degree of verbal or physical aggressiveness in the study, as rated and coded by two independent judges (i.e., judges who were kept blind to each other's ratings). Incidentally, you can assume that half the kids were randomly assigned to a condition in which they only had the opportunity to aggress verbally (because they could not touch the puppet), whereas the other half were randomly assigned to a condition in which they only had the opportunity to aggress physically (because they could shoot things at the puppet but thought the puppet could not hear them). You can find the hypothetical data in the file called [gender and aggression for interactions.sav].

Table 7.4 Hypothetical Data From a 2 × 2 Factorial Study of Gender and Two Kinds of Aggression

Type of Aggression	Gender	
	Male	Female
Verbal	4, 6, 7, 4, 8, 4, 5, 6	7, 9, 9, 8, 5, 7, 7, 9
Physical	8, 7, 4, 7, 8, 9, 5, 7	6, 7, 6, 7, 9, 5, 7, 8

QUESTION 7.6. Concisely summarize the results of this hypothetical study by reporting briefly whether you observed (a) any main effects and (b) an interaction. If you did observe an interaction, be sure to report the results of the follow-up tests (simple effects tests) you'd need to decode the exact nature of the interaction. Finally, as a methodological detour, provide at least two reasons why a critic of this study might like to see a follow-up study that focused on adults rather than kids. Even more important, explain why, in this study, the critic might insist on using alternate measures of aggression: (a) the delivery of electric shock as an indicator of physical aggression and (b) selection of insulting versus neutral feedback statements (chosen from a list and delivered as text messages) as a measure of verbal aggression.

Practice Study 2: A Lab Study of Self-Pay

A researcher interested in gender stereotypes gave men and women an experimental task (trying to unscramble a list of anagrams). The anagrams were easy for half of the participants but hard for the other half. Consistent with past work on *depressed entitlement*, the experimenter thought that women would pay themselves less than men would for their own work. However, she thought this effect might only occur under conditions of self-concept threat (when the task was difficult). The dependent measure was self-pay in dollars. Was the researcher correct? You can find the data in **[gender and self pay.sav]**.

	Gender	
Task Difficulty	Male	Female
Easy	7, 9, 9, 8, 5, 7, 7, 9	6, 7, 6, 7, 9, 8, 5, 8
Hard	8, 7, 4, 7, 8, 9, 5, 7	3, 4, 6, 3, 6, 3, 7, 5

QUESTION 7.7. Summarize the complete results of this hypothetical study in the same extremely concise way that you summarized the results of the study of aggression.

Three-Way ANOVAs and Beyond

Just as a two-way interaction means that a simple main effect is not the whole story, a three-way interaction in a study with three independent variables means that a two-way interaction is not the whole story. In other words, just as a two-way interaction qualifies a main effect, a three-way interaction qualifies a two-way interaction. The nature of such a two-way interaction "depends" on the level of a third independent variable. Let's begin with a two-way interaction that might be qualified by gender and thus become a three-way interaction. Research by Dov Cohen and colleagues (e.g., see D. Cohen, Nisbett, Bowdle, & Schwarz, 1996) has shown that insults and culture interact to predict aggression. Whereas Northerners show little increase in aggression following insults from strangers, Southerners often show huge increases. (Notice that insults affect one kind of person differently than they affect another, making this a two-way interaction.) However, suppose follow-up research showed that this interaction pattern held true only for men. More specifically, suppose that neither Northern nor Southern women became more aggressive after being insulted by a stranger. If this were true, then this would be reflected in a three-way interaction—the two-way Culture × Level of Insult interaction would apply only to men. Your final assignment in this chapter is to decide whether a different study of culture yields evidence of a three-way interaction. This study involves culture and self-evaluation. *Because this exercise is purely conceptual, you do not need to use SPSS.*

QUESTION 7.8. Research application question. It is widely believed that Westerners are more self-adoring than are Easterners. Furthermore, recent research in social cognition suggests that the tendency to be self-adoring (to view oneself favorably) is so automatic that self-relevant words such as *I* or *me* prime positive words more readily than they prime negative words. Thus, if an experimenter briefly flashes a word such as *I* or *me* in the middle of a computer screen, this will facilitate most people's recognition of positive words such as *love* or *good*. This sort of evaluative priming effect has been shown many times (Perdue, Dovidio, Gurtman, & Tyler, 1990). Imagine that a researcher interested in culture and self-evaluation believed that Westerners are only more "self-adoring" than Easterners when it comes to *individual* self-views. To test her idea, she created a list of pronouns that served to prime either individual or group identity. She then put American and Japanese participants through an experiment in which each kind of prime word sometimes preceded positive target words and sometimes preceded negative target words. She expected to see that Americans were more egocentric than Japanese participants only when it came to the individual target words. The following hypothetical results are based loosely on Hetts, Sakuma, and Pelham (1999). They reflect average response latencies to the target words in milliseconds. You may assume that between-condition differences of 70 ms or more are statistically significant. Did the researcher observe any support for her hypothesis? Did she observe a three-way interaction?

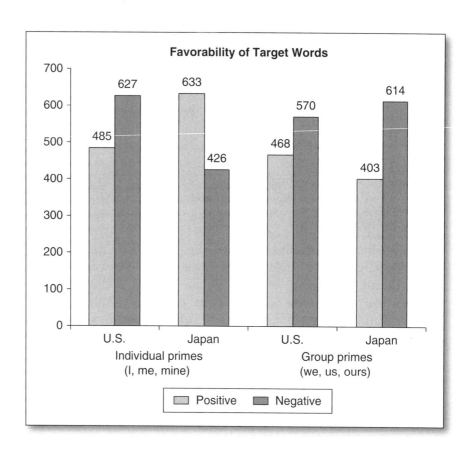

Putting It All Together

You might not be too surprised to learn that the statistician who formalized and popularized the use of factorial designs was R. A. Fisher. As a way of summarizing what we've said about factorial designs in this chapter, let's consider what R. A. Fisher had to say about the advantages of factorial designs over simpler, one-way designs. Fisher noted that two of the major selling points of factorial designs are that they are both more *efficient* and more *comprehensive* than one-way designs. Factorial designs are efficient in the sense that they allow us to look for more than one main effect at a time in a single study. They are comprehensive in the sense that, unlike one-way designs, they tell us more of the whole story behind a specific phenomenon—by allowing us to see how different variables may work together to influence the phenomenon. In addition to these two advantages of factorial designs, Fisher also noted that there is a third, less commonly appreciated, advantage of factorial designs. Although the term hadn't yet been coined in Fisher's day, modern methodologists would say that this third advantage has to do with external validity (i.e., the generalizability of a research finding). As Fisher (1935) put it,

> There is a third advantage... which, while less obvious than the former two... has an important bearing on the utility of the experimental results in their practical application. This is that any conclusion, such as that it is advantageous to increase the quantity of a given ingredient, has a wider inductive basis when inferred from an experiment in which the quantities of other ingredients have been varied, than it would have from any amount of experimentation, in which these had been kept strictly constant. The exact standardisation of experimental conditions, which is often thoughtlessly advocated as a panacea, always carries with it the real disadvantage that a highly standardised experiment supplies direct information only in respect of the narrow range of conditions achieved by standardisation. Standardisation, therefore, weakens rather than strengthens our ground for inferring a like result, when, as is invariably the case in practice, these conditions are somewhat varied.

In other words, a researcher who observes a robust main effect in a factorial study can be reasonably sure that the observed main effect will *generalize* across the two or more levels of the other independent variables that exist in the study. If these other independent variables were held at a single, constant value, as is necessarily the case in a one-way design, it would always be possible that the researcher had identified the one and only one set of conditions under which the observed effect usually occurs. Of course, main effects are sometimes qualified by interactions, but as we have noted repeatedly in this chapter, learning about the existence of interactions is something we can do only by using factorial designs.

Appendix 7.1: Results of a Unique Memory Study That Used Planned Contrasts

The main goal of this appendix is to provide you with a concrete example of how to write about a contrast analysis. Before we say anything more about this, however, please indulge us and read the following list of 15 words slowly—and out loud:

sour, candy, sugar, bitter, good, taste, tooth, nice, honey, soda, chocolate, heart, cake, tart, pie

Now if you would indulge us a little further, please put down that candy bar, pick up a pen or pencil, and write down the name of the *five U.S. states you would most enjoy visiting in the next year*. Are you done? It's very important that you complete this second task and *write down* the names of the five states. OK, now that you are done writing "Hawaii, California, Florida, New York, and Colorado," take a look at the five target words written at the end of this appendix and—*without consulting the list*—simply write down any of the five target words that you remember from the list of 15 words.

If you are like most other people (who haven't yet had a course in cognitive psychology), you probably had no trouble identifying the two words *sugar* and *sweet*. The problem, though, is that sweet is *not* on the list. Go ahead; check the list. If you *have* taken that course in cognitive psychology, you probably know that many people *think* they read the word *sweet* because *sweet* is a very close semantic associate of those other 15 words. It is impossible to read those words, that is, without activating the word *sweet* in memory. Apparently, we human beings are not as good as we might hope at knowing the difference between thinking something and experiencing that something. That turns out to be a very important point because it is related to a huge controversy about exactly how malleable human memory is.

This brings us to a very clever article that made use of planned contrasts. The gist of the article is that perhaps you should take it a little easier on your friend Kevin, who firmly believes that he was once abducted by space aliens. After all, Kevin may just be a little more susceptible than the rest of us to believing what his guts tell him about his memory. Unlike you, for example, Kevin might insist that he *vividly* remembers reading the word *sweet*. To test this sort of idea, Clancy, McNally, Schacter, Lenzenweger, and Pitman (2002) recruited three groups of participants for a study of memory and ran each group through a more sophisticated version of the memory task you just took yourself. The three groups were A, the *recovered memory group*, who believed that they had recovered repressed memories of alien abduction; B, the *repressed memory group*, who believed that they had been abducted by aliens but did *not* claim to have recovered any specific memories of the experience; and C, the *control group,* who did not believe they had ever been in contact with aliens. One of the predictions in this study can be succinctly summarized as follows: Group A should be the most susceptible to the memory bias, Group C should be the least susceptible, and Group B should fall between these two groups. It is worth noting that is was not exactly easy for these researchers to find a large group of people who felt they had been abducted by space aliens! So they had to make do with a small sample size of 9 to 13 participants per group. This meant that it was particularly important to conduct a statistical test with a lot of power. Here is a sample of one of their findings. You will notice that they converted their findings based on mean differences between the three groups to correlation coefficients to give readers an idea of how big their effects were

> According to the third hypothesis... the recovered memory group should exhibit the highest false recall and false recognition, followed by the repressed memory group, followed by the control group, respectively. Applying contrast weights of 1, 0, and −1 to the mean false recall and false recognition rates of the recovered, repressed, and control groups, respectively, we confirmed this hypothesis for false recall, $t(30) = 2.88$, $p = .01$, $r = .47$, and for false recognition, $t(28) = 3.51$, $p = .01$, $r = .59$. (Clancy et al., 2002, p. 458)

As you probably noticed from this example, when someone conducts a planned *linear* contrast, it is possible to express the result in the form of an F statistic, a t statistic, or (if you do a little conversion, as these authors did) even a correlation coefficient. Notice, for example, that a correlation coefficient is reasonable for the particular contrast these researchers were reporting because they always tested for a linear contrast. If these researchers had been testing for a quadratic effect in a planned contrast, they almost certainly would have reported some kind of F statistic.

We don't want to state the obvious by saying we think these results are interesting. So we will just say that we find them fascinating, intriguing, noteworthy, surprising, and attention grabbing. If you were to claim that we also said they are interesting, we could hardly blame you.

Five target words: *halo, sugar, table, tattoo, sweet.*

Within-Subjects and Mixed Model Analyses

8

Introduction: Controlling for Individual Differences

In the history of modern research methods, experimenters have faced only one central methodological challenge: to isolate the specific causes of the outcomes in which they are interested. One of the most important tools researchers have used to this end is **random assignment.** Random assignment occurs when all of the participants in an experiment have an equal chance of being assigned to any specific condition of the experiment. When you use this technique, you can almost always rest assured that your experimental and control groups differ in *one and only one way*. The experimental group received your treatment, and the control group did not. So if your experimental group behaved differently than your control group, you do not have to ask yourself whether the differing personalities of the people in your experimental and control groups were the real reason the two groups behaved differently. Random assignment eliminates confounds based on individual differences.

Random assignment is so useful and important that most researchers consider it the defining feature of the experimental method. However, there is an additional technique that researchers often use to control for individual differences, and in some ways, it is actually superior to random assignment. If we want to be sure that the people in our experimental group are identical to the people in our control group, why not just place the *very same people* in each of the two conditions? Research designs in which the same person serves in more than one experimental condition are referred to as **repeated measures** or **within-subjects** designs, and psychologists have been using within-subjects designs for more than 100 years (e.g., see Triplett, 1898). In fact, some cognitive and perceptual psychologists only use random assignment as a way of refining their use of repeated measures designs (e.g., by using random assignment to determine the exact order in which participants will receive their counterbalanced experimental treatments).

Some Bogus Within-Subjects Studies of Bogus Traits

In this chapter, we are going to compare between-subjects and within-subjects designs, with a focus on some of the advantages of within-subjects designs. You will do this by analyzing the data from some hypothetical between-subjects and within-subjects studies of self-enhancement. As you probably recall from Chapter 7, self-enhancement theories suggest that people have a strong desire to feel good about themselves and will typically do whatever they can to boost their own egos. To study self-enhancement processes in their purest form, imagine that a researcher decided to see if people ever develop positive beliefs about themselves in the absence of any objective, supporting information. More specifically, imagine that a researcher decided to examine people's tendencies to endorse *nonexistent* positive traits as self-descriptive. Pelham (1991) conducted a real study very much like the within-subjects version of this study. To get back to the statistical point, you will analyze the data from *one* between-subjects version and *three different* within-subjects versions of this study involving bogus traits. You will see that some of these four studies yield significant results, and some do not. It will be your job to figure out *exactly why* this is the case.

To begin with, suppose a researcher gave half of her 20 participants the following "personality measure" and asked them to rate the self-descriptiveness of each of the following 10 traits. Assume that the response scale ranged from 1 (*not at all like me*) to 9 (*exactly like me*). This means, of course, that higher scores indicate greater endorsement (i.e., greater claimed self-descriptiveness).

_____ immature   _____ considerate

_____ humorless   _____ hardworking

_____ dishonest   _____ friendly

_____ gamant   _____ creative

_____ boring   _____ talented

If you check out these 10 traits carefully, you'll notice that it looks like it's not so great to be gamant. After all, gamant is listed with all the bad traits rather than the good traits. But if you check out the following list, you'll notice that gamant is *now* listed with the positive traits. It now seems *much better* to be gamant. You've just seen each of the two between-subjects conditions of this simple study, and the hypothetical data for this simple study appear *in the first two columns* of the composite data file called [**bogus trait study for repeated measures.sav**].

Chapter 8 Within-Subjects and Mixed Model Analyses

_____ immature	_____ considerate
_____ humorless	_____ hardworking
_____ dishonest	_____ friendly
_____ disagreeable	_____ gamant
_____ boring	_____ talented

You'll want to begin by conducting an independent samples t test or a one-way analysis of variance (ANOVA) on these between-subjects data (from the first two columns of your SPSS data file). In case you'd prefer to run the one-way ANOVA, Image 1 offers you some reminders of how to do this. Don't forget, for example, to use the "Options" button to ask for "Descriptive" statistics so that you can see the means from this experiment rather than just the relevant F and p value.

QUESTION 8.1. Summarize the results of the between-subjects study. Be sure to offer a few sentences about what conclusions you'd draw from the study.

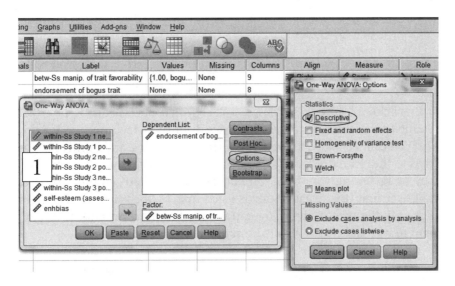

Examining Three Within-Subjects Versions of the Same Study

One good way to become familiar with the way within-subjects analyses work is to compare the statistical significance of the results of different within-subjects studies. Along these lines, we have included data from three different hypothetical within-subjects studies in the SPSS bogus traits data file. Each of the three studies is designed to see if people are more likely to endorse positive as opposed to negative bogus traits. For the purpose of this activity, we would like you to assume that the three within-subjects experiments all have the same research design. However, we will warn you that each study yields slightly different results. Furthermore, at least one of the studies does not yield a significant within-subject trait favorability effect. Your main job in this part of the chapter, then, will be to figure out exactly *why* some studies yield significant results and some don't. Before you begin to worry too much about this, however, let's take a quick peek at one way to do a repeated measures analysis on these data. For now, let's use the simplest possible test, which is the paired-samples *t* test. (We'll get to within-subjects ANOVA a little later.)

The screen capture that follows in Image 2 shows you where to find the paired-samples *t* test.

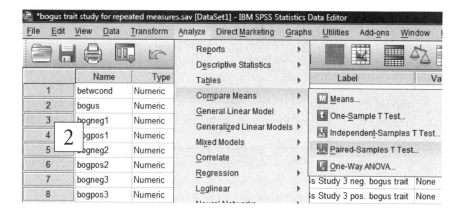

By clicking "Paired-Samples T test…" you'll open a dialog box that resembles the one that follows in Image 3. In the screen capture for this box, you can see that we have *already* sent the negative and positive bogus traits to the "Paired Variables:" box. To make it easy on the instructor who may be reviewing your SPSS output files, you should send the negative trait over first (the negative one is listed first, by the way). Make sure to pick only the two Study 1 bogus traits the first time around. You don't want to mix and match results from different studies!

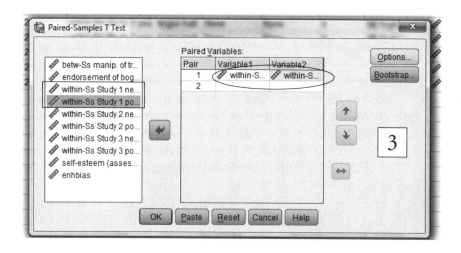

Chapter 8 Within-Subjects and Mixed Model Analyses

One nice thing about paired-samples *t* tests is that the SPSS default is to give you everything you need to interpret these tests, without needing to ask, for example, to see the means. So if you click on the "OK" button (let's not worry about pasting just yet), you should produce an output file that strongly resembles the one that appears in Image 4.

Paired Samples Statistics

		Mean	N	Std. Deviation	Std. Error Mean
Pair 1	within-Ss Study 1 neg. bogus trait	5.9000	20	2.24546	.50210
	within-Ss Study 1 pos. bogus trait	6.9000	20	1.58612	.35467

Paired Samples Correlations

		N	Correlation	Sig.
Pair 1	within-Ss Study 1 neg. bogus trait & within-Ss Study 1 pos. bogus trait	20	-.180	.447

Paired Samples Test

		Paired Differences					t	df	Sig. (2-tailed)
		Mean	Std. Deviation	Std. Error Mean	95% Confidence Interval of the Difference				
					Lower	Upper			
Pair 1	within-Ss Study 1 neg. bogus trait - within-Ss Study 1 pos. bogus trait	-1.00000	2.97357	.66491	-2.39167	.39167	-1.504	19	.149

The two means you see circled near the top of the screen capture tell you the degree to which these hypothetical participants said the two different bogus traits described them (on a 9-point scale). The section you see circled in the lower right-hand portion of the same screen capture tells you whether this mean difference (of exactly 1 point) is statistically significant. *Now all you need to do is to repeat this same repeated measures analysis for Study 2 and for Study 3.* Remember that these two additional studies are part of the very same data file (incidentally, researchers don't normally put more than one study in a single data file, but we didn't want you to have to open and manage four different data files for this activity. Just make sure you keep the studies straight, though, when analyzing them).

QUESTION 8.2. Summarize the results of the within-subjects studies. After you have conducted the three different paired-samples *t* tests (one for each within-subjects study), summarize concisely the results of each of the three within-subjects experiments. As we noted before, at least one of these studies will yield significant results, and at least one will not. It will be your job to figure out exactly why—based on what you know about the factors that affect power in a statistical test. This may be a little challenging, but a good hint is to take a careful look at either the raw data from each study or the patterns of correlations that are automatically generated when you conduct a paired-samples *t* test on each data set. Make sure that you address all the potential reasons why some studies might have yielded more significant results than the others (e.g., Could a difference in sample size have ever played a role?). Then focus on the one explanation that works best.

Combining Between-Subjects and Within-Subjects Designs: Mixed Model Designs

Now that you have had a little practice thinking about within-subjects designs, we would like to introduce you to a very clever research design that combines the positive features of within-subjects designs with the positive features of between-subjects designs. Designs that include at least one within-subjects variable and at least one between-subjects variable (whether measured or manipulated) are called **mixed model designs.**

Imagine that the researcher who conducted the third within-subjects study also assessed a **between-subjects** variable—namely, the global self-esteem levels of her 20 participants. How might you expect high self-esteem and low self-esteem people to differ in their reactions to positive versus negative bogus traits? More specifically, we'd like you to generate a prediction about whether people *high* versus *low* in self-esteem should be more likely to show the self-enhancing bias that was the focus of this original research. Then figure out the best way to test your hypothesis about whether the between-subjects variable of self-esteem is related to the size or nature of the within-subjects effect you have already observed. In the process of doing this, of course, you'll be graduating from a purely within-subjects design (and analysis) to a mixed model design (and analysis).

You will now see some step-by-step instructions for conducting a mixed model analysis, which will allow you to test your predictions about self-esteem and self-enhancement. You'll begin by (a) using the "General Linear Model" (ANOVA) analysis and (b) selecting "Repeated Measures. . . ."

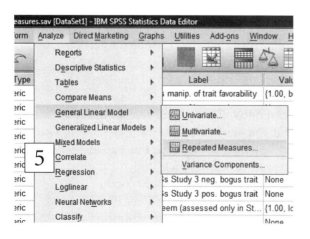
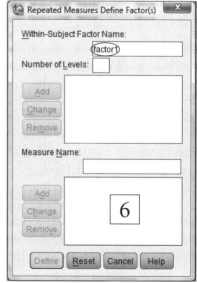

Chapter 8 Within-Subjects and Mixed Model Analyses

Once you click on the "Repeated Measures. . ." option, a dialog box like the one in Image 6 will appear (yours may be a bit smaller—we've blown it up so you can read it more easily). You'll now need to define any within-subjects variables you wish to make a part of your model:

a. Begin by replacing factor1 (circled in Image 6) with a more descriptive variable label such as "traitval" (valence of the bogus traits).
b. Then let SPSS know how many levels there are for this variable (a big hint: there are two levels of trait favorability in this particular study). Once you've entered a "2" in the little "Levels:" box, the "Add" button will be activated.
c. Click this "Add" button to let SPSS know that you want to create this within-subjects variable.
d. Then click "Define" so that you can let SPSS know exactly which scores from your data file will constitute the different levels of your within-subjects variable.

Now a new dialog box like the one you see in Image 7 will appear.

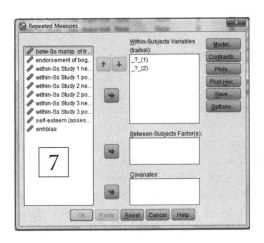
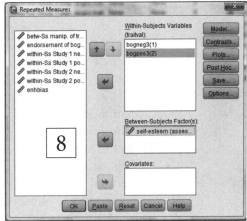

You'll want to (a) specify your within-subjects variable (the Study 3 negative vs. positive bogus traits), (b) specify your between-subjects variable (self-esteem, assessed only in Study 3), and (c) click on "Options. . ." to decide which specific descriptive statistics you want to see in your output.

From the "Options" dialog box (not shown here), you'll want to (a) send anything that looks like an *interaction term* to the "Display Means for:" box and (b) select "Descriptive statistics" from the "Display" options. Continue by pasting this command set into a syntax file, running the analysis, and deciding what kind of *follow-up tests* you should run if you observe a significant interaction.

If you'd like to check your work, your SPSS syntax file should now contain a mixed model ANOVA command that looks like the following (it may differ slightly if you have an older version of SPSS):

```
DATASET ACTIVATE DataSet1.
GLM bogneg3 bogpos3 BY esteem3
  /WSFACTOR=traitval 2 Polynomial
  /METHOD=SSTYPE(3)
  /EMMEANS=TABLES(esteem3*traitval)
  /PRINT=DESCRIPTIVE
  /CRITERIA=ALPHA(.05)
  /WSDESIGN=traitval
  /DESIGN=esteem3.
```

QUESTION 8.3. When you are done running this mixed model analysis, specify exactly what kind of analysis you conducted, what you found, and whether it supports your predictions. If you observe an interaction, you'll get to practice your "Split File..." skills a bit more and run a separate within-subjects test for people high versus low in self-esteem. If you'd like to see an example of how researchers describe the results of a mixed model analysis, you may refer to Appendix 8.1 at the end of this chapter.

QUESTION 8.4. After you have observed and interpreted the potential Self-Esteem × Trait Valence interaction effect (Oops! We weren't supposed to say that), it should be informative to compare this interaction effect with the results you would obtain by generating a *self-enhancing score* for each participant (a simple difference score that reflects the degree to which each participant endorses the positive trait to a stronger degree than he or she endorses the negative trait) and simply correlating this self-enhancement score with each person's self-esteem score. What is the correlation between self-esteem and this self-enhancement score? (This self-enhancement difference score should already be a part of your data file.) More important, what is the exact p value associated with this correlation? How does this p value compare with the p value for the interaction term you observed in your mixed model ANOVA (the one that involved self-esteem and the repeated measures variable of trait valence)? Does the similarity or dissimilarity of these two p values tell you anything?

A Repeated Measures Study of Optimism With Countries as the Unit of Analysis

Although the large majority of psychological research studies treat *research participants* as the basic unit of analysis, it is possible to treat other entities as the basic unit of analysis. For example, geographic units, rats, or crustaceans are sometimes a useful unit of analysis rather than people. You came very close to taking part in this kind of analysis in Chapter 2—when we compared the ethnic diversity of different U.S. states. Nothing would stop us, for example, from *correlating* the ethnic diversity scores of the 50 individual U.S. states with the percentage of people in each of the 50 states who voted for Barack Obama in 2008. Along somewhat different lines, macroeconomists might wish to know whether cities or countries that employ certain kinds of economic policies have better short- or long-term economic outcomes. Political leaders or policy makers might wish to know whether, on average, cities, states, or countries that differ in their average level of education also differ in their social or economic outcomes. Any time researchers measure (a) the same variable on several occasions in a set of cities, states, or nations or (b) measure two or more conceptually related variables in the same cities, states, or nations, they will usually be setting the stage for some kind of repeated measures analysis. We explore one such analysis here.

A great deal of research on North American and Western European populations has shown that most people possess unrealistically positive views of the self and maintain unrealistically rosy views of the future (e.g., Taylor & Brown, 1988; Weinstein, 1984, 1987). Most people think their future lives will be much better than their present lives. Furthermore, most people seem to think their future lives will be rosier than an objective account suggests they are likely to be. The health psychologist Neil Weinstein has argued that unrealistic optimism about the future can lead people to make some very risky health decisions (Why quit smoking when I am immune to lung cancer?). In contrast, Taylor and Brown have argued that, more often than not, optimistic biases make us happier, healthier, and more successful. The comedian and storyteller Garrison Keillor was wise enough not to take sides in the controversial issue of whether optimism is healthy or unhealthy in the long run. However, he did seem to believe that optimism is highly pervasive. Keillor is well known for his colorful stories of life in fictional Lake Wobegone, Minnesota, where "all the men are good-looking, all the women are strong, and all the children are above average."

Experts in cross-cultural psychology, however, have argued that many of the optimistic biases that have been documented so well in Western cultures may not exist (or may exist in muted form) in many non-Western cultures (e.g., Asian or African cultures). To see if optimism about the future is strictly a Western phenomenon, the first author capitalized on data from Gallup's ongoing World Poll and compared the present and predicted future *well-being* scores of representative samples of people in 154 countries worldwide. As you may recall from an earlier chapter, the Gallup World Poll covers well over 90% of the earth's population and surveys people in their native languages on topics ranging from the political and economic to the personal and spiritual. A core question that has been included in every annual World Poll survey since it began in late 2005 is based on the Cantril Self-Anchoring Striving Scale or "Ladder of Life." This ladder well-being question asks people to rate their present lives on a scale from 0 (*worst possible life*) to 10 (*best possible life*). Importantly, the World Poll survey also asks people to use the very same scale to report what they think their lives will be like in exactly 5 years. As you might guess, people who say their present lives are wonderful also tend to say that their future lives are wonderful, whereas those who say their present lives are sucky are also less upbeat about their futures. In other words, current and future well-being are positively correlated. This is true at both the respondent (micro-data) level and at the country level. So in countries, for example, where the average person says life is currently great, most people also say they think the future is going to be great. This makes a *repeated measures* approach an ideal way to compare present and perceived future well-being in different countries.

One very simple way to look at optimism is to see if people think their future lives will be better than their present lives and, if so, by how much. If we analyze World Poll data at the country level, this means treating individual countries as if they were people (making mean country well-being scores the basic data points, which means making individual countries the unit of analysis). Of course, treating individual counties as the unit of analysis is an extremely conservative way to analyze the data from a survey that included almost 600,000 people at the time the first author harvested these data, but it provides us with another opportunity to conduct a repeated measures analysis. The joys of within-subjects (or mixed model) analysis may well be worth the huge loss of statistical power. A partial screen capture for this worldwide study of well-being appears in Image 9. As you can see, the data file for this study is called [**present & future ladder w-being 4 re measures nov 3 2009.sav**].

In case you are wondering why Image 9 shows actual country names for variable wp5 whereas your SPSS data file (probably) shows only numbers for this wp5 (country) variable, check out Image 10 to learn about an

Chapter 8 Within-Subjects and Mixed Model Analyses

important feature of how SPSS data files appear to viewers. From the "View" tab, you always have the option of going to the "Value labels" button and toggling back and forth between showing your value labels or showing the numeric (or letter) codes for your variables in any given SPSS data file. We changed the view to the labels view because it is much more meaningful to see country names than to see 154 arbitrary country codes. Notice that this toggle switch also "turned on" the region labels as well, so you see "Africa" for a bunch of countries in Image 9 rather than "1."

Incidentally, the data in this file are based on every available World Poll data point as of late October 2009. For most of the countries you see in the file, country well-being scores are based on an average of either 3 or 4 years of data (usually involving at least 1,000 people per year). In a few countries, though, World Poll data were available only for one year. So long as at least one year of present and future well-being scores were available for a given country, we included that country in this data file. Finally, we should note that, if we really wanted to produce a worldwide assessment of the magnitude of the optimistic bias, we might want to weight countries by their populations. We did *not* do this here. This is partly because we wanted to keep things simple and partly because we decided that each country has its own unique culture. So be forewarned that, for better or worse, we are weighting huge countries such as China exactly like tiny countries such as Luxembourg.

Here are some screen captures that will remind you of how to conduct the repeated measures analysis to see if it is the case in most countries in the world that people believe their future lives will be better than their present lives. Image 12 tells you that we'd like you to name your repeated measures (time) variable "pres_v_future," and of course you'll need to let SPSS know that this variable has two levels, as we have done in Image 12. From this point, you'll be following the same procedures you followed earlier in this chapter in the third within-subjects bogus traits study. However, in this case, you will not (initially) add any kind of between-subjects variable to the analysis because initially we are just seeing if there is an average ("main") effect of the time period (now versus the future) people are asked to evaluate.

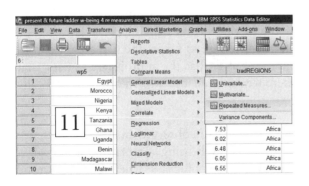
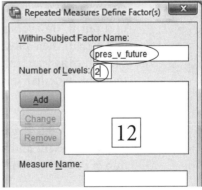

The short version of the remainder of these instructions (picking up in Image 12) is as follows: (a) Click the "Add" button, (b) click the "Define" button to specify which two within-subjects variables represent the "pres_v_future" variable, and (c) click the "Options..." button to open a new dialog box that will allow you to ask for means and descriptive statistics. The circled areas of Image 13 show you what things should look like as you are nearing the point where you can actually run the repeated measures analysis.

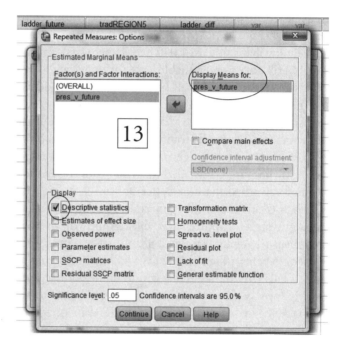

If you click "Continue," this will return you to the main dialog box where you can click either "OK" or "Paste" depending on your preference for using syntax files. "Paste" will send your command to a syntax file, where you can examine it and then run it; "OK" will simply run it.

QUESTION 8.5. Report the results of this repeated measures ANOVA to find out if, on average, people predicted that their future lives would be better than their present lives. Make sure to report the two separate well-being means (present and future) as well as the appropriate statistic that indicates whether any difference between the two means is statistically significant.

QUESTION 8.6. Regardless of whether you observed an average difference between present and perceived future well-being, a more important question of interest to most cross-cultural psychologists is whether the optimistic

Chapter 8 Within-Subjects and Mixed Model Analyses

bias occurs *to the same extent* in different world regions (sort of the way we wanted to know if the self-enhancement bias for bogus traits occurred *to the same extent* for people low versus high in self-esteem). It is possible, *in principle*, that there could be a large optimistic bias in many Western (i.e., independent) countries and a small pessimistic bias in many Eastern, Latin American, or African (i.e., collectivistic) countries. If this were the case, this would mean that the answer to Question 8.5, whatever it is, is a bit misleading because the magnitude or even the direction of the optimistic bias could vary greatly from region to region. If you look carefully at your data file, you will notice that we have carved the 154 countries into five distinct world regions. Although you might wish to quibble with a few of the assignments, let us take it for granted that this is one reasonable way to group the 154 countries. Your assignment in Question 8.6 is to expand the analysis you just ran by *adding* the region variable (tradREGION5) as a between-subjects variable to the same within-subjects analysis you just ran. Image 14 shows the crucial screen on which you will do most of this additional work. (Don't forget, however, to ask for the appropriate descriptive statistics you'll need to see what is really happening in these data.) Is there an interaction between the within-subjects (in this case, within-countries) present versus future well-being variable and the between-subjects (between-countries) region variable? If so, be sure to conduct and report some follow-up tests to specify the nature of the interaction. Although many different tests are possible, given the fact that there are five world regions, it will be sufficient for the purposes of this data analysis exercise to compare the most collectivistic region of the world (Latin America) with the least collectivistic (i.e., most individualistic) region of the world. Of course, this least collectivistic region is Europe, along with the four English-speaking countries that are all former British colonies. If you were thinking Asia is the most collectivistic region of the world, you are not alone, but this is a common misconception (see Hofstede & Hofstede, 2005).

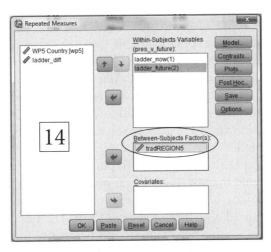

A Mixed Model Study of Implicit Political Attitudes

Keith Payne and his colleagues (Payne, Cheng, Govorun, & Stewart, 2005) have developed a technique called the affect misattribution procedure (AMP) to assess implicit (i.e., unconscious) attitudes. The logic of the AMP is that if you have just been exposed to a photo of something you like or dislike, the pleasant or unpleasant gut reaction you have to that thing might influence your ratings of a highly ambiguous stimulus that appears immediately *after* that thing. Moreover, because Payne et al. felt that this process runs its course *unconsciously,* they typically give participants a clear warning prior to the beginning of their judgment tasks—asking them to try not to allow their feelings about the first things they see to influence their ratings of the pleasantness of some neutral stimuli that follow. In one of their first experiments using the AMP, Payne et al. (2005) briefly flashed photos of political candidates (George Bush or John Kerry) in the middle of a computer screen. (They also included neutral primes as a control, but we are simplifying their design a bit by ignoring this third within-subject condition.)

These photos of Bush and Kerry were presented multiple times, and each photo was always followed quickly by an image of a different Chinese character, carefully pretested to be both neutral and unfamiliar to people who cannot read Chinese. Participants' task was to rate each individual *Chinese character* for whether it was more pleasant or more unpleasant than average (for all of the many characters). Notice that because all of the participants made pleasantness ratings following Bush primes and following Kerry primes, this study included a within-subjects variable. Because Payne et al. (2005) expected participants to have very different gut reactions to Bush and Kerry based on whatever was driving their participants' voting preferences, the researchers categorized people based on whether they said they would have voted for Bush or Kerry if the election had been held the day of the study. Of course, this voting preference variable had to be a between-subjects variable.

QUESTION 8.7. We've created an SPSS data file [**Payne et al AMP approximation data.sav**] that contains a very close approximation of the original Payne et al. (2005) AMP election data. *Your job is to analyze these data and decide what conclusions the data suggest about the validity of the AMP as an implicit measure of attitudes.* For the purpose of this exercise, you may assume that each participant saw 12 neutral Chinese characters after a Bush photo and 12 neutral characters after a Kerry photo. The two variables that follow the "preference" variable in your data file indicate *how many* of the 12 stimuli (characters) in each set each participant rated as more favorable than average. Thus, you can see that the first participant rated 11 of the 12 neutral characters as more favorable than average when they followed a picture of George Bush. The next two variables in the file are simply the two "stimuli" ratings converted to proportions (thus, 11/12 becomes .917 and 6/12 become .500). We'll work with the

Chapter 8 Within-Subjects and Mixed Model Analyses

raw numbers (that will thus range from 0–12) rather than the proportions. It will thus be useful to remember that 6.0 is a meaningful (chance) midpoint for these scores because it reflects no bias in either a favorable or unfavorable direction.

Because you've already had a couple of chances to conduct a mixed model ANOVA, we now provide you with some suggestions and reminders rather than with a set of detailed instructions. First, to make it easy for your instructor to see if your analysis is correct, please label the within-subject variable "typeofprime" (Bush images vs. Kerry images). Second, don't forget to click the "Options..." button from the main analysis screen so that you can (a) ask for "Descriptive statistics" and (b) send "preference*typeofprime" to the "Display Means" box. Images 15 through 17 should also help you check your work.

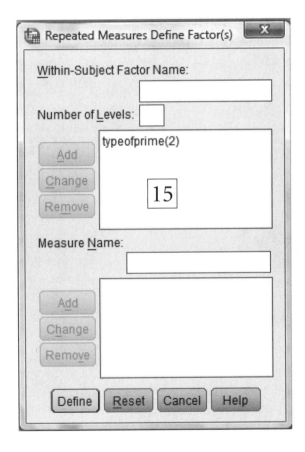
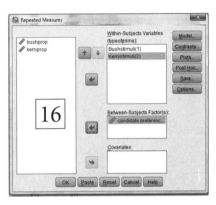
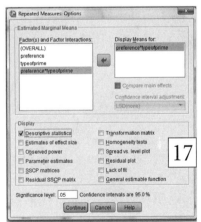

Be sure to report whether you observed any main effects as well as whether you observed an interaction. If you did observe an interaction, you'll want to follow it up with some simple effects tests (separate within-subjects tests for those who planned to vote for Bush and those who planned to vote for Kerry).

QUESTION 8.8. Limitations of within-subjects designs. OK, we hate to take a step backward, but remember that we began this chapter by discussing within-subjects designs. Given the incredible power and precision of within-subjects designs, we'd like you to give some thought to the question of why anyone ever goes to the trouble of conducting a purely between-subjects (or mixed model) design. Answer this question in at least two ways. First, what serious methodological problem can easily occur in within-subjects—but not in between-subjects—designs? (Make sure that you briefly discuss how you usually fix this problem.) Second, identify at least two studies, or classes of studies, in which it is simply impossible to use a within-subjects design. One big clue to this question is "reversibility." Another clue is "manipulatability." Are there some experimental manipulations that can't be reversed? Are there other manipulations that can't reasonably, or ethically, be done in the first place?

Appendix 8.1: Sample Results of a Study Using a Mixed Model Design

This appendix contains the results of a mixed model study taken from a research article that radically changed the way many researchers think about stereotypes and prejudice. Prior to the publication of this paper (Lowery, Hardin, & Sinclair, 2001), there was a great deal of evidence that stereotypes and prejudice have a substantial automatic (i.e., unconscious) component (Devine, 1989). In other words, stereotypes and prejudice become so familiar and well practiced early in most people's lives that they live on in unconscious form even after certain open-minded adults put the conscious part of their stereotypes and prejudice behind them. Research showed, for example, that highly prejudiced and less prejudiced people differed greatly in the opinions that they consciously endorsed but seemed to differ very little in the degree to which they automatically associated certain stereotypic things together. For example, research showed that both prejudiced and nonprejudiced people alike found it easier to pair Black names or faces with negative words than to pair White names or faces with negative words (and the reverse occurs for positive words; it's easier for most people to pair them with White names or faces). These disturbing results suggested that—at some very basic level—prejudice and stereotypes are with us all—and perhaps that they are here to stay.

Lowery et al. (2001) did not doubt the many pioneering studies that showed such effects, but they did strongly suspect that stereotypes and prejudice—even automatic ones—might be dynamic and flexible rather than rigidly cast in stone. To see if they could push automatic prejudice around a little, they adapted a now-classic research design intended to reveal automatic prejudice, but they added one (between-subjects) manipulation that no one had ever thought to add before: the race of the likable, professional experimenter who was carrying out the study in the first place. The key prediction of this experiment, based on a theory called shared reality theory, is that even people's unconscious gut associations about Blacks and Whites would vary depending on whether the guy who ran the experiment happened to be Black or White.

The first experiment in the Lowery et al. (2001) paper was a mixed model design, by the way, because the researchers manipulated the *race of the experimenter* on a between-subjects basis but manipulated the *nature of some names and words* people had to categorize quickly (e.g., Is Tyrone a stereotypically Black name or a stereotypically White name?) on a within-subject basis. We won't cover all the details of their methodology, but we do need to tell you that the researchers labeled stimuli as *congruent* if they involved more negative associations about Blacks than about Whites and *incongruent* if they involved more negative associations about Whites than about Blacks. So think of *congruent* as prejudicial if you like, and think of *incongruent* as counter-prejudicial if you like. They key finding from past research is that people will make more correct categorizations with congruent than with incongruent stimuli (e.g., it'll be easier for me to associate names like Tyrone with a word such as *fear* than with a word such as *sunrise*).

Here, then, are the results of Experiment 1 in the Lowery et al. (2001) paper (our only edits were to remove a couple of footnotes):

> Error rates were small—on average less than 5%—and hence comparable to those reported by Greenwald et al. (1998).
>
> To test the hypothesis that automatic prejudice is subject to social influence, we examined automatic prejudice as a function of experimenter race in a 2 (critical phase: congruent vs. incongruent) × 2 (experimenter race: Black vs. White) mixed model analysis of variance (ANOVA), with critical phase as the within-subjects variable. Results replicated previous demonstrations of anti-Black prejudice by showing that participants categorized more items in the congruent phase (Black-negative/White positive) than in the incongruent phase (White negative/Black-positive), as indicated by a significant main effect of critical phase, $F(1, 31) = 31.18$, $p < .001$, $\eta^2 = .501$. As predicted, however, experimenter race moderated this effect. As shown in Figure 1, participants exhibited more automatic anti-Black prejudice in the presence of a White experimenter than a Black experimenter, as indicated by a significant Critical Phase × Experimenter Race interaction, $F(1, 31) = 8.44$, $p < .01$, $\eta^2 = .214$. Although participants made more congruent than incongruent categorizations in the presence of the White experimenter, as indicated by a significant simple main effect of critical phase, $F(1,18) = 53.19$, $p < .001$, $\eta^2 = .747$, this difference disappeared in the presence of a Black experimenter, as indicated by a nonsignificant simple main effect of critical phase, $F(1,13) = 2.48$, $p = .143$, $\eta^2 = .158$. (pp. 844–845)

If you work your way carefully through these results, you should get a good feel for how to report the results of a study with a mixed model design. You may also learn a little bit about the social nature of automatic thoughts and feelings in the process.

Multiple Regression 9

Introduction: Ceteris Paribus

There are many times when a researcher wishes to evaluate which of several predictors of a phenomenon is the best predictor of the phenomenon. Although it is almost always risky to draw causal conclusions from correlational data, another way to put this is that researchers often wish to determine which of several potential causes of a phenomenon is the best candidate for the *true* cause. For instance, if you learned that the number of dentists per capita in a large, representative sample of American cities is a good predictor of crime rates in these cities, you probably would *not* conclude that dentistry causes crime. Instead you would look for a more plausible cause that happens to be correlated with the popularity of dentistry. Perhaps, the number of liquor stores, gun shops, or candy manufacturers per capita is the true reason for the spurious connection between acts of dentistry and acts of crime. Or perhaps it is just a matter of capita. That is, perhaps both dentistry and crime occur more often in cities with greater population densities. Determining which of several competing predictors of an outcome is the best predictor of the outcome is often an important first step in figuring out what the phenomenon is all about. Multiple regression analyses are the most common analytic tool for achieving this goal with passive observational data. In the example just given, a researcher who predicted crime rates from (a) population densities, (b) the number of liquor stores per capita, and (c) the number of gun shops per capita could use multiple regression techniques to evaluate the unique association between each of these three predictors and crime rates. That is, he or she could make a statement about whether each of these three variables predicted crime rates *while holding levels of the other two variables constant.* For instance, if we hold both population densities and gun store ownership at some constant mean level, is there *still* an association between the popularity of liquor stores and the popularity of crime?

If we had the right kind of data at our disposal, we could ask the same kind of question about almost anything. *In principle,* multiple regression techniques could be used to see (a) if damage to the ventromedial hypothalamus or the lateral hypothalamus was more closely associated with a particular feeding disorder in rats, (b) which of several specific TV commercials was associated with the greatest increases in sales of Lay's Potato Chips (e.g., during a test market), or (c) which variable (e.g., gender, political party affiliation, sex-role attitudes) best predicts people's attitudes toward electing a female U.S. president. In practice, of course, multiple regression techniques are used to address some of these kinds of research questions much more often than others. Behavioral neuroscientists don't usually ask questions about the brain using multiple regression analyses because they can selectively destroy specific areas of the brain (and only those areas) to answer such questions. (That being said, however, the increasing popularity of passive brain imaging techniques such as functional magnetic resonance imaging [fMRI] has increased the use of regression techniques in some areas of neuroscience.) If we could selectively destroy some gun shops in some cities and force hordes of people to relocate—so as to hold population densities constant—we wouldn't need multiple regression to get a better handle on what best predicts crime rates. We could just see if the cities that had more gun shops after our massive experimental intervention came to have higher crime rates. Of course, we cannot usually enact massive interventions like this one (although some test markets come pretty close). This is why we often need covariance techniques such as multiple regression analysis.

What all these techniques (e.g., regression, analysis of covariance [ANCOVA], path analysis, structural equation modeling) have in common is that they allow us to disentangle the associations between several different variables to assess each variable's unique association with the variable we would like to predict. Whenever you hear the Latin term *ceteris paribus* (KAY-ter-us PAIR-uh-bus: "all else being equal"), you should think of multiple regression because multiple regression analysis attempts to examine the association between one variable and another ceteris paribus—that is, all else being equal (with all other variables held constant).

Predictor Variables and Criterion Variables

Conveniently enough, the potential variables that may predict an outcome of interest are referred to as **predictor** variables. *Inconveniently* enough, the outcome we wish to predict is called the **criterion** variable (by more common use, the word *criterion* means deciding factor, standard, or reason). Just remember that statisticians were nice enough to use the more familiar term *predictor* to indicate just what they meant. You might also remember that multiple regression is very PC. **P**redictor (i.e., cause) comes before **C**riterion (i.e., effect). Nonetheless, you should also remember that in a regression analysis, it is sometimes arbitrary which variables you call independent variables (presumed causes) and which you call dependent variables (presumed outcomes). If gun shops are associated with crime, then gun shops *could* cause crime. However, as the late Charlton Heston and his associates at the National Rifle Association might be quick to note, crime could also cause gun shops. That is, gun shops might spring up *in response* to crime—as decent, peace-loving people like you and me go out and purchase automatic assault weapons so we can shoot back at all those bad people who shot at us first. To be a little fairer to the NRA, nonexperimental data regarding gun shops and crime rates cannot definitively answer this causal question. To be fairer still, Levitt and Dubner (2005), the authors of the popular book *Freakonomics,* argue that careful multiple regression analyses have shown that gun ownership rates are *not* uniquely associated with crime rates. Although it is possible we would interpret their data differently than they do, we applaud their efforts to use a rigorous statistical technique to try to answer a thorny social question.

The Logic of Multiple Regression Analysis

But how does multiple regression analysis work? By what means can we decide which potential predictors of a criterion variable are uniquely associated with that variable? By means of plain and simple logic, as it turns out. Multiple regression techniques do—in a very powerful and sophisticated way—what you yourself could do logically if you could keep track of observations as well as computers can. To acquaint you with the logic of multiple regression analysis, let's look at a simple data set in which it is safe to make some assumptions about causality. If you analyze this situation logically, you should see that a multiple regression analysis suggests the same conclusions about data that a logician or a juror would if he or she were making careful judgments about likely causes.

Another way to put this is that one good way to see how multiple regression works is to compute the results of a multiple regression analysis *in your head* and compare your conclusions with those you'd draw if you performed a regression analysis on the same data! We're willing to bet that this is easier than you might think. The data you will evaluate come from an unobtrusive observational study of cookie theft (we'll let you guess whether the data are real). To test our suspicion that our acquaintance Lisa was stealing cookies from Brett's kitchen cookie jar, we made observations on 12 different days. At the end of each day, we recorded (a) whether Lisa had visited the kitchen and (b) whether any cookies were missing from the cookie jar. To convert the data to numbers, we coded them as follows: For Lisa's presence in the kitchen, a "0" means that she had not visited, and a "1" means that she had visited on a given day. For cookie thefts, a "0" means that cookies had not been stolen, and a "1" means that cookies had been stolen on a given day. The data are presented in Table 9.1.

Table 9.1 Hypothetical Data From an Observational Study of Cookie Thefts

	Lisa Visited?	Cookies Stolen?
Day 1	No (0)	No (0)
Day 2	No (0)	No (0)
Day 3	No (0)	No (0)
Day 4	No (0)	No (0)
Day 5	No (0)	No (0)
Day 6	No (0)	Yes (1)
Day 7	Yes (1)	No (0)
Day 8	Yes (1)	Yes (1)
Day 9	Yes (1)	Yes (1)
Day 10	Yes (1)	Yes (1)
Day 11	Yes (1)	Yes (1)
Day 12	Yes (1)	Yes (1)

We conducted an analysis of these data and recorded the correlation. In the unusual case of these data, you can calculate the correlation yourself by hand using the following extremely simple formula (which, as far as he can recall, was derived by your primary author):

$$r = (\text{matches} - \text{mismatches})/n.$$

In this super simple formula, *matches* refers to observations for which the two variables take on the same value, and *mismatches* refers to observations for which the two variables take on different values. For example, if there were 12 matches and 4 mismatches in a set of 16 observations, the correlation would be $(12 - 4)/16$, which is 8/16 or .50. Be forewarned that this super simple formula only works perfectly when you are correlating two dichotomous variables, both of whose outcomes are equally likely—as they happen to be here. If cookies disappeared on about 60% of all days rather than exactly 50% of all days, this super simple formula would be slightly inaccurate. If Lisa appeared in the kitchen only about 10% of the time (about 1 day in 10), this super simple formula could be super inaccurate!

QUESTION 9.1. Who's the cookie monster? Assume that these are the only data you have at your disposal. What is the correlation between Lisa's presence and cookie thefts? Based on this correlation, what conclusion would you draw? That is, how likely is it that Lisa is guilty of at least some of the cookie thefts? (She need not be the only thief in the world to be considered guilty.)

Chapter 9 Multiple Regression

Considering More Data

Now suppose that we had conducted a more thorough study—one in which we also observed the behavior of Lisa's brother Bart on the same 12 days. Because Lisa and Bart are both children of our friend Homer, they tend to come as a package (i.e., their presence in the kitchen is *confounded* or positively correlated). Because this is the case, you could think of Bart as a "third variable." It is obviously possible that Bart is the true culprit. Statistically, it is possible to conduct a regression analysis to get a better idea of who is responsible for the cookie thefts. First, we could compute a regression coefficient (a lot like a correlation coefficient) to describe the unique association between Lisa's presence and cookie thefts (controlling for Bart's presence). Second, and just as important, we could also compute a regression coefficient between Bart's presence and cookie thefts (controlling for Lisa's presence). In other words, we could try to see who is the true cause of the cookie thefts. The complete data for this expanded study appear in Table 9.2.

QUESTION 9.2a. Who's seems more guilty, Bart or Lisa? If we gave these data to a computer, the computer would give us a very clear idea of who is responsible for the cookie thefts (in the form of a couple of beta weights or regression coefficients). Based on the additional data, who would you blame? In answering this, be sure to report whether you feel you now have a better grip on who was responsible for the cookie thefts.

Table 9.2 Some Additional Hypothetical Data From an Observational Study of Cookie Thefts

	Lisa Visited?	Bart Visited?	Cookies Stolen?
Day 1	No (0)	No (0)	No (0)
Day 2	No (0)	No (0)	No (0)
Day 3	No (0)	No (0)	No (0)
Day 4	No (0)	No (0)	No (0)
Day 5	No (0)	No (0)	No (0)
Day 6	No (0)	Yes (1)	Yes (1)
Day 7	Yes (1)	No (0)	No (0)
Day 8	Yes (1)	Yes (1)	Yes (1)
Day 9	Yes (1)	Yes (1)	Yes (1)
Day 10	Yes (1)	Yes (1)	Yes (1)
Day 11	Yes (1)	Yes (1)	Yes (1)
Day 12	Yes (1)	Yes (1)	Yes (1)

QUESTION 9.2b. To be more specific, provide your estimate of the unique association between Bart's presence and cookie thefts. A good way to think about this is to ask yourself what portion of the blame for the cookie thefts should logically be placed on Bart—given everything you know about what is going on. If Bart seemed to deserve about half of the blame, for example, you might estimate his standardized regression coefficient at about .50 (or about .71, which becomes .50 when squared). You do not necessarily have to hit the answer to the regression coefficient right on the head, but you must logically defend your answer.

QUESTION 9.2c. Now provide your estimate of the standardized regression coefficient representing the unique relation between Lisa's presence and cookie thefts. That is, how uniquely responsible is Lisa for the cookie thefts? (Be sure to defend your answer.)

Checking Your Answers in SPSS

Perhaps you wish you could check your work by conducting a multiple regression analysis in SPSS. Let's do that. The SPSS data file you will use is called [**lisa bart regression demo.sav**], and it will allow you to check your logic while also serving as a useful introduction to the nuts and bolts of multiple regression analysis in SPSS. Before you run a multiple regression analysis on these data, however, you should take a quick peek at the value labels for the three variables in your SPSS data file to confirm that we coded the data in a manner that is consistent with what you see in Tables 9.1 and 9.2. You should see, for example, that we coded the data so that 0 means cookies are not missing and 1 means cookies are missing. Once you've confirmed this, check out Images 1 and 2, which show you where to point and click to begin a multiple regression analysis on these hypothetical cookie theft data.

Once you click "Linear. . ." (see Image 2), you'll open the dialog box you see in Image 3.

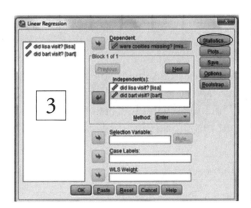
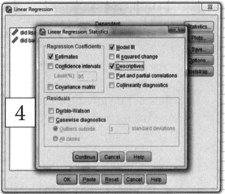

By now, you should be pretty familiar with how to send variables from your variable list to the appropriate place in an analysis box, so we're not going to belabor this. However, make sure you remember (as shown in Image 3) that "missing" is your dependent variable and "bart" and "lisa" are your two independent variables. You might also notice that SPSS labels variables using the language of experiments and ANOVA rather than using the more traditional statistical terms *predictor* (independent variable) and *criterion* (dependent variable). Once you've entered your independent and dependent variables, click on "Statistics. . ." (circled in Image 3) to generate the Statistics box you see in Image 4.

The default statistic "Estimates" will tell you what your regression coefficients are for your predictor variables. This will include both **Bs** or unstandardized regression coefficients and **beta weights** or standardized regression coefficients. The other default statistic, "Model fit," tells you whether your overall model was significant (i.e., collectively, did your predictors significantly predict anything?). In addition to accepting these two very useful defaults, you should also check "Descriptives." Doing so will give you some very useful descriptive information, such as means, standard deviations, and all of the simple correlations between all of the variables that went into your multiple regression analysis.

When you click "Continue," you will be whisked back to the main "Linear Regression" box. Once you get there, click "Paste" to create a syntax file that includes your multiple regression commands. The crucial part of your syntax file should read as follows:

REGRESSION

/DESCRIPTIVES MEAN STDDEV CORR SIG N

/MISSING LISTWISE

/STATISTICS COEFF OUTS R ANOVA

/CRITERIA=PIN(.05) POUT(.10)

/NOORIGIN

/DEPENDENT missing

/METHOD=ENTER lisa bart.

As usual, click anywhere in the analysis statement, or highlight the entire statement. Then click the "Run" arrow to run the analysis. Doing so will generate an output file that will include several important things. The most important section of the output file appears in Image 5.

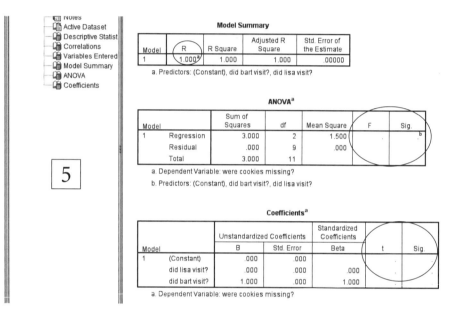

In comparison with any output files you are ever likely to see based on analyses of real data, this output file is extremely unusual in three ways. First, it is virtually unheard of to obtain a perfect multiple R as we have done here. A multiple R of 1.00 means that your predictors perfectly account for the scores on your criterion (cookie thefts). Needless to say, this will practically never happen with real data. The other two unusual features of this output file are both a direct result of this perfect multiple R.

In particular, SPSS didn't generate an F value or a p value to correspond to your perfect multiple R. This is because an F ratio (i.e., F fraction) is always a function of the size of your error term (with the error term serving as the denominator). When prediction is perfect, there is 0 error, and we thus have division by zero, which is impossible (although some of us keep trying). A more intuitive way to put this is that the F ratio that corresponds to your overall regression model is so large that it's impossible to calculate! The same thing holds for the t value that is normally listed in the "Coefficients" box as a test for the significance of each individual predictor (each regression coefficient) in the regression equation. Needless to say, it'd be nice to ever have this problem with real data. We should also add that it was easier than usual to experience this problem because we analyzed dichotomous data using a technique designed to deal with continuous data. Although multiple regression is pretty robust to some violations of the assumption that your predictor and criterion variables are normally distributed, the dichotomous nature of these data did contribute to this glitch in the output. With a sample size as small as ours, it's much easier to observe a perfect correlation when you're dealing with dichotomous as opposed to continuous, normally distributed variables (a point we'll reinforce later in this chapter).

Finally, if you're wondering why we broke the usual rules in the first place and used dichotomous variables here, remember that the goal of this exercise was to let you explore the logic of how regression works. It's much easier to think about dichotomous rather than continuous scores. So to make a long story a little longer, we hope you can see that multiple regression analysis is based on a very logical analysis.

QUESTION 9.3. The most helpful days. Just to be sure you appreciate the logic of multiple regression analysis, check back on the raw data and ask yourself which 2 of the 12 days proved to be the most informative with regard to who was responsible for the cookie thefts. Which 2 days were these (which 2 were most informative) and why? (Be concise.)

Now if all research questions were as simple as our example involving cookie thefts, we might be able to abandon multiple regression analyses altogether in favor of simple logic. However, imagine how hard it would be to figure out who the real cookie culprit was if there were four or five potential suspects. It might require hundreds of days of careful observations to get a clear handle on each person's precise degree of guilt. But with hundreds of days of observations to keep track of, the kind of mental arithmetic required to logically deduce any answers would be difficult if not impossible. The same thing is often the case when you collect data involving several potential predictors of romantic love, employee satisfaction, or consumer purchase intent. You need a powerful tool to do the logical analysis for you. That tool is multiple regression analysis (and its many variations). Let's apply that tool to a couple of more realistic data sets.

Correlation, Multiple Regression, and Multiple Predictor Variables

Let's begin by revisiting a data set that you've already seen before. As you may recall, your first SPSS activity in Chapter 3 involved a data set that focused on variables such as height, gender, self-esteem, education, and income (age and weight were also thrown in for good measure). Previously, you ran some correlations on these data. Then you speculated about what some of these correlations might mean. Finally, you offered criticisms for your own speculations. As a means of further exploring multiple regression, let's focus on one of the most interesting correlations in that data set and see if we can rule out some of the alternative explanations critics might offer for the correlation. As you may recall, these data revealed a substantial positive correlation between height and personal income. (The bigger they come, they more money they bring with them, you might say.) The most interesting interpretation for this finding is grounded in stereotypes regarding height and success or leadership. Perhaps people value tall people more or expect them to be better leaders. As a result, people (such as bosses) might be more likely to offer tall people nice starting salaries, to succumb to taller people's requests for raises, and so on. As you probably noted, however, several less interesting alternate explanations could explain the zero-order correlation (the simple correlation) between height and income. Let's consider two of these alternate explanations. Each alternate explanation will require a different regression analysis.

The first problem is that young people tend to be taller than old people. This empirical fact about height is reflected in a joke about great grandmothers: Fully half of them never die; they just get so short you can't find them anymore. And young people *might* also tend to make more money than older people—either because young people are more highly educated or because they are much less likely to be retired. So in other words, height could be confounded with age. Is height still positively associated with income when you statistically hold the ages of tall and short people constant in a multiple regression analysis? To find out, you would need to follow the model we followed for the cookie theft example and enter both (a) height and (b) age as predictors of income. The results of this multiple regression analysis should give you a better idea of whether height still predicts income when you statistically control for age.

The second problem is that women tend to be shorter than men, and it is well-established that women generally earn less money than men. That is, gender is **confounded** with height. In a second multiple regression analysis, we'd like you to predict income from (a) height, (b) age, and (c) gender. You'll enter all three independent variables at once and let them all compete as predictors of income.

The file in Image 6 may look a little different than your data file, but this image should remind you of where to point and click to run your regression analyses. This image should also remind you that the file you will use is called [**hypothetical data for correlation activity 1.sav**].

Chapter 9 Multiple Regression 241

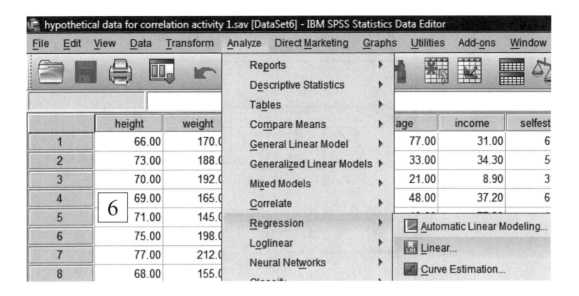

When you're ready to run the first multiple regression analysis, your main analysis box should look like Image 7. And when you're ready to run the second regression analysis, your main analysis box should look like Image 8.

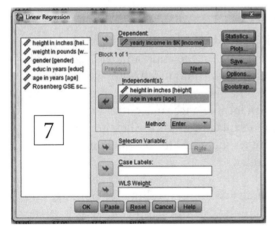
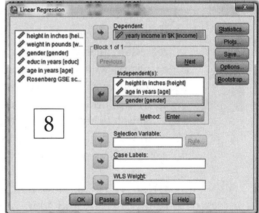

Clicking "Paste" from either of these screens will send your regression command to your syntax file, and from there you can run both of these two regression analyses. When you run the first multiple regression command, it should generate an output statement that looks a lot like the one you see in Image 9.

Model Summary

Model	R	R Square	Adjusted R Square	Std. Error of the Estimate
1	.536	.287	.234	19.09383

a. Predictors: (Constant), age in years, height in inches

ANOVA[a]

Model		Sum of Squares	df	Mean Square	F	Sig.
1	Regression	3965.678	2	1982.839	5.439	.010[b]
	Residual	9843.511	27	364.574		
	Total	13809.190	29			

a. Dependent Variable: yearly income in $K
b. Predictors: (Constant), age in years, height in inches

Coefficients[a]

Model		Unstandardized Coefficients B	Std. Error	Standardized Coefficients Beta	t	Sig.
1	(Constant)	-100.498	66.601		-1.509	.143
	height in inches	2.167	.908	.436	2.388	.024
	age in years	-.191	.203	-.172	-.940	.355

a. Dependent Variable: yearly income in $K

9

We've circled four important things in Image 9 to which you should pay attention. However, by far, the most important parts of your output file are the beta weights and the *p* values associated with those beta weights. We've placed boxes around the beta weights and the *p* values that correspond to (a) the unique association between height and income as well as (b) the unique association between age and income.

QUESTION 9.4. What conclusion would you draw from the results of this regression analysis? Do these results support the criticism that height is only correlated with income because height is confounded with age? That is, is height *uniquely* associated with income once you control for age? What about age? Is age uniquely associated with income once you control for height?

Now, let's subject our preferred explanation for the height-income correlation to an even more stringent test designed to rule out concerns about age as well as concerns about gender.

QUESTION 9.5. Now, do the same analysis controlling for both age and gender. Specify and address the unique role of each predictor as it relates to income.

R-Square, Adjusted *R*-Square, and Standard Errors in Multiple Regression

Recall that an *R*-square statistic is literally the squared value of the multiple *R* that summarizes the total ability of a specific set of predictors to predict the criterion variable of interest. A multiple *R*, then, is the multiple regression equivalent of the lowercase r that signifies a regular (i.e., simple, zero-order) Pearson correlation coefficient. Furthermore, just as r^2 (the coefficient of determination) tells you the percentage of variance in y you can account for by knowing people's scores on a single x variable, R^2 tells you the percentage of variance in the criterion you can account for by knowing people's scores on the total *set* of predictors in a multiple regression equation. To emphasize the close connection between correlation and multiple regression, consider what you'd expect to happen in a multiple regression analysis with *two completely uncorrelated predictor variables, each of which was moderately correlated with your criterion variable.* Specifically, consider the data you'll find in the SPSS file [**multiple R meaning demo.sav**]. This file contains hypothetical data on only three variables. The first, income_sat, refers to people's satisfaction with their personal income. The second, friendship_qual, refers to an interviewer's objective rating of the quality of people's friendships. The third, life_sat, refers to people's overall ratings of life satisfaction. As you can guess, we'd expect both income satisfaction and friendship quality to be positively associated with life satisfaction. However, for the sake of this next activity, assume that there is absolutely no correlation at all between income satisfaction and friendship quality. People who are satisfied versus dissatisfied with their income do not differ at all in the overall quality of their friendships. To be even more specific, assume that (a) both income satisfaction and friendship quality correlate $r = .50$ with life satisfaction and (b) income satisfaction and friendship quality correlate $r = .00$ with one another. On the basis of this information, an expert in multiple regression analysis could tell you, without even running a regression analysis, exactly what the value of your multiple R would be.

QUESTION 9.6. In case you have not quite yet achieved that level of expertise, run a simultaneous multiple regression analysis predicting life satisfaction from both income satisfaction and friendship quality and report (a) the beta weights for both income satisfaction and friendship quality as well as (b) your multiple R and your multiple R-square. Based on what you know about the three zero-order correlations noted above, explain why it makes perfect sense that these two predictors would collectively allow you to account for exactly 50% of the variance in life satisfaction.

Adjusted R-square. If you have looked carefully at the output files for the last couple of regression analyses you've conducted, you may have noticed that SPSS gave you two different versions of your R-square value—the basic version and the "adjusted" version. Many years ago, statisticians who worked out the details of how to calculate multiple R realized that as you increase the number of predictors in a multiple regression equation, each additional predictor poses an additional risk of producing an *artificially high* multiple R. This risk is particularly high when you have a small number of observations (i.e., a small sample size) relative to the number of predictors. Let's examine why this is the case.

We hope you agree that there's no reason to expect a correlation between a person's Social Security number and his or her height. However, in *any* sample of only $n = 2$ in which x and y each take on two different values (as they arbitrarily do in both Set A and Set B of Table 9.3), there will always be a **perfect correlation** (either +1 or −1) between x and y. In other words, there is no possible place you can put two pairs of x, y scores on a plane in which you won't be able to draw a straight line that includes both pairs of scores.

The example that follows illustrates this by showing you that if you take the heights in Set A and simply switch them (without changing the Social Security numbers) as in Set B, the perfect *positive* correlation between Social Security numbers and height simply becomes a perfect *negative* correlation.

Table 9.3 The Perfect Correlation Between Height and Social Security Number (SSN) When $n = 2$

Set A Scores		Set B Scores	
SSN	Height, in.	SSN	Height, in.
223565678	74.0	223565678	73.0
567314067	73.0	567314067	74.0

Figure 9.1 shows this graphically (as you could confirm in SPSS, if you wished to create a couple of tiny data sets based on Table 9.3).

If you remember that the absolute value of r approaches 1.0 to the degree that a scatterplot of a complete set of x, y scores comes closer to falling perfectly on a straight line, you can see why this is a problem. The situation wouldn't be quite this dramatic if you had sampled even three or four cases instead of only two. However, assuming the null hypothesis is true, it's still a lot easier to get pairs of scores from 3 or 4 cases to fit pretty close to a straight line than it is to get pairs of scores from 100 cases to do so. Furthermore, this problem is magnified when you have multiple predictors. In the case of two predictors rather than one, for example, it is rarely possible to get three arbitrary pairs of scores to line up perfectly in a straight line, but it is *always* possible to fit them perfectly on an arbitrary three-dimensional plane. From a physical perspective, this is exactly what you are doing when you find a best-fit equation for two predictors in a multiple regression analysis. Add a few additional predictors, and those who can visualize six-dimensional space have reassured the rest of us that

Figure 9.1 Scatterplots for the Two Sets of Scores in Table 9.3

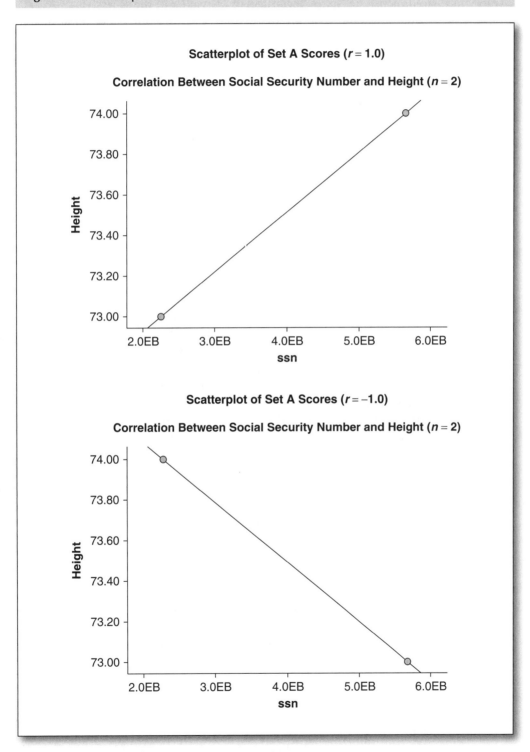

you can always fit any $n = 6$ sets of six predictor-criterion scores to such a six-dimensional space, regardless of the values of the scores!

To put this more simply, there is a very real sense in which a multiple R, if not corrected for the ratio of predictor variables to participants, will reflect the operation of arbitrary model fitting rather than a true multivariate association between a set of predictors and a criterion. An adjusted R-square value corrects for this arbitrary model-fitting bias to produce an estimate of what the multiple R is likely to be in the population from which you sampled *given the limitations of your exact sample size and the exact number of predictors in your regression analysis.* This means, for example, that if you hold n constant and increase the number of predictors in a multiple regression equation, the discrepancy between your R-square and your adjusted R-square values will get larger. Holding n constant, the adjusted R-square will get smaller—relative to the multiple R-square—as the number of predictors in your analysis increases. Conversely, if you hold the number of predictors constant, increasing your sample size will decrease the discrepancy between R-square and adjusted R-square. In fact, unless you have a ridiculously high number of predictors, the discrepancy between R-square and adjusted R-square in samples of even a few hundred or more is negligible. This means that if you happen to be working with a small sample, it is particularly important to pay attention to (and report) your *adjusted* R-square value rather than your simple R-square value, especially if there is a big reason to care about total variance accounted for.

Now that you have digested this discussion of six-dimensional space, compare the "R Square" value and the "Adjusted R Square" values from the Model Summary portion of the results of the multiple regression analysis you saw back in Image 9 (see the second and third small, circled items in the top portion of Image 9). Now you should understand very well why the adjusted R-square (.234) is smaller than the R-square (.287)—despite the fact that both statistics came from the same analysis of the same data. With two predictors and only $n = 30$ cases, there is a need to make a modest adjustment downward in R-square.

Small samples in a multiple regression analysis are also worrisome in other ways. The most important worry you should have when trying to interpret the results of a small-sample multiple regression analysis is the same worry you have when conducting *any* statistic on a small sample. The smaller the sample, the less confidence you have in the exact values of any observed statistical result. As you may recall from our discussion of t tests, for example, the standard error of the mean is closely tethered to sample size. And this standard error of the mean comes in very handy when we wish to offer sophisticated estimates of the population values of any mean (or mean difference) we may have observed. When you conduct a multiple regression analysis in almost any software package (SPSS included), the same output file that yields regression coefficients and p values for each predictor in your model also yields standard errors for each of your regression coefficients.

Chapter 9 Multiple Regression

This means, for example, that if you wish to know the 95% confidence interval in the population for a specific unstandardized regression coefficient (B), you merely need to multiply your standard error by the familiar value 1.96 (the critical value for z with alpha set at .05 for a two-tailed test). By subtracting this product from your B value, you'll have the lower end of your 95% confidence interval. By adding this same product to your B value, you'll have the upper end of the same confidence interval. Let's examine this in the case of the specific SPSS output we just presented to help you answer Question 9.4—involving height, age, and income. Image 10 contains a snippet of the same SPSS output file you saw earlier. Because unstandardized regression coefficients are expressed in the original units in which your predictor and criterion variables are measured, the value of 2.167 that you see circled in Image 10 means that for every one-unit (1-inch) increase in height in these data, you should expect an increase of $2,167 in yearly income. (That's just the B of 2.167 times the $1,000 units in which annual income was reported.)

Coefficientsᵃ

Model		Unstandardized Coefficients		Standardized Coefficients	t	Sig.
		B	Std. Error	Beta		
1	(Constant)	-100.498	66.601		-1.509	.143
	height in inches	2.167	.908	.436	2.388	.024
	age in years	-.191	.203	-.172	-.940	.355

a. Dependent Variable: yearly income in $K

We hope it is clear, though, that $2,167 is our best estimate based on this particular sample and this specific regression equation. The 95% confidence interval for this regression coefficient should be

$$2.167 \pm (1.96 * .908) = 2.167 \pm (1.780) = 0.387 \text{ to } 3.947.$$

On the basis of a sample this small, then, we can only say that we are 95% sure that each additional inch in height is associated with between $387 and $3,947 per year in additional income. If this sounds awfully imprecise, it should be. Remember that we only had a sample size of $n = 30$ people! If you had greatly increased your sample size, the standard errors for the regression coefficients would have gotten a lot smaller, and you could be a lot more precise about your predictions. By the way, there is also an obvious sense in which our estimates of regression coefficients are even more imprecise than they might appear here. The confidence intervals you can easily calculate for B usually only apply to the population of people from whom you randomly (i.e., representatively) sampled. If you have a convenience sample of college students, it would usually be pretty risky to assume that your 95% confidence intervals would apply to the general U.S. population. In fact, unless you truly sampled your college students randomly, your confidence intervals might not even apply very well to other college students.

A Real-World Multiple Regression Application

Now that you have had some practice with multiple regression analysis, we'd like you to try your hand at using a simultaneous multiple regression analysis to analyze some real marketing data that appear in a popular marketing research textbook. The data for this project focus on annual sales figures for 20 convenience stores. For each of the 20 stores in the sample, a research team assessed (a) annual sales in thousands of dollars, (b) average daily traffic (based on vehicles counted passing the store for a month), (c) population within a 2-mile radius of the store (assessed using 1990 census data), and (d) median family income for households within 2 miles of the store (based on 2000 census data). We added a hypothetical variable (employee satisfaction) to these real data to give you another interesting variable to consider. For this final variable, you may assume that the researchers assessed the average satisfaction level for all of the employees at each convenience store by means of an interview and used a 7-point scale to record the mean (the anchors for the scale appear as "Values" that you can observe in the variable mode from the SPSS data window).

The data file for this question is called [**mcdaniel & gates convenience store july 2007 w errors data.sav**]. We also included a screen capture of the SPSS data file in Image 11. Because a secondary goal of this SPSS question is to give you some practice at data cleaning, *we intentionally entered two erroneous scores in these data.* Thus, before you do anything else, you should find and correct these errors. The first error is based on the fact that we seem to have had a sticky "8" key when we were entering the population data for Store 10. Here, you will want to replace the huge population score you see for Store 10 with the correct score of 18814.00. To find the second error, you will need to think about the range of possible scores for employee satisfaction and find a score that is off the scale (i.e., lower than 1 or higher than 7). You should assume that you don't have access to a hard copy of the data. Thus, the only reasonable solution to this second error will be to erase it—that is, to replace this score with nothing. When you do this, the erroneous score will become a period, and SPSS will be forced to ignore the data from the store that had an invalid employee satisfaction score. This means the loss of an observation. Your *n* will effectively become 19 rather than 20. However, this is a lot better than using invalid data. (Yes, this kind of thing just happens sometimes.)

Chapter 9   Multiple Regression

storeid	annual_sales	avgtraffic	pop2miles	avgincome	employeesatisf
1.00	1121.00	61655.00	17880.00	28991.00	2.00
2.00	766.00	35236.00	13742.00	14731.00	2.00
3.00	595.00	35403.00	19741.00	8114.00	1.00
4.00	899.00	52832.00	23246.00	15324.00	3.00
5.00	915.00	40809.00	24485.00	11438.00	4.00
6.00	782.00	40820.00	20410.00	11730.00	4.00
7.00	833.00	49147.00	28997.00	10589.00	5.00
8.00	571.00	24953.00	9981.00	10706.00	1.00
9.00	692.00	40828.00	8982.00	23591.00	3.00
10.00	1005.00	39195.00	1888814.00	15703.00	5.00
11.00	589.00	34574.00	16941.00	9015.00	5.00
12.00	671.00	26639.00	13319.00	10065.00	4.00
13.00	903.00	55083.00	21482.00	17365.00	6.00
14.00	703.00	37892.00	26524.00	7532.00	2.00
15.00	556.00	24019.00	14412.00	6950.00	33.00
16.00	657.00	27791.00	13896.00	9855.00	5.00
17.00	1209.00	53438.00	22444.00	21589.00	5.00
18.00	997.00	54835.00	18096.00	22659.00	4.00
19.00	844.00	32916.00	16458.00	12660.00	7.00
20.00	883.00	29139.00	16609.00	11618.00	7.00

QUESTION 9.7. Report the "storeid" numbers for the two stores with errors, and explain exactly how you know (from what is documented in the "Values" of your SPSS data file) why the second error must be an error. Justify your decision to enter a missing data point for the second error (rather than making your best guess about what the data point was supposed to have been). By the way, your instructor will know you have actually corrected these errors in your data file because your results for the questions that follow will be different from the results with the errors included.

QUESTION 9.8. Run a simultaneous multiple regression analysis to predict annual sales from all four of the predictors that you have available in these data. Your most important results will be beta weights and p values for each of the four predictors. Summarize your results for each predictor, paying attention not only to whether each predictor has a significant unique association with sales but also to what kind of association you observe. (Recall that just as correlation coefficients can be positive or negative, beta weights can be negative or positive.) Offer at least one plausible explanation and practical marketing implication for each observed effect.

QUESTION 9.9. Summarize briefly the results you would have obtained if you had failed to correct these two data entry errors. (To do so, you'll need to run the same regression analysis you just ran using the original—flawed—data.) How could a failure to correct these errors have led to serious problems for a manager wanting to make good future decisions about where to open new convenience markets?

QUESTION 9.10. Imagine that a researcher for a competing marketing research firm conducted a simple but otherwise careful study very much like the one we are currently discussing. However, imagine that the competing researcher measured only one predictor of annual sales—namely, daily traffic flow. In his study, he found that daily traffic flow was strongly and significantly correlated $r = .77$ with annual sales. He argued that this very high correlation strongly indicates that managers should choose high-traffic areas as locations for new stores. Based on the results of your more sophisticated regression analysis, explain why he was wrong.

QUESTION 9.11. The power of multiple regression analysis might lead some people to conclude that we no longer need tools such as lab experiments or test markets to assess causal relations between variables. Attack this position. Here are some hints based on the cookie thief example: (a) What if Lisa had a great number of siblings and you could only observe one or two of them at once? (b) In the cookie example, it is obvious that causality can run in only one direction (i.e., it seems unlikely that disappearing cookies can cause Bart to appear in the kitchen), but what if you were trying to assess the relation between income and education controlling for a personality variable such as achievement motivation? Would multiple regression analysis be any more or less informative here than it was in the case of missing cookies? (c) Multiple regression analysis controls for any variable(s) you were sophisticated enough to measure and add to the analysis. What does random assignment (the defining feature of experiments) control for?

QUESTION 9.12. If your instructor wishes to expand your understanding of multiple regression even further, he or she may ask you to explore an additional data set that will give you some extra practice conducting multiple regression analyses. More specifically, you may have the chance to analyze some data from the [**World95.sav**] SPSS demonstration data set. The research question you will address in these data is whether women's life spans are shorter in countries (a) where fewer women are literate and (b) where women have more children. However, to rule out some worrisome confounds, you will want to add at least one predictor to your model to account for the fact that in countries where, for example, literacy rates are lower, wealth and access to quality health care are also lower. In short, your multiple regression analysis predicting female life span should include at least three predictors, two having to do with women's social status or treatment and at least one having something to do with the quality of health care (or quality of life) that exists in a given country. Of course, you'll have to rely heavily on the "Label" and "Values" entries that are part of your World95 SPSS data file to be able to do this. Remember that you can gain access to these values and labels by using the toggle button in the lower left-hand corner of your SPSS data file. See Image 12 for a quick reminder.

Because understanding multiple regression can be considered a foundation for understanding all multivariate statistics, your instructor may wish to discuss or review not only (a) some of the terms tossed around in this chapter so far but also (b) some new terms we didn't yet cover. These terms are listed in Appendix 9.1 at the end of this chapter.

Logistic Regression: Multiple Regression Analysis With Categorical Criterion Variables

Now that you have become pretty familiar with how to run and interpret multiple regression analysis, we have a serious confession to make. This chapter has encouraged you, our devoted and trusting reader, to break the law. And we're not talking about the moderately important laws that keep judges and trial lawyers in business. Instead, we are talking about the hallowed, sacred, iron-clad laws of statistics. To be more specific, the conceptual foundation of this chapter was built on a multiple regression analysis involving cookie thefts. We won't belabor the details of this analysis because we assume you already sweated these details. But we must tell you that it is not statistically kosher to run a multiple regression analysis of the sort that we instructed you to run using the cookie theft data. This is because one of the assumptions of multiple regression analysis is that both your predictor variables and your criterion variables are continuous, normally distributed variables. This includes variables such as height, weight, and liking for Britney Spears (or Brittany Spaniels) on a 15-point Likert scale. The folks, like Karl Pearson, who invented multiple regression analysis designed it with such normally distributed variables in mind.

But the distribution of scores for whether cookies disappeared on a given day is decidedly *non*normal. Instead, a variable such as cookie theft—as we're sure you'll remember—is categorical. Categorical variables take on discrete values that do not indicate amount or ranking (e.g., political party, favorite brand of Ben & Jerry's ice cream, whether you play the zither). The simplest possible kind of categorical variables are *dichotomous* categorical variables. Dichotomous variables may take on one of only *two* values (e.g., success vs. failure, yes or no). For example, the two values could be whether someone is able to tell if another person is lying about where he spent the night after winning a major golf tournament or whether cookies disappeared on a given day.

When your criterion variable is categorical, you really should run a *logistic regression* rather than a simple multiple regression analysis to see which of several variables best accounts for variation in your categorical outcome. Logistic multiple regression, just like traditional linear multiple regression, assesses whether, and how strongly, your predictor variables are uniquely related to the categorical criterion variable. But logistic regression makes some special adjustments for the fact that your criterion is categorical rather than continuous. Without belaboring exactly how those adjustments are made, we should note that these adjustments are not trivial. For example, if you conduct a simple linear multiple regression analysis in which you predict, for example, the chances that a student will pass a licensing exam based on the student's GPA in grad school, you might see that a person's GPA in grad school predicts the chances of passing the exam even after controlling for (a) a person's college GPA and (b) whether the person took a licensing exam preparation course. So far, so good. The trouble might begin, though, when you got serious about prediction and tried to use your linear multiple regression equation to predict who would pass the exam. If the correlation between professional school GPA and whether people pass the exam is substantial, and if a person had a very high GPA, the predicted probability that the person will pass the exam (a person's predicted y, or y', score) could be greater than 1.0! As you well know, probabilities can never be greater than 1.0. It is because of problems such as this one that statisticians developed and refined logistic multiple regression, which is specifically designed to deal with categorical criterion variables.

Having said all that, we hasten to add that we have *not* lied to you in this text on the occasions when we suggested that, as long as you have a large sample size, most parametric statistical analyses (such as ANOVA and regression) are pretty robust to violations of the assumption of normality (e.g., see Lindman, 1974; Winer, 1971; and cf. Zumbo & Zimmerman, 1993). In fact, linear multiple regression *is* pretty robust (valid) in the face of significant violations of the assumption that criterion variables are normally distributed. (As we have said before, it is much more worrisome, for example, to have a continuous variable that is highly skewed or has extreme outliers than to have a dichotomous variable whose platykurtic outcomes are about equally likely.) In other words, if you were to observe

Chapter 9 Multiple Regression

Bart and Lisa on a great number of days, you would be able to trust the basic linear regression model pretty well to tell you which predictor (which kid) matters and which doesn't. You just couldn't trust it to produce predicted scores (y' scores) that always fall between 0 and 1.0. In many applied settings (e.g., when you are interested in predicting the actual rather than relative probability of an outcome), producing predicted scores that fall between 0 and 1.0 (i.e., that do not break the laws of probability) is very important.

Back to Missing Cookies

Now that you know *when* logistic regression should be used, let's look at *how* to use it. To do this, you will analyze another set of data involving cookie thefts. This time around, though, you'll have to assume that we made some observations of Bart and Fred rather than Bart and Lisa (who was attending a prestigious conference on female entrepreneurship in the European Union at the time that we conducted the newer study). In case you don't know about Fred, we should mention that he is *not* one of Bart's friends. But he does frequently pop by the first author's kitchen to ask for advice about making homemade taralli or penne á la vodka with pancetta. Because Fred is extremely honest, we have little reason to suspect he would ever steal cookies. These things are important to know because when you run a logistic regression you are going to see (a) that the correlation between Fred's presence and Bart's presence on a given day is $r = .00$ and (b) that the correlation between Fred's presence and the disappearance of cookies is $r = .00$. Fred doesn't avoid Bart, but he also doesn't seek out Bart's company. Cookies do sometimes disappear when Fred has visited the kitchen, but they disappear with equal likelihood when Fred has *not* visited the kitchen. We could provide you with some similar statistics for Bart, of course, but that would largely defeat the point of this delicious exercise on logistic regression.

Logistic Regression Analysis of Cookie Thefts: Disentangling Bart and Fred

To get an idea of how logistic regression works, open the data file called [**bart fred missing cookies for logistic regression.sav**]. Using these data, you will essentially repeat our Lisa and Bart cookie theft analysis, but you will use *logistic* regression on these *new* data involving Bart and Fred. You have 16 days of such data. As before, the presence of someone in the kitchen is coded as 1 for present and 0 for absent. "Cookies gone" is coded 1 for yes, they disappeared and 0 for no, they did not.

Open these data and analyze them using logistic regression. You'll use this analysis to answer some questions later. To get you started, though, remember that "bart and fred" are the two competing predictor variables. If you have any doubts about what the criterion variable is, check out the box called "Dependent:" in Image 13.

To run a logistic regression, click "Analyze" and then click the same "Regression" button you've already clicked several times to run linear regressions. To run a logistic regression, however, you must choose "Binary logistic" from the "Regression" menu. *Binary,* meaning "two levels," refers to the fact that we have a dichotomous criterion variable—one that has two and only two possible values.

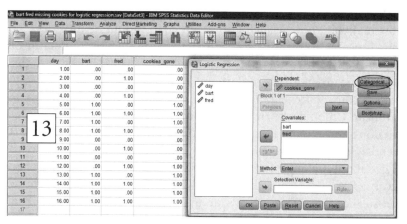

From the Logistic Regression box, it's very important to click "Categorical..." (circled in Image 13) to be able to designate your predictors (bart & fred) as categorical variables. In this particular example, our criterion *and* both of our predictors are categorical (you can also conduct a logistic regression analysis when your criterion is categorical but some, or even all, of your predictors are continuous).

Once you click the "Categorical..." button, you'll see a dialog window like the one we've captured in Image 14.

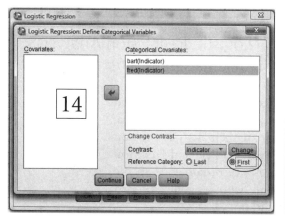
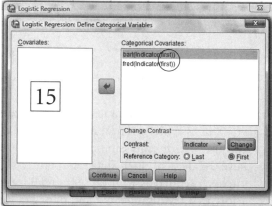

Chapter 9 Multiple Regression

From the "Define Categorical Variables" box you see in Image 14, you should send both "bart" and "fred" to the "Categorical Covariates:" box because both predictors (aka "Covariates") are, in fact, categorical. Then from within the "Change Contrast" box, click the "Reference Category:" option "First"—which is circled for you in Image 14. *This crucial change will make 0 the point of reference for this analysis rather than 1.* Assuming that you've highlighted the "fred" variable (as we did in Image 14), clicking "Change" *after* you've clicked "First" (see the button just above the "First" button) will actually enact this change in reference point for the "fred" variable. In other words, clicking "First" won't do what you need to do unless you *also* click "Change." (Apparently, true "Change" really is hard.) Once you've done this, you'll also need to highlight the "bart" variable and repeat the same procedure (although you may not have to click "First" again, you *will* need to click "Change"). You'll know that you've done all this correctly if SPSS has added the word *first* in parentheses after both "bart" and "fred" (see Image 15). You are not ready to continue until you've made these crucial changes. If you don't set things up this way, it will be very hard to figure out your logistic regression output—and even harder to compare it with an odds ratio that you can calculate pretty easily by hand.

Once you click "Continue" to return to the main "Logistic Regression" analysis box, make sure to click "Options" so you can ask for the 95% confidence interval for the **logistic regression odds ratio.** It's not super obvious which box that is, so we have reproduced the screen for you in Image 16, with the appropriate box checked.

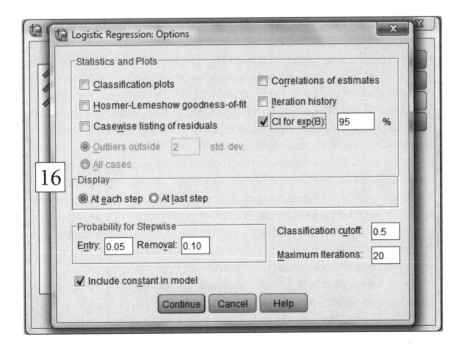

Recall that confidence intervals just give you a range of likely values for an observed statistic in the total population of the people (or the days, in this case) in which you are ultimately interested. In case your memory for odds ratios is a little rusty, we'll say more about them in a moment.

But first, click "Continue" from the Options box you see in Image 16 to get back to the main Logistic Regression box. Now you can run this logistic regression analysis (with or without pasting). Image 17 shows the most important part of the output:

Classification Table[a]

		Predicted		
		cookies_gone		Percentage Correct
Observed		.00	1.00	
Step 1 cookies_gone	.00	7	1	87.5
	1.00	1	7	87.5
Overall Percentage				87.5

a. The cut value is .500

Variables in the Equation

		B	S.E.	Wald	df	Sig.	Exp(B)	95% C.I.for EXP(B)	
								Lower	Upper
Step 1[a]	bart(1)	3.892	1.512	6.626	1	.010	49.000	2.531	948.619
	fred(1)	.000	1.512	.000	1	1.000	1.000	.052	19.360
	Constant	-1.946	1.309	2.209	1	.137	.143		

a. Variable(s) entered on step 1: bart, fred.

Understanding Odds Ratios in Logistic Regression

Remember that, as we implied earlier, SPSS uses the predictor variable(s) in a logistic regression to generate a best guess (prediction) for the criterion variable. In this case, using Bart and Fred as predictors, SPSS creates a model to best predict whether a cookie went missing on a given day. In this particular model, this prediction is based on who was present on a given day and on what tends to happen when each specific person is present. Much of this logistic regression output will probably look familiar to you. However, there is an important way in which logistic regression outputs differ from linear regression outputs. The *B*s in a logistic regression output differ from the *B*s in a linear regression output. In logistic regression, associations between variables are typically summarized as **odds ratios**. Yes, that's a ratio of odds, one set of odds divided by another set of odds. In this case, it is the odds of a cookie being stolen when Bart is present *compared with* (divided by) the odds of a cookie being stolen when Bart is *not* present (controlling, of course, for Fred's presence).

If you look carefully at the top half of Image 17, where it says "Classification Table," you'll see that when the model predicted a cookie would be stolen, the model got things right seven times out of eight

Chapter 9 Multiple Regression

(87.5%). The same is true when the model predicted no cookie would be stolen—seven of eight correct. To get back to odds, you could say it's seven times more likely (.875/.125) that a cookie is stolen when the model predicts a theft *and* seven times more likely that no cookie is stolen when the model predicts a nontheft. Now, looking at the lower half of the Classification Table, you can see which of the model's predictors are significant—only Bart (even after controlling for Fred). Fred's presence doesn't help predict cookie thefts one bit beyond what Bart's presence predicts—so Fred's Exp (B) is 1.0 (remember that an odds ratio of 1.0 is like a correlation coefficient of zero!). This makes Bart's presence the only thing that matters in the model. Everything the model can predict is because of Bart.

Putting all this together, it's seven times more likely that a cookie is stolen than not stolen when Bart is present *and* seven times more likely that a cookie is *not* stolen than stolen when Bart is *not* present. Because we have a dichotomous criterion variable, we may use these data to compute an odds ratio. Refer again to the Classification Table that you see in the upper half of Image 17. The odds ratio is the product of the matching diagonals divided by the product of the mismatching diagonals. In the table in Image 17, we have placed circles around the matching diagonals and squares around the mismatching diagonals. So the odds ratio is the product of the two circled numbers (matching diagonals) divided by the product of the two "squared" numbers (mismatching diagonals). That is, $(7*7)/(1*1)$ = odds ratio = 49:1. Notice that's this is *exactly* what Exp(B) is. Exp(B) refers to exponentiating the estimate of B to get an odds ratio. Don't fret about the math on this. But what does a 49:1 odds ratio mean? In our example, it means it's very, very likely that Bart is a cookie thief, even after we control statistically for the possibility that Fred could have also been the culprit.

Misunderstanding Odds Ratios in Logistic Regression

Incidentally, it is a very common mistake for budding statisticians to interpret an odds ratio as if it were simply a ratio rather than a *ratio of two ratios*. That is, people sometimes interpret odds ratios as if they were simply odds. In the particular example we just discussed, it would be a big mistake to say that cookies are 49 times more likely to disappear when Bart has visited the kitchen than when he has not. If you check out the Classification Table in the output screen capture one more time, you'll see that in this case, the odds ratio of 49:1 means that it is exactly *seven* times more likely that a cookie will disappear when Bart has recently visited the kitchen than when he has not. So if you were to run a logistic regression predicting whether people like the color purple from their gender and you observed an odds ratio of exactly 2:1 (with women preferring purple more often than do men), you would want to be careful *not* to say that women are twice as likely as men to prefer the color purple. If this were the case (if women *were* twice as likely as men to prefer purple), you'd expect an odds ratio of about 4:1 rather than 2:1. The 4:1 ratio would up hold exactly, for example, if exactly 1/3 of 300 men preferred purple but exactly 2/3 of 300 women preferred purple. If you want to confirm this for yourself very easily, just reproduce the Classification Table in our Bart-Fred example and replace each of the 7s with a 2. $(2 \times 2)/(1 \times 1) = 4.0$. However, if men preferred purple exactly 50% of the time, you'd only get an odds ratio of exactly 4:1 if women preferred purple 80% of the time. You may wish to play around with some 2×2 classification tables a bit to better appreciate what an odds ratio really means. One thing you'll quickly see is that because an odds ratio is a ratio of two ratios, there are many different ways to get the same odds ratio. If 4 out of 5 men (80%) preferred purple, for example, 16 out of 17 women (94.12%) would have to prefer purple to yield an odds ratio of 4:1!

In case that caveat hasn't made you think hard enough, allow us to add a caveat to that caveat. There are occasionally times when you can *sort of* interpret an odds ratio literally, but they are exceptions rather than the rule. For example, if you developed a special way of tossing coins that allowed you to toss 7 heads for every 1 tail you tossed, we could compare the odds of this event (7:1) with the even odds (1:1) we should observe for someone who didn't have your special coin-tossing talent. We'd get an odds ratio of 7:1. In this rare case, we *could* say that you are *seven times* more likely to toss a head than to toss a tail, and we could remind people that normal mortals are *equally* likely to toss heads and tails. However, we could *not* say that you're seven times more likely to toss heads than is a normal person. If you tossed a coin 800 times, we would expect 700 heads, but if Little Jimmy did the same, we'd expect 400 heads—not 100. Finally, in case you're on a high caveat diet and your taste for complexity is *still* not filled, you may wish to check out an excellent online article by Davies, Crombie, and Tavakoli (1998).

Back to Missing Cookies

OK, getting back to missing cookies, we hope you can see why Bart's Exp(*B*) value was exactly 49:1, which is exactly the value you calculated by hand from the logistic regression classification table. We'd now like you to repeat the very same logistic regression analysis we just covered but to *remove Fred as a predictor in the model*. But in case you're still not feeling completely comfortable running a logistic regression analysis, we'll tell you that when you do run the new model, you should find that Bart's Exp(*B*) value didn't change one bit. Go ahead and run it, though, to confirm this.

QUESTION 9.13. That may seem a little puzzling. Remember that the whole goal of regression is to *control for* one or more competing predictors while assessing the association between a given predictor and your criterion variable. If we included Fred in the original logistic regression model but left him out of the second model, why didn't Bart's Exp(B) value change at all? Why was Bart's unique association the same as Bart's zero-order (simple) association? Answer this question in a way that makes it clear that you understand the logic of multiple regression analysis, logistic or otherwise.

Confidence Intervals in Logistic Regression

Now look back one more time at the lower right-hand section of your original output (the one captured for you in Image 17). Always the optimists, we feel sure that you'll fondly remember what confidence intervals are. Notice the "95.0% C.I. for EXP(B)" column. This is the 95% confidence interval for the odds ratio (or, EXP(B)). For bart, then, you see there's a 95% chance that the true (population) odds ratio is between 2.53 and 948.62. Notice also that even though the "bart" effect is significant at $p = .01$, there's a *humungous* range for the 95% confidence interval. This is because $n = 16$ (a very small sample), and *especially with dichotomous outcomes*, sampling error can play a big role. That is, changing just a few cookie theft scores could radically affect the observed odds ratio and its associated confidence interval.

Because we feel that knowing how much confidence you can place in odds ratios in logistic regression is very important, we'd like you to see what would happen if we increased our sample size by a factor of 5 in these data (leaving all the data points, and thus the effect sizes, absolutely unchanged). How much would this narrow your confidence interval for Bart? To find out, we'd like you to analyze the data from a new file called **[bart fred missing cookies for logistic regression TIMES FIVE.sav]**. This file is simply the original $n = 16$ data file with $n = 80$ cases. We simply copied the original 16 data points (16 days) five times each. None of the data points are changed at all.

QUESTION 9.14. Repeat the original logistic regression analysis (with both bart and fred as predictors) using this new data set. Summarize the results. What did you expect to happen to the confidence interval for "bart" with five times as many observations? Did the confidence interval for Bart's Exp(B) change any more or less than you expected?

It Sure Is Messy Out There: Multivariate Data Cleaning

Because the first author has two children at home, he has had lots of practice saying things such as "You can't play with that; it's filthy" or "You can do that as soon as you've finished cleaning your room." So here it goes: You can't safely analyze any freshly collected data using (continuous) multiple regression analyses until you've cleaned them (the data, that is). You may recall that when we discussed the correlation coefficient in Chapter 3, we noted that one or two extreme outliers can wreak havoc on a correlation coefficient, either by making it look like two variables are correlated when they really are *not* or by making it look like two variables are not correlated when they really *are*. In a sense, problems of this general sort become worse when you conduct a multiple regression analysis rather than a simple correlation because more variables means more ways for things to go wrong. In more technical terms, one of the most important underlying assumptions of multiple regression analysis is "multivariate normality." This means not only that every predictor in your regression should be approximately normally distributed (without any whacky outliers) but also that there shouldn't be any *combinations* of the kind of unusual scores on the various predictor variables that are highly inconsistent with the general pattern of associations between your predictors. We discuss data cleaning in great detail in Chapter 14, and this includes issues such as how to deal with multivariate outliers. For now, however, we just wanted to remind you that you can only trust a multiple regression analysis when you are sure that all of the variables you are analyzing are pretty well behaved.

Appendix 9.1: Terms for Further Reading or Discussion

Adjusted R (or R-square)

B

Beta (β)

Constant

Criterion

Hierarchical multiple regression

Multicollinearity (see the upcoming chapters on suppression and data cleaning for discussions)

Multiple R

Predictor

R-squared change

Residual

Restriction of range problem

Reverse causality

Standard error of the regression coefficient

Tolerance (inverse of the variance inflation factor)

Variance inflation factor (inverse of tolerance)

Zero-order correlation

10 Examining Interactions in Multiple Regression Analysis

Introduction: Type of Variable Determines Type of Analysis

In the last chapter, we covered some of the fundamental concepts behind multiple regression analysis. As you now know, multiple regression techniques are designed to tell you which, if any, of several competing predictors of an outcome are the strongest predictors of this outcome ceteris paribus (all else being held equal). Thus, multiple regression techniques allow you to sort out the *unique associations* between your predictors and your criterion. In a very important sense, multiple regression analysis is a lot like the factorial (two-way and higher) analysis of variance (ANOVA). In fact, based on what you have learned so far about factorial ANOVAs and multiple regression analysis, there are only two differences between these two techniques. First, researchers typically conduct ANOVAs when their independent variables are *categorical* (e.g., drug group or placebo control group, cued recall task or free recall task, hot room or cool room). In contrast, researchers typically conduct multiple regression analyses when their independent variables are *continuous* (years of education, height, level of neuroticism, etc.). Second, researchers who conduct ANOVAs are frequently interested in both main effects and interactions, whereas researchers who conduct the most basic multiple regression analyses often focus exclusively on main effects. Thinking back to the example of Bart and Lisa, for instance, we simply wanted to know who was more responsible for cookie thefts. We didn't discuss the issue of whether the two kids together might have a different effect than you'd expect based on either kid alone.

To put this a little differently, recall that interactions exist when the relation between one of your independent variables and your dependent variable *is different* at different levels of the *other* independent variable (or variables). That is, the relation between one independent variable and your dependent variable "depends" on the level of *another* independent variable. To cite another familiar example from a previous chapter, people high in self-esteem might prefer a positive evaluator over a negative evaluator. People low in self-esteem might show the opposite preference.

Moderators and Interactions in Multiple Regression

Could you test this same kind of idea if your independent variables (e.g., self-esteem and current level of job stress) were continuous rather than categorical? That is, is it possible to test for interactions in multiple regression analyses? If you read the title of this chapter, you probably guessed that the answer is a resounding yes. In fact, *you can do anything with multiple regression analyses that you can do with ANOVA*. The only real difference between these two data-analytic techniques is whether your independent variables are categorical or continuous. Furthermore, if you don't mind dummy-coding your categorical independent variables, you can conduct a multiple regression analysis on the same data you might normally be inclined to analyze using ANOVA. However, as you will soon see, to conduct a multiple regression analysis in which you test for an interaction, you have to take a few extra steps in your regression analysis that would otherwise be unnecessary. By the way, in the language of multiple regression analysis, testing for interactions is referred to as testing for **moderator** variables. If one independent (predictor) variable *moderates* the association between another independent variable and your dependent (criterion) variable, you have an interaction—the nature of the association "depends" on the level of the moderator variable. In this chapter, we'll focus on cases in which you have only *two* independent variables, but just as you can conduct three-way ANOVAs and test for three-way interactions, you can also conduct moderator analyses in multiple regression (with three independent variables) that allow you to test for three-way interactions. Once you have mastered the procedures needed to test for two-way interactions in multiple regression, you will be ready to make the transition to testing for three-way (or higher) interactions. Appendix 10.1 covers three-way interactions in multiple regression analysis in great detail (see also Aiken & West, 1991).

In mathematical terms, you always test for two-way interactions in multiple regression analysis by examining a *cross-product* term that represents the joint effect of your two predictor variables. This interaction term is literally the cross-product ($A \times B$) of people's scores on the two predictor variables. If LJ's self-esteem score were 30 and the feedback score she received

Chapter 10 Examining Interactions in Multiple Regression Analysis

from a coworker were 10 (on a 12-point scale), her score for the interaction term in a regression analysis involving feedback and self-esteem would be 30 × 10 or 300. If you are the curious type, you may be wondering why a *cross-product* does the trick. Why not a sum? Why not a log-transformed ratio score involving the two predictors? (If you are *not* the curious type, you should be. Curiosity may have killed the cat, but it's the common thread that unites Thomas Edison, Edmund Hillary, and Eddie Vetter.) To get back to the point, in case you're wondering why a cross-product does the trick, let's stick with the familiar example of feedback, self-esteem, and liking and see how a cross-product can account for variance that you couldn't account for if you only focused on the main effects of your two predictor variables. Specifically, let's consider a highly simplified example involving four hypothetical participants whose z scores for self-esteem, feedback, and liking for a confederate are summarized in Table 10.1 (along with the cross-products that serve as their scores for the interaction term).

If you compute the simple correlation between self-esteem and liking, you'll find that it is zero. Needless to say, when simple correlations are zero, the predictor variables behind these correlations don't usually account for much unique variance in a regression equation. The same thing is true for the correlation between feedback and liking. It's also zero. Because you are probably thoughtful as well as curious, it may have dawned on you that this is just another way of saying that there is *no main effect* of self-esteem (or feedback) on liking in this tiny, highly simplified data set. You may have also realized that our R-square so far for both of the predictors taken together would have an embarrassingly low value of zero! But cheer up. All is not lost. As you can see, the correlation between the Self-esteem × Feedback *interaction term* and people's liking scores is a perfect 1.0. Mathematically, people's liking scores are more than just the sum of their self-esteem and feedback scores (instead, they're the *product* of these scores). But enough of this talk of curiosity and thoughtfulness. We can sense that you're probably getting pretty hungry—and would like to see

Table 10.1 Why Cross-Products Are Appropriate for Detecting Statistical Interactions

Participant	Self-Esteem	Feedback	Self-Esteem × Feedback	Liking
Therma	−1	−1	+1	+1
Kelvin	−1	+1	−1	−1
Centi	+1	−1	−1	−1
Farhen	+1	+1	+1	+1

another delicious example involving cookies. Let's look at one. Of course, this new example follows the same familiar format you observed in the last chapter with Bart and Lisa (see Table 10.2).

Table 10.2 Yet Another Observational Study of Cookie Thefts (When a Statistical Interaction Is at Work)

	Homer Visited?	Marge Visited?	Cookies Stolen?
Day 1	No (0)	No (0)	No (0)
Day 2	No (0)	No (0)	No (0)
Day 3	No (0)	No (0)	No (0)
Day 4	Yes (1)	No (0)	Yes (1)
Day 5	Yes (1)	No (0)	Yes (1)
Day 6	Yes (1)	No (0)	Yes (1)
Day 7	No (0)	Yes (1)	No (0)
Day 8	No (0)	Yes (1)	No (0)
Day 9	No (0)	Yes (1)	No (0)
Day 10	Yes (1)	Yes (1)	No (0)
Day 11	Yes (1)	Yes (1)	No (0)
Day 12	Yes (1)	Yes (1)	No (0)

Image 1 comes from an SPSS data file that contains all of these data, along with a variable (for which all the data are missing) for an interaction term that will consist of the cross-product of scores for "homer" and "marge" on each day of the observational study. The data file is called [**homer marge interaction example.sav**]. By now, you should have no trouble

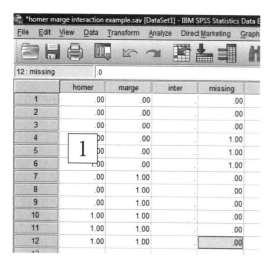

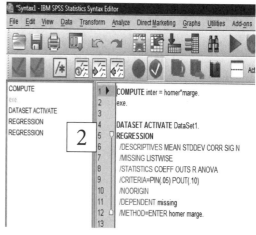

getting SPSS to compute the simple correlations between (1) "homer" and missing cookies and (2) "marge" and missing cookies. When you've done this, you should see that (a) Homer is *positively* associated with disappearing cookies, $r(10) = .577$, $p < .01$,[1] whereas (b) Marge is *negatively* associated with disappearing cookies, $r(10) = -.577$, $p < .01$. This is useful to know because prior to testing for an interaction in these data, it will be useful to see if these two significant correlations translate into significant main effects in a simultaneous multiple regression analysis predicting missing cookies from both "homer" and "marge."

QUESTION 10.1a. Calculate the main effects. Run a multiple regression analysis and then report whether you observed (a) a main effect of Homer's presence and (b) a main effect of Marge's presence. While you're at it, report (c) the value of your adjusted multiple R (the square root of your adjusted multiple R-square) for this regression analysis. By the way, for this analysis, be sure to paste the multiple regression command into a syntax file, so that you can later edit the command. The "REGRESSION" command in Image 2 shows what your multiple regression command should look like.

As impressive as your observed multiple R is, it is not the whole story. If you inspect the data very carefully, you should be able to see that the relation between Homer's presence and cookie thefts appears to *depend upon* whether Marge is present. That sounds a lot like an interaction. To see if this is the case, go to the top of your syntax file and carefully add the compute statement that you see at the top of Image 2 (in lines 1 and 2). Once you run this compute statement (making sure to execute it as well), you should see that your SPSS data file now contains cross-product scores for the variable "inter."

Now that you have added the interaction term to your data file, you can test for the apparent interaction between homer and marge. The easiest way to do this, by the way, is to simply paste a copy of the entire REGRESSION command that should already be part of your syntax file below the original REGRESSION command. Your new REGRESSION command should read as follows:

REGRESSION

/DESCRIPTIVES MEAN STDDEV CORR SIG N

/MISSING LISTWISE

/STATISTICS COEFF OUTS R ANOVA

/CRITERIA=PIN(.05) POUT(.10)

/NOORIGIN

/DEPENDENT missing

/METHOD=ENTER homer marge inter.

The one and only difference between this REGRESSION command and your original is that you added "inter" at the very end.

QUESTION 10.1b. Run and report the results of this simple multiple regression analysis with three predictors. Make sure to report several important things. First, report whether your adjusted multiple R changed at all. Second, report the beta weights and p values for all three of your predictors in the new model ("homer," "marge," and your interaction term "inter"). By the way, don't worry if the beta weight for "homer" seems *impossibly* large—that's a result of this particular and highly unusual situation, and it's a problem we'll fix once we get to more realistic situations.

The fact that you observed an interaction means that the magnitude of the "homer–missing cookie" correlation depends dramatically on whether Marge is present. To specify the nature of this interaction effect as precisely as possible, you should offer your readers the regression equivalent of **simple effects tests** in ANOVA. In this highly simplified case, the easiest way to do this is to compute the correlation between Homer's presence and missing cookies (a) when Marge is absent and (b) when Marge is present. That is your next task.

QUESTION 10.1c. Who's the cookie cop? Do this task and then report these two correlations. (Recall that you can use the "Split File" command from the "Data" window to split up your data file.) Then offer a very brief interpretation for why you might observe these two different correlations (e.g., do they suggest that someone is playing the role of cookie police officer?). Finally, taken together with the simplified example involving self-esteem and feedback, this example involving missing cookies should also illustrate that there are *different kinds* of interactions in multiple regression (i.e., spreading vs. crossover interactions). Comment on which kind of interaction you seem to have observed here.

A More Realistic Example: Centering and Simple Slopes Tests in Multiple Regression Analysis

This greatly simplified—and thus unrealistic—example involving cookies can also be used to illustrate two important problems that often come up when you wish to test for interactions involving real, continuous predictors. The first problem is that when you are dealing with real data, your predictor variables (the conceptual equivalents of homer and marge) are often correlated with one other. This can make it *much* harder than it was in the cookie example (a) to separate the unique main effects of your independent variables and (b) to separate your interaction effect(s) from these main effects. This problem is worsened by the fact that when your two predictors are correlated with each other, your interaction term will

Chapter 10 Examining Interactions in Multiple Regression Analysis

virtually always—by mathematical necessity—be correlated with both of your predictors. If you are a big fan of experiments, it has probably occurred to you by now that experiments allow you to create a situation in which your independent variables are completely uncorrelated with one another. This makes it easy to identify the unique association between different independent variables (aka predictors) and your dependent variable (aka criterion). Unfortunately, reality isn't usually as cooperative as experiments are. Many things that are correlated with a criterion variable are also correlated with one another.

Before discussing what we can we do about this first problem, let's discuss a second, even trickier, problem that arises when dealing with continuous, nonexperimental predictors of the sort that abound outside the laboratory. This second problem has to do with figuring out exactly what *kind* of interaction you have when you conduct tests of moderation using multiple regression analysis. Because the hypothetical cookie data involved categorical predictors (present vs. absent), it was very easy to conduct the regression equivalent of simple effects tests by just looking at two separate correlations between "homer" and "missing"—one in the condition in which Marge was present and one in the condition in which she was absent. But as you recall, we normally conduct multiple regression analyses when we have *continuous* predictors. How could you conduct the equivalent of simple effects tests if you were dealing with continuous data? In the old days, people often solved this problem by performing **median splits** on one of their two continuous predictors. Everyone who scored below the median on one predictor might be placed in a "low" group, and everyone who scored at or above the median on this same predictor might be placed in a "high" group. Then you could conduct *two separate correlations* between the *other* predictor and your criterion variable. If an interaction were at work, these two correlations should be different (as they were in the homer–missing cookie example). Unfortunately, however, there are a lot of problems with this highly intuitive approach. In addition to the fact that it greatly reduces your *power* by cutting your total sample size in half for each subgroup, it can also create problems involving things such as restriction of range and tarnished validity. For instance, the median split approach often compromises the validity of the variable that you artificially dichotomized by treating people who are *barely* above or below the median the same as people who are *way* above or below the median. As an extreme example, suppose you were a sports psychologist whose sample of four runners included (1) the first author of this text, (2) the comedian Drew Carey, (3) a 452-pound Sumo wrestler, and (4) Olympic sprint champion Usain Bolt. If you performed a median split based on running speed in 200 meters, the first author, slow though he may be, would probably end up in the same "fast" category as Usain Bolt. This would clearly be less than ideal. So how do you solve this problem? By performing the regression equivalent of simple effects tests—namely, **simple slopes tests.** But before we get to that solution, let's back up and solve the first problem. As it turns out, you have to solve the first problem (involving intercorrelated predictors) before you can easily solve the second.

So how can we solve the problem of intercorrelated predictors when testing for interactions in multiple regression analysis? The bad news is that we *can't*, at least not completely. All we can do is minimize it. But wait! That's not *all* the bad news. There's more. The best way to minimize the problem is to run a very large number of participants, and that's not always possible. Now for the good news. Once you've done whatever you can to get a large sample, you can perform a very simple procedure on your raw data that will not only help a lot with this problem of intercorrelated predictors but also get you on track to solve the simple slopes problem. *The first thing you always do when you plan to conduct tests for interactions in multiple regression is to perform a simple but extremely important transformation on each of your continuous predictor variables.* This simple but important transformation is called **centering.** Centering is a variation on a mathematical procedure that was probably invented by an ancient Phoenician and put to great use by Thomas Edison. This procedure is called **subtraction.** That is, to center a variable is to simply *subtract the sample mean* from the variable. This is a lot like *standardizing* a score (i.e., converting it to a *z* score) except that you keep the standard deviation of the score in its original units (you don't bother to divide your difference score by the standard deviation of the predictor).

As a concrete example, suppose you wanted to center people's self-esteem scores. If the mean self-esteem score in your sample were 2.9 on a scale with a theoretical range of 1 to 4, you'd just subtract 2.9 from each person's raw self-esteem score to get a centered self-esteem score. You'd then repeat this centering procedure for your *other* predictor variable—say, number of negative life events experienced in the past month (using the sample mean score for negative life events, of course). Then you'd do just what we did in the cookie example above. You'd compute an interaction term involving your two centered predictors (you'd take the crossproduct of the two *centered* predictors). Finally, you'd enter all three of these terms (two centered main effects and a corresponding interaction term based on the centered terms) as predictors in a **simultaneous multiple regression.** By the way, you always run *simultaneous* multiple regressions when testing for interactions. This ensures that all main effects are fully accounted for when evaluating the interaction term. With all three predictors entered in a simultaneous multiple regression, you can see if you observed a main effect of either centered predictor (in this case, self-esteem and negative life events) and an interaction (in this case, a Self-Esteem × Life Events interaction).

That's it so far. To date, that's the best solution we have to the first problem involving correlated predictors. Centering your scores is the beginning of all tests for interactions in multiple regression analyses. One of the advantages of centering, for example, is that when you have centered your predictors, you can safely interpret the main effects of these predictors in a regression analysis that also included your interaction term (otherwise, entering an interaction term seriously messes up your estimates for main effects). So we've made a lot of progress, but we still have the second problem. Assuming you observe some kind of interaction in a regression involving continuous predictors, how do you figure out exactly what *kind* of interaction you have? The answer is simple slopes tests (see Aiken & West, 1991, for a much more detailed treatment of this topic).

The *logic* of simple slopes tests is exactly like the logic of simple effects tests. These tests are done to see if the correlation between variables *A* and *C* is different at different levels of variable *B*. But the procedure is very different than what you might expect. First, when you conduct simple slopes tests, you simply repeat a variation of your original multiple regression analysis two additional times. The goal of *one* of these repeated multiple regression analyses is to get an estimate of the association between one of your independent variables and your dependent variable for people with *low* scores on the other independent variable. The goal of *the other* repeated multiple regression analysis is to get an estimate of the association between one of your independent variables and your dependent variable for people with *high* scores on the other independent variable. For example, you might want to know the association between age and upper body strength among people who *do* exercise regularly versus those who do *not* exercise regularly. You might expect to see that age is only associated with large reductions in upper body strength among the couch potatoes. The way you do this kind of thing with continuous predictors (like age and exercise level) is to perform a couple of transformations on your variables that will still allow you to analyze all of your data (no median splitting allowed) while focusing on some participants much more than others.

You could think of this as a **weighting** process in which the weight given to a person's data in estimating a beta weight for predictor variable x depends on how close this person scores to some high or low value of predictor variable z. Stop and read that last sentence again, just the way Edmund Hillary would have read a map or Eddie Vetter would have read a musical score. That's all there is to it. *You still analyze all of your data* (so you can capitalize on the power of your entire sample size) *but you weight some observations more heavily in one simple slopes test* (e.g., an estimate that focuses on an *x-y* association for people low on variable *z*) *and you weight other observations more heavily in the other simple slopes test* (e.g., an estimate that focuses on an *x-y* association for people high on variable *z*).

Beyond Median Splits: Isolating and Analyzing Subgroups in Multiple Regression

The tricky part is that the way you do this is counterintuitive. The way to get your regression equation to focus on people who score *low* on predictor variable z is to convert their original, centered score on variable z to a score from which you have *subtracted* the centered score that you've decided best represents a low score. For example, if one of your predictor variables were self-esteem, and if the standard deviation for self-esteem in your sample were 8.42, you could compute a new self-esteem *stand-in* variable that would let you estimate the association between variable x and your criterion for people whose self-esteem scores placed them exactly one standard deviation below the mean in self-esteem. That is, you could estimate the association between variable x and your outcome variable y for people low in self-esteem. Go back and read that least sentence again, just the way Thomas Edison would have read a formula for computing the energy efficiency of a new kind of lightbulb. Suppose variable x is job stress and variable y is physical illness. You could use simple slopes to focus on people low in self-esteem and see, for example, if the positive correlation between job stress and illness is stronger than usual for people lower than usual (i.e., lower than average) in self-esteem. If you're like Eddy Vetter, you may be wondering, how low is low? As low as you like. By convention, researchers usually generate simple slope estimates for people whose scores place them exactly one standard deviation from the mean (both above and below). However, if you had a good legal, economic, clinical, or theoretical reason to want an estimate corresponding to a specific score (e.g., people with specific SAT, TAT, or IAT scores that have an important meaning), you could use that specific score as well. So how would you do this in our example involving self-esteem? To answer this in the language of syntax commands, assume that the standard deviation for self-esteem in your sample is 8.42. In this case, your stand-in self-esteem variable to get an estimate of the association between, say, job stress and illness for people low in self-esteem would look like this:

COMPUTE LOWESTM = CENTESTM − (−8.42).

EXECUTE.

(where −8.42 is thus the centered self-esteem score of participants who are 1 standard deviation below the mean in self-esteem). Your new interaction term would now consist of the cross-product of this stand-in variable for people low in self-esteem and your original, centered score for your other predictor (job stress). Your new interaction term might look something like this:

COMPUTE LOINTER = LOWESTM * CENTJOBSTR.

EXECUTE.

Chapter 10 Examining Interactions in Multiple Regression Analysis

Thus, your simultaneous multiple regression command to examine the association between job stress and illness for people low in self-esteem would include three predictors: (1) your "LOWESTM" stand-in term, (2) your centered job stress term, and (3) the new, special interaction term (LOINTER) that was the cross-product of these two terms.

But that's only half the story. That would only generate the simple slopes analysis for people low in self-esteem. To do the same kind of thing for people high in self-esteem, you'd just change the score you subtracted out of your original centered self-esteem score. Now you want this score to correspond to people high in self-esteem. So you'd need to do this somewhere in your syntax file:

COMPUTE HIESTM = CENTESTM – (8.42).

EXECUTE.

And then, of course, you'd want to add your new interaction term.

COMPUTE HIINTER = HIESTM * CENTJOBSTR.

EXECUTE.

Now you'd be ready to run your simultaneous multiple regression analysis to get estimates of the association between job stress and illness for people high in self-esteem. Finally, if you're wondering where you'd look to see these estimates, you'd focus on the beta weights for centered *life events*. After all, what you want to know about is the strength of the association between life events and illness for these two different estimate groups (people low in self-esteem and people high in self-esteem). If your interaction is significant, the beta weights for life events should look pretty different in the two different "stand-in" analyses.

Some Practice With Real Data

OK, enough of this abstract stuff. Let's look at an example involving real data. We're going to introduce you to the data set, but given that you now have some training under your belt, we are only going to give you (a) a list of logical procedures that you will need to convert into syntax statements and (b) some analysis commands to complete this task. The data set you will analyze for this assignment is a subset of a real data set the first author collected in the late 1990s from attendees at a seminar for health care professionals. In an effort to replicate a study he had previously conducted with college students, he measured three things in this sample: (1) implicit sociotropy (people's automatic, presumably unconscious, self-relevant associations to interpersonal loss or rejection), (2) negative life events (using a well-validated life event checklist), and (3) self-reported depressive symptoms (using the popular Beck Depression Inventory). *We expected that the association between negative life events and depressive symptoms would be stronger for people who received higher scores on the implicit measure of sociotropy (because higher scores indicate greater vulnerability).* Thus, in clinical terms, we wanted to know if there was an *interaction* between the "diathesis" variable (implicit sociotropy) and the stress variable (negative life events). The data file for this study is called [**subset of implicit depression data for interactions.sav**]. Image 3 shows you the first few lines of your SPSS data file. Just to be sure you're clear on the variables, remember that higher scores on "mplicit" reflect more vulnerability (higher scores are worse), higher scores on the "bdi" indicate more self-reported depressive symptoms, and higher scores on "lifevnts" indicate a greater number of self-reported negative life events. In short, everything important is coded so that higher numbers are worse.

This file contains complete data from 92 participants. The three crucial variables you care about most are those in the last three columns. It is important to note that because you have complete data on all three crucial variables, these data make it easier than usual to determine the means and standard deviations for your two predictor variables. If this were *not* the case, you would need to identify the subset of participants who did have the crucial scores on all of the variables in your regression analysis before calculating the appropriate means and standard deviations (and you'd want to focus exclusively on the subset of participants with complete data on all of the variables you plan to analyze).

After you open a syntax file, we recommend that you do the following:

1. Center your two main predictors (an easy way to get means and standard deviations for your variables is to conduct a correlation with them and ask for "Descriptives"). To simplify the worlds of those who will grade your work, please call the centered predictors "clife" and "cimpl." Remember, *C* is for centered. (And, yes, *E* is for Eddie.)

2. Compute an interaction term for the centered variables. Call it "cinter."

3. Run a test for an Implicit Sociotropy × Life Events interaction. Make sure to include all three predictors.

Chapter 10 Examining Interactions in Multiple Regression Analysis

Assuming that you observe an interaction, you'll need to do simple slopes tests. Remember that there are two of these. Thus, you'll need to do the following:

4. Conduct a simple slopes test for people low in implicit sociotropy (this will require you to make two compute statements, one involving a variable you'll call "loimpl" and one involving a variable you'll call "lointer").

5. Conduct a simple slopes test for people high in implicit sociotropy (this will require you to make two more compute statements, one involving "hiimpl" and one involving "hiinter").

Because the steps involved in testing interactions is painstaking, it should go without saying that we recommend marching through this process one step at a time and checking your work carefully as you go. Then after you develop each syntax statement, you can run it, save the data file with all your newly created variables, run the appropriate analyses, and write up a summary of all of your results. Don't forget to save your syntax file, too.

QUESTION 10.2. Report the complete results of these three different multiple regression analyses. Make sure you begin by reporting whether there were any main effects of your two predictors and then report whether there was an interaction (the preceding Test 3). Assuming you got one, report the results of your two separate simple slopes tests (Tests 4 and 5), and provide a concise and insightful interpretation of all of your simple slope findings.

More Real Practice Data

QUESTION 10.3. To give you some more practice testing for interactions in multiple regression, we'd like you to examine an additional data set that will require you to test for an interaction using some correlational survey data. Specifically, we will introduce you to a pilot study in which Mitsuru Shimizu and the first author assessed (a) explicit self-esteem (people's conscious thoughts and feelings about their value or worth), (b) implicit self-esteem (people's *non*conscious thoughts and feelings about their value or worth), and (c) a measure of intuitiveness, which we define as the tendency to rely on hunches or gut feelings when evaluating things (see Shimizu & Pelham, 2011, for more details). The simple idea you will want to test has to do with how strongly implicit and explicit self-esteem are related. We wanted to see if *the relation between implicit and explicit self-esteem would be higher among people who are more intuitive*. The data file you'll need to test this idea is called [**intuitiveness pilot for regression interactions.sav**]. Analyze these data to test the prediction that implicit and explicit self-esteem are more strongly correlated among people who score high as opposed to low on this new measure of intuitiveness. Be sure to include two separate simple slopes tests if you do see any evidence of the predicted interaction.

Important Moderator Effects Sometimes Add Minimally to R-Square Values

If you were pretty familiar with multiple regression analysis before reading about it in this book, there is a very good chance that you learned that a goal of multiple regression analysis is to maximize the value of R-square (or adjusted R-square). There is an obvious sense in which this is true. Mathematically speaking, the stronger a predictor's unique association with some criterion variable, the more that predictor contributes to the size of the overall multiple R-square in a multiple regression analysis. Furthermore, if you put several strong predictors of a criterion in a regression equation, you will usually get a much larger R-square value than if you put only one strong predictor (unless the strong predictors in the first case all happen to be highly redundant with one another). Finally, the strength of the unique association between a predictor and a criterion variable often tells you something very important about the practical or theoretical importance of that predictor. For example, if my theory predicts that agreeableness and extraversion are superior predictors of driver safety but your theory predicts that neuroticism and conscientiousness are superior predictors of driver safety, one way to evaluate our competing theories is to run a multiple regression analysis predicting driver safety from all four of these predictors. We can then see how much each of these four personality traits contributes to our overall R-square value (controlling for the other three variables, as well as any other important control variables such as driver age or average neighborhood traffic density). If your two pet variables contribute more to the overall R-square value than my two pet variables, then this supports your theory. Let's say that your two variables collectively account for 40% of the variance in driver safety, whereas my two variables collectively account for only 10% of the variance. This is very good news indeed for your theory. So far, so good. However, researchers sometimes become overly fixated on exactly how much of the variance in a criterion variable different predictors account for. Being too fixated on how much a specific predictor variable adds to your overall R-square value can be especially unwise when statistical interactions are at work. This is because an interaction term that adds only minimally to one's overall R-square value can often be of tremendous practical and theoretical importance. To explore this important but often overlooked idea, let's consider a research question adapted from one of the first author's favorite papers—namely, Pelham and Swann (1989; see also Pelham, 1995a, 1995b).

Chapter 10 Examining Interactions in Multiple Regression Analysis

More than 100 years ago, William James argued that when people are good at things they value (even if they are not so good at things they *don't* value), they are likely to enjoy high self-esteem. As James (1890) put it, identities on which people have "staked their salvation" should play a much bigger role in self-esteem than identities about which people don't care very much. Usain Bolt would presumably be pretty devastated if he suddenly lost his phenomenal sprinting abilities. However, we suspect that it wouldn't bother him too much to learn that he has very little talent as a poker player or a polka dancer. For the first 99 years after James suggested this idea, research examining people's specific self-views (e.g., are you artistic? are you physically attractive?) and how much importance people attribute to these specific self-views (does being artistic really matter to you?) yielded little support for James's hypothesis. However, Pelham and Swann (1989) argued that researchers had not looked as carefully as one possibly could for evidence in support of James's hypothesis. Without sweating the details, let's simply note that Pelham and Swann developed a measure of "differential importance," which is the degree to which people believe that their self-perceived strengths across many different self-views are more important than their self-perceived weaknesses across the same set of self-views. People who score high in differential importance tend to have high global self-esteem, even after controlling statistically for how favorably they evaluate themselves on a long list of specific traits such as intelligence, athletic ability, or physical attractiveness. We think that's mildly interesting. However, we think it's a little more interesting—and a little more relevant to the theme of this chapter—that differential importance is a much better predictor of self-esteem for some people than for others. Specifically, differential importance is much more strongly related to global self-esteem for people who do not report having a lot of positive specific self-views than it is for people who do. So if Clark believes that he is überathletic, superintelligent, very handsome, and pretty artistic, his self-esteem might not vary much depending on whether he values his super strengths more than his so-so strengths. However, if Wally believes that he is only moderately athletic, slightly intelligent, not so handsome, and abysmally unartistic, whether he values his moderate strengths more than his abysmal weaknesses usually tells us quite a lot about whether Wally is high or low in self-esteem. Statistically speaking, there is an interaction between the overall level of people's specific self-views and what Pelham and Swann called differential importance.

The hypothetical data set [**Jamesian self-worth hypothetical data.sav**] illustrates just such an interaction. To confirm this, take a look at Image 4. It contains some selective screen captures of the results from an SPSS simultaneous multiple regression analysis of these data. The regression model includes three predictors: (a) the centered main effect of people's specific self-views, (b) the centered main effect of differential importance, and (c) the Self-Views × Differential Importance interaction term based on these two centered scores. As you can see in Image 4, the overall adjusted *R*-square value is a highly respectable .606, both main effect terms are significant, and the interaction term is significant ($p = .012$).

Model Summary

Model	R	R Square	Adjusted R Square	Std. Error of the Estimate
1	.794ᵃ	.630	.606	6.76710

a. Predictors: (Constant), centinter, centself_views, centdiff_import

Coefficientsᵃ

Model		Unstandardized Coefficients		Standardized Coefficients	t	Sig.	Collinearity Statistics	
		B	Std. Error	Beta			Tolerance	VIF
1	(Constant)	54.185	1.002		54.053	.000		
	centself_views	.298	.049	.572	6.061	.000	.902	1.109
	centdiff_import	13.752	3.852	.343	3.570	.001	.869	1.151
	centinter	-.505	.192	-.243	-2.632	.012	.940	1.063

a. Dependent Variable: esteem

But what would the *R*-square value have been if we had only looked at the two main effects that were a part of this more complex three-predictor model? Some selective screen captures from the output of a simultaneous multiple regression analysis that *excluded* the interaction term appear in Image 5. As you can see from the top of Image 5, if we had completely ignored the interaction term, we still would have been able to account for more than half of the variance in people's global self-esteem scores. The adjusted *R*-square value for the two-predictor model is .557. On the basis of the difference in the two adjusted *R*-square values (.606 − .557 = .049), we can account for only 4.9% more of the variance in global self-esteem by including the interaction term.

Model Summary

Model	R	R Square	Adjusted R Square	Std. Error of the Estimate
1	.758ᵃ	.575	.557	7.18131

a. Predictors: (Constant), centdiff_import, centself_views

Coefficientsᵃ

Model		Unstandardized Coefficients		Standardized Coefficients	t	Sig.	Collinearity Statistics	
		B	Std. Error	Beta			Tolerance	VIF
1	(Constant)	53.400	1.016		52.580	.000		
	centself_views	.280	.052	.538	5.421	.000	.919	1.088
	centdiff_import	16.123	3.974	.403	4.057	.000	.919	1.088

a. Dependent Variable: esteem

On the basis of the modest contribution of the interaction to the overall *R*-square value, you might be tempted to question the practical and/or theoretical importance of this interaction. Perhaps differential importance has only *slightly* more to do with self-esteem for people with negative as compared with positive self-views. That is, perhaps the simple slopes that reflect the association between differential importance and global self-esteem are not all that different for people with negative versus positive specific self-views. To put it differently, maybe Clark and Wally are not very different after all, at least when it comes to the role differential importance plays in their self-esteem. To find out exactly how different these two simple slopes are, you'll need to follow the usual procedures we've outlined in this chapter and conduct simple slopes tests for people with negative versus positive specific self-views.

QUESTION 10.4a. Conduct and report a simple slopes test that produces the estimate of the association between differential importance and global self-esteem in these hypothetical data for people 1 standard deviation *above* the mean in specific self-views. Now conduct and report the same simple slopes test for people 1 standard deviation *below* the mean in specific self-views. Be sure to comment briefly on the meaning of these two simple slopes tests. By the way, this is not nearly as difficult as it might sound because we have already created all of the "stand-in" variables you need to conduct both sets of simple slopes tests (e.g., "low_sviews," "low_inter").

QUESTION 10.4b. What do these simple slopes tests tell you about the importance of the interaction involving differential importance? Specifically, in what sense is it misleading to point out that the interaction term in this analysis only allows one to account for an additional 4.9% of the total variance in self-esteem? As a big hint, remember that the squared beta weight (squared standardized regression coefficient) for a predictor tells you roughly, sometimes exactly, how much of the variance in the criterion you can account for by knowing a person's score on the predictor in question (see Aiken & West, 1991, p. 115, Note 1).

Examining Interactions Between Categorical and Continuous Predictors in Multiple Regression

As the social sciences have become more complex and more integrative over the past few decades, researchers have increasingly begun to test for statistical interactions between one or more *continuous* predictors and one or more *categorical* predictors. For example, personality and social psychologists often conduct *person-by-situation quasi-experiments* by combining categorical experimental manipulations with measured individual difference variables (see Pelham & Blanton, 2013). They might do this, for example, to see if degree of extraversion (a continuous measure) predicts friendliness better under some manipulated laboratory conditions than others (e.g., during a staged social activity rather than a staged work-related activity). Moving from the lab to the field, some kinds of attitudes and behaviors are categorical (e.g., your favorite color, whether you were born in Chile), whereas others are continuous (e.g., exactly how much you like the *Bone* series of graphic novels, how many episodes of *Psych* you have ever watched). The simplest possible case of a moderated regression analysis with both a categorical predictor and a continuous predictor has (a) a single categorical predictor with only two levels (e.g., gender) and (b) a single continuous predictor (e.g., the number of siblings you have).

An easy way to test for moderation in the simplest possible case of a single, two-level categorical predictor combined with a single continuous predictor is to (a) dummy code the categorical predictor (e.g., 0 vs. 1); (b) center this categorical, dummy-coded predictor as well as the continuous predictor; and then (c) proceed through the remaining testing process pretty much as you would if the dummy-coded categorical variable were a continuous variable, usually focusing the two simple slopes tests on the two different groups that are part of the categorical variable (e.g., women vs. men). The only exception to the procedures you've already learned in this chapter is in how you set up your simples slopes tests. The key is to determine exactly what centered score corresponds to each of the two discrete groups (e.g., women vs. men, platypuses vs. echidnas) on which you wish to focus for your two different simple slopes tests. It is extremely unlikely that the "stand-in" score that allows you to focus separately on each of the two levels of your categorical predictor will happen to be exactly +1 and −1 standard deviation from the centered mean. In fact, scores of exactly +1 and −1 standard deviation from the mean won't usually exist for a categorical predictor. For example, if your categorical variable is coded 0 versus 1 (prior to centering), zero will almost never be exactly 1 standard deviation below the mean, and 1 will almost never be exactly 1 standard deviation above the mean (although they may both be pretty close if scores are about equally split between 0 and 1).

To see exactly how this works, let's analyze a subset of a real (and very large) data set. In these data, the criterion is self-reported health status (scored from 1–5, where 1 is "excellent" health and 5 is "poor" health). The categorical predictor is gender, and the continuous predictor is BMI (body mass index). Roughly speaking, BMI is a person's weight relative to his or her height, and a very high BMI score means that a person is considered obese.[2] Needless to say, it is well documented that obesity is associated with a wide range of health problems. The hypothesis we'd like you to test, however, is a bit more complex than this. The hypothesis is that *the relation between BMI and self-reported health status is stronger for women than for men*. As we said before, the data you will use are real. They were sampled randomly from a very large-scale health survey of a representative sample of Americans. (The original data file contained more than 100,000 cases. We thought this was a bit much, and so we randomly sampled only 2% of the cases.) The SPSS data file is called [**big health 2 percent sample 32408**

Chapter 10 Examining Interactions in Multiple Regression Analysis

.sav]. For this question, let's test our hypothesis using "health," which is the primary self-report measure of health status (this variable name is a little misleading, by the way, because on this "health" variable, higher scores mean *poorer* health).

Let's begin by centering gender and BMI for this analysis. To do so, we'll want to focus only on the participants who answered (a) the gender question, (b) the height and weight questions necessary to calculate BMI, and (c) the self-reported health status question. Although there were 2,404 respondents in this subset of the larger data file, only 1,715 of them answered all of the crucial questions we need to see if there is an interaction between gender and BMI predicting self-reported health. In case your "Select Cases..." skills have gotten a little rusty, let us remind you that the SPSS Data commands include the "Select Cases..." option, which is designed to allow users to specify that SPSS focus only on the subset of cases in a data file that fulfills certain requirements. Images 6 through 8 contain sequential screen captures showing how we focused on the 1,715 respondents who had complete data on all the measures we care about. By the way, the reason we need to focus on this subset of 1,715 people is that all of our centering and simple slopes tests require very precise values for the means and standard deviations of the predictors. If we centered BMI using the mean BMI score for the total sample of 2,404 people, this would almost certainly yield a different centered BMI value than the true value for the subset of people who happened to answer all of the crucial questions—and who can thus be used in the regression analysis. So before you begin any analyses involving moderation effects with real data files, you almost always have a little data preparation to be sure you consistently focus on exactly the right people (those who have no missing data on the variables of interest).

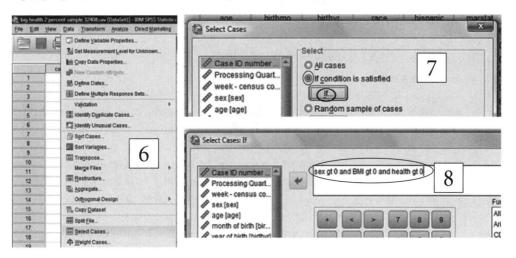

Once you click "Continue" (which has been cropped away in Image 8) and then "OK" (cropped away in Image 7) from the "Select Cases...." function, you'll have instructed SPSS to focus only on the 1,715 participants who have complete data for this analysis. This means you'll be ready

to get the information you need to center gender and BMI. Remember that a very easy way to find out the means and standard deviations for a set of variables is to correlate the variables with one another—making sure to ask for "Means and standard deviations" as a Statistics "Option" in the "Correlate" command. Image 9 is a quick reminder of what to request when you run the correlation in SPSS.

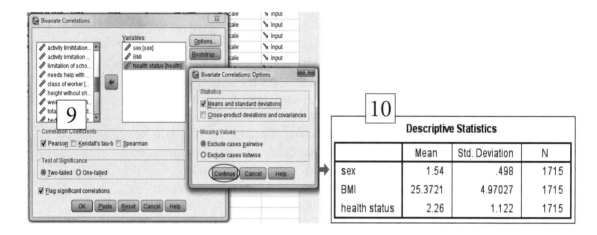

If you click "Continue" from the pop-up "Options" dialog window (circled in Image 9) and then click "OK" from the main "Bivariate Correlations" window, you should generate an SPSS output file whose contents include the important descriptive information you see in Image 10. To center gender, then, you'll simply run the following compute command from a syntax file:

COMPUTE cent_sex = sex − 1.54.

EXE.

You can write a similar COMPUTE command to create a centered BMI score. Once you've created both of these centered scores by running both compute statements, check out the last two columns of your original SPSS data file to see what scores correspond to being male versus female. If you recall that for the sex variable, 2 = female and 1 = male, it should make sense that for the variable "cent_sex," women always have a score of +0.46 (2 − 1.54 = +0.46), whereas men always have a score of −0.54 (1 − 1.54 = −0.54). This will be very important to know when it comes time to conduct separate simple slopes tests for women and men. But let's make sure there is a reason to do those simple slopes tests in the first place. That is, let's first test for the gender-by-BMI interaction effect. And recall that to test for this interaction, you'll need to create a cross-product term that is based on the two centered predictors.

Chapter 10 Examining Interactions in Multiple Regression Analysis

COMPUTE cent_inter = cent_sex*cent_BMI.

EXE.

Of course, if you observe a significant interaction, you'll want to use the following syntax commands to create variables that will allow you to assess the strength of the association between BMI and health separately for women and men. Notice, for example, that in the COMPUTE statement that allows us to focus on women, the score that is subtracted from "cent_sex" is simply the exact centered score that corresponds to being female. This is very much like subtracting a score from a centered self-esteem variable that corresponds to being low (or high) in self-esteem.

COMPUTE women = cent_sex − (.46).

COMPUTE women_inter = women*cent_BMI.

COMPUTE men = cent_sex − (−.54).

COMPUTE men_inter = men*cent_BMI.

exe.

QUESTION 10.5a. Report the results for the regression analysis testing for the gender-by-BMI interaction, including the relevant beta weights and p values for the two centered main effects as well as the interaction term. If there is a significant interaction, follow it up with two simple slopes tests, one for women and one for men.

QUESTION 10.5b. Now, for purposes of comparison, conduct the rough logical equivalent of these same simple slopes tests by splitting the file by gender and simply correlating BMI and health separately for women and for men. How do the results of this simpler approach compare with those based on your more sophisticated simple slopes tests? Finally, make sure to offer some guesses, however cautious, about *why* you observed these findings. A potential clue is that you should remember that your criterion is *self-reported* health status, which is positively but imperfectly correlated with most physiological indicators of true health status.

Why Does This Technique for Estimating Simple Slopes Work?

If it seems a little odd to you that a bit of subtraction from centered variable z can radically change the size of the regression coefficients for the x-y association when an interaction is at work, you are not alone. It's pretty counterintuitive to most people, for example, that subtracting a standard deviation worth of self-esteem from the centered self-esteem variable (and then using this new score to create a new interaction term) should make the regression analysis focus on people high in self-esteem when it estimates, say, the association between job stress and illness. Upon close inspection, however, the transformations that are part of simple slopes estimates are quite logical—and not very complex. To see exactly why simple slopes tests do what they do, let's examine some hypothetical survey data on the association between height and weight. We hope it is obvious that height is a pretty strong predictor of weight. In representative samples of Americans, the correlation between the two is usually somewhere in the neighborhood of $r = .70$ to $.80$. But imagine that on a distant planet where beings have telekinetic control of their bodies, there is little or no association between height and weight. In fact, the only rule about weight on this planet is that it's considered attractive to have a weight that ends in double digits (e.g., 122 and 355 are both highly desirable weights). Let's look at a highly simplified data file involving five earthlings and five aliens (Telekinetians) to see exactly what happens when we conduct a routine simple slopes analysis after observing a Planetary Origin × Height interaction in predicting body weight.

Let's begin by examining the SPSS data file [**alien height weight demo of why simple slopes work.sav**] that includes all the variables we need to assess the interaction and follow it up with simple slopes tests. As documented in Image 11, the thoughtful and industrious alien research assistants who conducted this survey have already completed all the transformations we need to analyze these data. In the screen capture of the SPSS data file you see in Image 11, please notice four important things: (1) The arbitrarily coded alien status variable is centered from the very beginning, (2) height is included in the file in both raw score form and in centered form, (3) a centered interaction term has already been calculated for you, and (4) the stand-in variables for both simple slopes tests (including both special interaction terms) have already been calculated for you. Your alien research assistants were also thoughtful enough to provide you with the syntax commands they used to create all of the transformed variables.

First, they centered height and created an interaction term based on centered alien status and centered height.

COMPUTE cent_height = height – 3.

COMPUTE inter = alien*cent_height.

EXE.

Second, they created (a) a stand-in variable that will focus a simple slopes analysis on the aliens (the variable is appropriately called "is_alien") and (b) the interaction term that goes along with it.

COMPUTE is_alien=alien – (1).

COMPUTE inter_is_alien = is_alien*cent_height.

EXE.

Third, they created the comparable stand-in scores that will focus a simple slopes analysis on the earthlings ("isn't_alien" and "inter_isnt_alien").

COMPUTE isnt_alien = alien − (−1).

COMPUTE inter_isnt_alien = isnt_alien*cent_height.

EXE.

human beings ↓

	alien	height	weight	cent_height	inter	is_alien	inter_is_alien	isnt_alien	inter_isnt_alien
1	-1.00	1.00	150.00	-2.00	2.00	-2.00	4.00	.00	.00
2	-1.00	2.00	160.00	-1.00	1.00	-2.00	2.00	.00	.00
3	-1.00	3.00	245.00	.00	.00	-2.00	.00	.00	.00
4	-1.00	4.00	240.00	1.00	-1.00	-2.00	-2.00	.00	.00
5	-1.00	5.00	255.00	2.00	-2.00	-2.00	-4.00	.00	.00
6	1.00	1.00	211.00	-2.00	-2.00	.00	.00	2.00	-4.00
7	1.00	2.00	244.00	-1.00	-1.00	.00	.00	2.00	-2.00
8	1.00	3.00	255.00	.00	.00	.00	.00	2.00	.00
9	1.00	4.00	177.00	1.00	1.00	.00	.00	2.00	2.00
10	1.00	5.00	122.00	2.00	2.00	.00	.00	2.00	4.00

aliens ↗

Before we go any further, notice that there is clearly a positive correlation between height and weight for the human beings (participants with "alien" scores of −1). The shortest earthling is the lightest (150 pounds), and the tallest earthling is the heaviest (255 pounds). This familiar correlation clearly does not apply to the Telekinetians. In fact, the tallest Telekinetian happens to be the lightest, and the heaviest Telekinetian is exactly average in height. Thus, it probably won't surprise you much to learn that when we conducted a test for an Alien × Height interaction predicting body weight, there was a significant interaction, despite our tiny sample size. That's all as it should be, but what we'd really like to know is how and why the simple slopes work the way they do.

Let's begin by focusing on the aliens. Notice that we've drawn a dashed squircle around both the stand-in variable for being an alien ("is_alien") and the special interaction term that goes with this stand-in variable. If you remember that a regression analysis is simply a mathematical way of analyzing what goes hand in hand with what, we hope you can see that there is an obvious sense in which "is_alien" and "inter_is_alien" are being held constant in the transformed scores. If you'll forgive the personification, it's as if SPSS is saying, "Well, I only have three predictors, and two of the predictors (is_alien and inter_is_alien) aren't varying at all for these five cases. So here's my chance to pay special attention to height to see how well it predicts weight in these specific cases." This is very much like our familiar cookie theft situation except that we'd have 5 days to pay close attention to Bart's presence or absence in the kitchen when both Lisa and

Homer's presence in the kitchen were held constant. (We added Homer to the picture because there are three rather than two predictors at work in this analysis.) Of course, SPSS does not completely ignore the data from the five earthlings, but if you look carefully at the is_alien and inter_is_alien score for the earthlings, you should be able to see that although is_alien takes on a constant value of −2, the interaction term inter_is_alien has a perfect negative correlation with height. Thus, just as SPSS didn't put much stock in days when Bart and Lisa were both present (or both absent), SPSS would essentially ignore the earthling data when drawing conclusions (for the Telekinetians) about the strength of the height-weight association.

Of course, the same kind of logic characterizes what happens when SPSS estimates the simple slopes for the earthlings. The special stand-in variables, isn't_alien and inter_isnt_alien, are both held at a constant value of zero for the earthlings (see the five sets of scores in the squircle drawn with a solid line), causing SPSS to focus disproportionately on the height-weight association for these five participants. Furthermore, when it comes to the aliens, it is impossible from SPSS's perspective to tell whether the height variable or the interaction term that is perfectly (and in this case positively) correlated with the height variable is responsible for any possible observed association between height and weight. So there you have it. These simple but counterintuitive transformations do the trick because they set up a logical situation in which SPSS focuses disproportionately on the association between x and y at some specified level of the other predictor z. Of course, with two continuous predictors, you won't see too many cases in which SPSS *completely* ignores the stand-in variable and the stand-in interaction term because the stand-in variables will often take on values that are very close to zero rather than zero. However, the closer a participant scores to the score on which you wish to focus for the simple slopes tests, the closer that person's stand-in scores will be to the zero values that would make SPSS focus completely on that person when assessing the x-y association. To be more concrete, if a person scored 0.93 standard deviations below the mean in self-esteem, then this person's stand-in self-esteem score and stand-in interaction term wouldn't both be zero for the simple slopes test focusing on people low in self-esteem, but they'd both be very close to zero. So this person's x scores would be weighted very heavily, although not quite as heavily as the x scores of a person whose centered self-esteem score was *exactly* −1.0.

Finally, we hasten to remind you that the situation we have described here applies only when the association between one of your predictors and the criterion differs at different levels of the other predictor. For example, if there were no interaction whatsoever between self-esteem and job stress in predicting illness, then it wouldn't matter at all whether you focused disproportionately on one self-esteem group or the other. The association between job stress and illness would be about the same for both self-esteem groups, and your two simple slopes tests would thus be highly similar.

It's Not Easy Studying Green: Dealing With Interactions Involving Categorical Predictors With More Than Two Levels

If categorical predictors always had only two levels, it would be pretty easy to assess interactions involving a mixture of categorical and continuous predictors. However, many categorical variables have multiple levels. Ask 50 people their favorite color, or even their political party, and you will usually get more than two different answers (unless you happen to limit your sample to attendees at the National Alliance of Republicans Favoring Teal). Furthermore, the different answers you get can't usually be lined up on some meaningful scale involving amounts. If they could, after all, you'd be dealing with a continuous variable. This means that testing for interactions involving categorical variables requires some statistical juggling of variable levels. This juggling is crucial. For example, it allows you to choose exactly which of the three levels of a three-level categorical variable you wish to compare with one of the other two specific levels of the variable. It can thus tell you, for example, whether your Political Party × Extraversion interaction is more of a Republican versus Democrat × Extraversion interaction or more of a Democrat versus Marxist × Extraversion interaction.

Speaking of politics, political party identification is a great example of a very important categorical variable that has more than two levels. If you are a Democrat, you may be happy to learn that according to Gallup polls taken over the past decade or so, only about a third of Americans identify themselves as Republican (e.g., see Jones, 2009). If you are a Republican, there is good news for you, too. Only about a third of Americans identify themselves as Democrats. If you are an Independent, there is both good and bad news. The good news is that about one third of Americans identify themselves as Independents. The bad news is that very few people seem to recognize that there are so many of you. Getting back to our statistical point, let's consider a hypothetical data set involving political parties. These data are hypothetical but are based loosely on a real data set analyzed by Pelham and Hardin (unpublished data, 2012).

Pelham and Hardin (unpublished data, 2012) were interested in the contact hypothesis, which is an idea that goes back at least as far as Gordon Allport (1954). According to the contact hypothesis, one of the reasons why people often harbor prejudice toward social groups other than their own is because they often have very little personal contact with members of social or ethnic groups other than their own (e.g., for a review, see Pettigrew & Tropp, 2008). If people do have regular contact, especially equal status contact, with the members of other social groups, their attitudes about members of this group should become more favorable (and perhaps also less stereotypical). In the hypothetical data we'd like you to analyze, you should assume that all of the respondents are White. The criterion variable is the favorability of these White respondents' attitudes about Barack Obama (their approval of the job they felt he was doing as president). The categorical predictor is people's political party affiliation (Democratic, Independent, or Republican), and the continuous predictor (based on census data) is the proportion of people's neighbors who are Black. The hypothesis we'll want to test in these hypothetical data is that White people who have a higher proportion of Black neighbors will only have more positive attitudes toward Barack Obama if they happen to be Democrats. One could even hypothesize that among Independents or Republicans, having more Black neighbors might lead to a "boomerang effect" or a reversal of the usual contact effect (so that those who have a higher percentage of Black neighbors are *less* likely to be pro-Obama, perhaps because they feel extra pressure to support the worldview of their political party). For our purposes, however, let's assume that the only a priori prediction is that Democrats will show stronger evidence for a contact effect than will Republicans.

Getting back to the statistical point, when a categorical variable has more than two levels, it is no longer possible to account for the effects of this categorical variable in a multiple regression analysis with only one variable. As a general rule, it takes $k-1$ variables in a regression analysis to code effectively for a categorical variable with k levels. Thus, for a categorical variable with only two levels, $k-1$ is simply 1. This is why, for example, you only needed one variable to deal with gender in the moderated multiple regression analysis we covered earlier. In the case of political party, though, there are three very popular categories, and you may assume that we intentionally sampled only members of these three parties. This means we'll need two variables to deal with political party, and Table 10.3 illustrates three possible dummy-coding schemes for the political party variable. These three schemes all necessarily involve two variables, but the focal group differs for each of the three specific schemes. Following the example of Aiken and West (1991, p. 117), we refer to the two dummy variables as D_1 and D_2 although we add the additional subscripts d, i, and r to remind readers that each D_1 and D_2 variable would be part of a separate focus and thus a separate statistical analysis. It's also important to remember that variables such as D_{1d} and D_{2d} are not two different levels of the same variable but represent *two different variables* whose scores are always 0 or 1.[3]

It is important to note that it is always the two dummy-coded variables *working together* (i.e., entered into the same simultaneous multiple regression analysis) that allow you to focus on a specific level of the categorical variable of interest. Notice, for example, that in the case of the dummy codes focusing on Democrats, a person who is a Democrat would be coded as a 0 on both D_{1d} and D_{2d}. For the variable D_{1d} Independents are

Table 10.3 Three Dummy-Coding Options for Representing Contrasts Between Three Different Political Parties

	Democrats as Focal Group		Independents as Focal Group		Republicans as Focal Group	
	D_{1d}	D_{2d}	D_{1i}	D_{2i}	D_{1r}	D_{2r}
Democrat	0	0	0	1	0	1
Independent	0	1	0	0	1	0
Republican	1	0	1	0	0	0

also coded 0, whereas only Republicans are coded 1. However, for D_{2d} this state of affairs is reversed. Independents are now coded as 1, whereas Republicans are coded as 0. In short, taken together, (a) variable D_{1d} discriminates Democrats from Republicans and (b) variable D_{2d} discriminates Democrats from Independents. This need to have two dummy-coded variables to cover one categorical, three-level predictor means that if you are interested in interactions, you also need two separate interaction terms, one for D_{1d} and one for D_{2d}. If you are as likely to think ahead as Edmund Hillary must have been, it may have already occurred to you that it is no accident that when a pair of dummy codes is intended to focus on a specific group, the dummy codes are both set at *zero* for that group. On the basis of our recent discussion of why simple slopes work as they do, we hope you can see that for all of the Democrats in our sample, D_{1d} and D_{2d} will always take on values of zero. Likewise, for all of the Democrats in our sample, the two *interactions terms* that involve D_{1d} and D_{2d} will also take on a value of 0 (because anything multiplied by zero is zero). Thus, we hope you can see that for the purposes of main effects as well as interactions, the first pair of dummy codes in Table 10.3 will make Democrats the focal group. As you'll soon see, this simplifies things a bit when it comes to simple slopes tests.

Because of your extreme patience with this challenging chapter, we are going to save you quite a bit of trouble and provide you with a data file that has already been coded for the first set of dummy codes in Table 10.3. The data file is called [**Obama approval by political party and neighborhood ethnicity.sav**]. As you can see in the following, it was necessary to do some recoding of the original political party variable ("pol_party") to translate the wonderful state of affairs you see summarized in the left-hand portion of Table 10.3 into an SPSS data file. Here, for example, are the syntax commands we used to convert the original "pol_party" variable to the useful dummy-coded variables we called "dem_v_rep" (D_{1d}) and "dem_v_ind." (D_{2d}).

IF (pol_party eq 1) dem_v_rep = 0.

IF (pol_party eq 2) dem_v_rep = 0.

IF (pol_party eq 3) dem_v_rep = 1.

IF (pol_party eq 1) dem_v_ind = 0.

IF (pol_party eq 2) dem_v_ind = 1.

IF (pol_party eq 3) dem_v_ind = 0.

> As documented in the variable values in the SPSS data file, for pol_party, 1 = Democrat, 2 = Independent, and 3 = Republican. Notice that the three dummy codes highlighted here correspond to the circled values in Table 10.3.

And here are the syntax commands that create the two necessary *interaction terms* that build on the above terms.

COMPUTE interdem_v_rep = cent_perc_Black*dem_v_rep.

COMPUTE interdem_v_ind = cent_perc_Black*dem_v_ind.

EXE.

Image 12 contains a peek at the data file sorted by political party, with Democrats at the top. This allows you to see the scores of 0 that Democrats always get for the two dummy-coded variables *and* the two dummy-coded interaction terms.

	id	pol_party	perc_Black	cent_perc_Black	approve_Obama	dem_v_ind	dem_v_rep	interdem_v_ind	interdem_v_rep
1	3.00	1.00	.182	-.01	2.00	0	0	.0	.0
	9.00	1.00	.233	.04	4.00	0	0	.0	.0
12	10.00	1.00	.170	-.02	2.00	0	0	.0	.0
	17.00	1.00	.212	.02	3.00	0	0	.0	.0
5	18.00	1.00	.346	.15	4.00	0	0	.0	.0

Finally, here is the syntax version of the multiple regression analysis you'd need to conduct to see if contact effects are stronger among Democrats than among Independents or Republicans.

REGRESSION

/MISSING LISTWISE

/STATISTICS COEFF OUTS R ANOVA

/CRITERIA=PIN(.05) POUT(.10)

/NOORIGIN

/DEPENDENT approve_Obama

/METHOD=ENTER cent_perc_Black dem_v_ind dem_v_rep interdem_v_ind interdem_v_rep.

You'll notice that this simultaneous regression analysis has *five* rather than three predictors. It took two variables to represent the dummy-coded main effects of political party and two variables to represent the two pieces of the focal interaction involving the dummy-coded political party. Assuming you and Barack are both dying to know whether we observed an interaction, Image 13 contains the most important part of the SPSS output file for the multiple regression model we just presented.

Coefficients[a]

Model		Unstandardized Coefficients		Standardized Coefficients	t	Sig.
		B	Std. Error	Beta		
1	(Constant)	3.414	.291		11.721	.000
	cent_perc_Black	3.521	1.548	.464	2.274	.029
	dem_v_ind	-.606	.405	-.226	-1.494	.143
	dem_v_rep	-1.698	.464	-.633	-3.660	.001
	interdem_v_ind	-3.857	2.194	-.319	-1.758	.087
	interdem_v_rep	-8.926	3.342	-.463	-2.671	.011

a. Dependent Variable: Obama approval

Beginning with the main effects for the two dummy-coded variables, this model illustrates that, on average, Democrats and Independents didn't differ significantly in their degree of approval for Obama, $\beta = -.226$, ns. Not surprisingly, however, Democrats and Republicans did differ quite dramatically in their approval of Obama, $\beta = -.633$, $p = .001$. Remembering that Democrats were coded 0 (and Republicans 1) for this variable, this merely supports the well-documented fact that Democrats like Obama better than Republicans do. There was also a marginally significant trend ($p = .087$) toward a Democrat versus Independent by percent Black neighbors interaction. For now, we'll ignore that ambiguous finding and focus on the key predicted finding. There was, in fact, a significant *Democrat versus Republican* by percent Black neighbors interaction, $\beta = -.463$, $t(39) = -2.67$, $p = .011$.

Now, of course, we'd like to know two things. First, what's the association between the percentage of one's neighbors who are Black and support for Obama among Democrats? Second, what's the association between the percentage of one's neighbors who are Black and support for Obama among Republicans? We'll answer the easy part for you. Among Democrats, the more Black neighbors people had, the more favorable their attitudes toward Obama, $\beta = .464$, $p = .029$. We didn't have to conduct any extra simple slopes tests to know this because this particular way of setting up these dummy codes already focused on Democrats. Remember that this is why Democrats got a score of 0 on both of the dummy-coded variables and both of the interaction terms. They are the focal group, and the results for "cent_perc_Black" (the centered percentage of one's neighbors who are Black) in the portion of the SPSS output file you've just been examining gives us the answer to our first simple slopes question.

QUESTION 10.6. Now for the slightly more difficult part. Using both Table 10.3 and the syntax commands we recently showed you as a guide, create a set of two new dummy-coded variables that focus on the Republicans rather than the Democrats. Then run a *new* regression analysis and first confirm that the Republican versus Democrat by percentage Black neighbors interaction term is identical to the comparable one we showed you when we focused on Democrats. Second, report and briefly comment on the results of the simple slopes test for Republicans that is the equivalent of the one we just discussed for Democrats. Copy your new syntax commands into your answer (at the end) to document your setup work.

We hasten to add that, although we will not ask you to conduct any more simple slopes test on these data, it is also possible to conduct a set of simple slopes tests that focuses on Independents. Such a test could tell you whether this specific kind of contact effect for the Independents more closely resembled what was happening for the Democrats or what was happening for the Republicans.

QUESTION 10.7. We noted that for a categorical variable with n levels, you need $n - 1$ variables to code appropriately for that categorical predictor. With that rule in mind, imagine that you needed to analyze some data involving a potential interaction between a continuous predictor (e.g., level of social support) and a categorical predictor with four levels—namely, which group of people whose lives involve prison you are focusing on: (a) wardens, (b) guards, (c) medical staff, or (d) prisoners. Create a table patterned directly after Table 10.3 in which you fill out all of the dummy codes necessary to look for moderator effects in such a research design.

Before we conclude this chapter, we should reinforce a point we have made elsewhere in this text, which is that there is often more than one correct way to analyze a data set. In the case of studies with (a) a continuous, normally distributed dependent variable, (b) at least one categorical predictor, and (c) at least one continuous predictor, it is just as appropriate to analyze the data using analysis of covariance (ANCOVA) as it is to analyze them using multiple regression. One of the tricks you must know if you are doing so in SPSS, though, is that SPSS needs a little coaxing to create an interaction term involving the categorical predictor and the covariate (the term for a continuous predictor variable in ANCOVA). The way categorical variables are coded in ANOVA or ANCOVA is also somewhat different than the dummy-coding procedure we have emphasized here (ANOVA uses the equivalent of unweighted effects coding rather than dummy coding; see Aiken & West, 1991, pp. 127–130, for some details). Nonetheless, moderated multiple regression analysis involving a categorical variable and a continuous variable is a very, very close cousin of ANCOVA, and in most ways, the two techniques are identical.

In this chapter, we have covered two closely related advanced multiple regression techniques: (1) how to test for and interpret interactions

Chapter 10 Examining Interactions in Multiple Regression Analysis

involving two continuous predictors and (2) how to test for and interpret interactions involving one continuous predictor and one categorical predictor (with either two or three levels). As Question 10.7 suggests, however, we hope it is clear that the logic we have laid out here generalizes to more complex research questions. For example, the same centering procedures and simple slopes tests used to assess and interpret two-way interactions in multiple regression are readily generalized to the assessment and interpretation of three-way interactions. Along these lines, Appendix 10.1 extends this chapter a bit by providing an example of how you might test for (and follow up on) a three-way interaction involving three continuous predictors. Appendix 10.2 complements this chapter in a very different way. This second appendix provides an example (borrowed from Shimizu & Pelham, 2004) of how to write about interactions in multiple regression in an APA-style research report.

Appendix 10.1: Testing for and Interpreting Three-Way Interactions in Multiple Regression

Just as higher order interactions (e.g., three-way or four-way interactions) are possible when all of your multiple independent variables are categorical—and when you thus analyze your data using ANOVA—higher order interactions are also possible when one or more of your multiple independent variables are continuous. In this appendix, we briefly discuss procedures for testing for and interpreting a three-way interaction when you have three continuous predictors in a multiple regression analysis. Because of our deep and abiding love of soft drinks, we decided to focus this analysis on soft drink research.

On Monday October 10, 2011, Dr. Pepper rolled out a new kind of Dr. Pepper, which they called Dr. Pepper Ten. This cola, obviously designed to be attractive to men who think diet colas are wimpy, came out in a can that looked like it may have contained equal parts cola and gunpowder. Moreover, the ad campaign for the new cola—apparently borrowed from one of the "He-Man Woman Hater's Club" episodes of *The Little Rascals*—proclaimed unapologetically, "It's not for women." Who knows if this cola will prove to be the iPad or the Edsel of diet colas targeted at men, but let's assume that Dr. Pepper would love to know more about exactly who is likely to respond favorably to this new diet soft drink. Assuming that you don't mind contributing a bit to the basest aspects of male socialization toward unfettered machismo, imagine that Dr. Pepper wanted you to figure out exactly what kind of macho man will find this macho cola most attractive. As an experienced colaologist, you believe that three predictors are crucial. First, men who are overweight (operationally defined as having a high BMI) will be more interested than will men who are thin. Second, men who report that they have a strong desire to lose weight (regardless of BMI) might find the cola more attractive than others. Third, the more men like the classic, socially insightful TV show *The Three Stooges,* the more they'll like this cola. However, these are all just main effects. You want to test the more interesting idea that men will disproportionately love this new cola if they happen to be obese *and* wish to lose weight *and* love *The Three Stooges.* To be a little more specific, suppose you predicted that there would be a two-way interaction such that whether men strongly wish to lose weight will matter more for the obese men than for men who are not so heavy. Crucially, however, you believe that this two-way interaction will be even stronger for men who love *The Three Stooges* (presumably because this is a decent proxy for loving all things stupidly and/or humorously masculine). This means you'd be expecting a three-way interaction. The two-way BMI × Weight Loss Desire interaction would be qualified by the strength of men's liking for *The Three Stooges.*

To test for this three-way interaction, we need to generalize what we already learned about testing two-way interactions in multiple regression.

Chapter 10 Examining Interactions in Multiple Regression Analysis

Image 14 contains a screen capture of an SPSS file [**Bubba Pepper.sav**] whose data the Dr. Pepper Snapple Group desperately wishes were real. All of the transformed variable names are circled for you, and there are three important things to note about the transformed data. First, all three predictors are centered (and labeled in such a way as to make this obvious). Second, there are three different two-way interactions terms, which collectively cover every possible kind of two-way interaction for these three predictors (in generic form, the three two-way interaction terms are always $A \times B$, $A \times C$, and $B \times C$). Third, there is also a three-way interaction term, which is simply the triple cross-product of the three centered predictors. An appropriate test for the three-way interaction requires that we include all three of the main effects, all three two-way interaction terms, and the three-way interaction term.

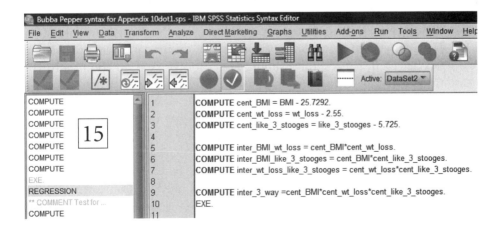

Image 15 contains a screen capture of the syntax commands we used to create all seven of the transformed variables you see circled in the data file. In case you wish to consult this syntax file yourself, it is called [**Bubba Pepper syntax for Appendix 10dot1.sps**]. This syntax file may prove useful if you ever need to test for a three-way interaction when you have three continuous predictors because it contains all of the compute statements and analyses we cover in this appendix.

Coefficientsᵃ

Model		Unstandardized Coefficients		Standardized Coefficients	t	Sig.
		B	Std. Error	Beta		
1	(Constant)	5.180	.162		31.993	.000
	cent_BMI	-.037	.044	-.064	-.834	.406
	cent_wt_loss	.235	.130	.129	1.813	.072
	cent_like_3_stooges	.413	.114	.254	3.616	.000
	inter_BMI_wt_loss	.139	.032	.393	4.391	.000
	inter_BMI_like_3_stooges	-.127	.030	-.349	-4.208	.000
	inter_wt_loss_like_3_stooges	.143	.080	.125	1.788	.076
	inter_3_way	.041	.019	.224	2.160	.032

a. Dependent Variable: like_Dr_Pepper10

Image 16 contains the most important part of the SPSS output file from the multiple regression analysis that included these seven variables as predictors of liking for Dr. Pepper Ten.

We hope you can see from the three main effects tests enclosed in the rectangle in Image 16 that there was no significant main effect of BMI on liking for Dr. Pepper Ten. Furthermore, although there was a trend for men who reported a stronger desire to lose weight to report greater than average liking for Dr. Pepper Ten, this potential main effect was only marginally significant ($p = .072$). The only significant main effect was for liking for *The Three Stooges*. On average, men who liked the Stooges more also liked Dr. Pepper Ten more. But remember that main effects aren't all that informative when significant interactions are at work.

Speaking of interactions, there were two significant two-way interactions. The positive sign of the BMI × Weight Loss Desire interaction ($\beta = .393$, $p < .001$) suggests that men who scored high on both variables (heavier men who strongly wished to lose weight) seem to have liked Dr. Pepper Ten more than other men. The negative sign of the BMI × Liking for *The Three Stooges* interaction ($\beta = -.349$, $p < .001$) suggests that the positive correlation between liking for the Stooges and liking for Dr. Pepper Ten is probably stronger for thinner men as opposed to heavier men. However, we shouldn't get too worked up about either of these two-way interactions because both interactions are qualified by a significant three-way interaction, $p = .032$.

One useful way to break down this three-way interaction is to look at the two-way BMI × Weight Loss Desire interaction separately for men who *like* and men who *dislike The Three Stooges*. To do that, we must change the "cent_like_3_stooges" variables twice. First, we'll focus on men who dislike *The Three Stooges*. We'll need to transform (a) *The Three Stooges* variable as well as (b) both two-way interactions that include this variable and (c) the three-way interaction that includes this variable. Here are the syntax commands for so doing, along with the revised multiple regression command that includes all of the newly transformed variables—along with the original variables that didn't need to be changed.

COMPUTE dislike_3_stooges = cent_like_3_stooges − (−1.445).

COMPUTE inter_BMI_dislike_3_stooges = cent_BMI*dislike_3_stooges.

COMPUTE inter_wt_loss_dislike_3_stooges = cent_wt_loss*dislike_3_stooges.

COMPUTE inter_3_way_dislike_3_stooges =cent_BMI*cent_wt_loss*dislike_3_stooges.

EXE.

REGRESSION

/DESCRIPTIVES MEAN STDDEV

/MISSING LISTWISE

/STATISTICS COEFF OUTS R ANOVA

/CRITERIA=PIN(.05) POUT(.10)

/NOORIGIN

/DEPENDENT like_Dr_Pepper10

/METHOD=ENTER cent_BMI cent_wt_loss dislike_3_stooges inter_BMI_wt_loss inter_BMI_dislike_3_stooges inter_wt_loss_dislike_3_stooges inter_3_way_dislike_3_stooges.

If you were to run these syntax commands on the Bubba Pepper data file and then rerun the test for the three-way interaction, you'd see that when we focus on men who dislike *The Three Stooges* (i.e., those scoring 1 standard deviation below the mean in liking for the Stooges), the two-way BMI × Weight Loss Desire interaction is no longer significant ($p = .132$). See the circled part of the left-hand portion of the SPSS output screen capture in Image 17. You might also notice, for example, that the significance of the three-way interaction did not change at all compared with the original analysis in which all the predictors were centered.

Coefficients[a]

Model		Unstandardized Coefficients		Standardized Coefficients	t	Sig.
		B	Std. Error	Beta		
1	(Constant)	4.583	.245		18.718	.000
	cent_BMI	.146	.068	.254	2.146	.033
	cent_wt_loss	.029	.186	.016	.156	.877
	dislike_3_stooges	.413	.114	.254	3.616	.000
	inter_BMI_wt_loss	.079	.052	.225	1.513	.132
	inter_BMI_dislike_3_stooges	-.127	.030	-.541	-4.208	.000
	inter_wt_loss_dislike_3_stooges	.143	.080	.181	1.788	.076
	inter_3_way_dislike_3_stooges	.041	.019	.364	2.160	.032

a. Dependent Variable: like Dr Pepper10

Coefficients[a]

Model		Unstandardized Coefficients		Standardized Coefficients	t	Sig.
		B	Std. Error	Beta		
1	(Constant)	5.776	.217		26.674	.000
	cent_BMI	-.220	.056	-.382	-3.953	.000
	cent_wt_loss	.442	.161	.242	2.751	.007
	love_3_stooges	.413	.114	.254	3.616	.000
	inter_BMI_wt_loss	.198	.028	.562	7.129	.000
	inter_BMI_love_3_stooges	-.127	.030	-.393	-4.208	.000
	inter_wt_loss_love_3_stooges	.143	.080	.155	1.788	.076
	inter_3_way_love_3_stooges	.041	.019	.160	2.160	.032

a. Dependent Variable: like Dr Pepper10

When we repeated this regression analysis using the stand-in variable that focuses the analysis on men who *do* like *The Three Stooges*, the two-way BMI × Weight Loss Desire interaction was clearly significant, as you can see in the right-hand half of Image 17 ($p < .001$). Because this process is pretty painstaking, we've also included the syntax commands and multiple regression analysis commands that allowed us to focus on the men who do, in fact, adore *The Three Stooges* (those scoring 1 standard deviation above the mean in Stooge liking). Here are those commands, along with the edited command to rerun the multiple regression analysis (and remember that these commands also exist in the Bubba Pepper syntax file).

COMPUTE love_3_stooges = cent_like_3_stooges − (1.445).

COMPUTE inter_BMI_love_3_stooges = cent_BMI*love_3_stooges.

COMPUTE inter_wt_loss_love_3_stooges = cent_wt_loss*love_3_stooges.

COMPUTE inter_3_way_love_3_stooges = cent_BMI*cent_wt_loss*love_3_stooges.

EXE.

REGRESSION

/DESCRIPTIVES MEAN STDDEV CORR SIG N

/MISSING LISTWISE

/STATISTICS COEFF OUTS R ANOVA

/CRITERIA=PIN(.05) POUT(.10)

/NOORIGIN

/DEPENDENT like_Dr_Pepper10

/METHOD=ENTER cent_BMI cent_wt_loss love_3_stooges inter_BMI_wt_loss inter_BMI_love_3_stooges inter_wt_loss_love_3_stooges inter_3_way_love_3_stooges.

Of course, even after doing all of this, we are still not done. To truly understand the two-way interaction that holds only for lovers of *The Three Stooges,* you'd need to run two more regression analyses, both of which are slight variations on the one you see above. In the first, of course, you might choose to replace the centered BMI scores with a stand-in variable that focuses on men who score 1 standard deviation below the mean in BMI. In the second, you might choose to replace the centered BMI scores with a stand-in variable that focuses on men who score 1 standard deviation above the mean in BMI.[4]

Notice that, at this point, you are doing almost the same thing you have already done several times when conducting simple slopes tests to follow up on two-way interactions. The only differences are (a) that your regression model now has seven predictors rather than three and (b) you need to remember that when you create a stand-in variable for, say, men who are thin, you need to include this stand-in variable consistently when calculating both of the two-way interaction terms that include BMI as well as the three-way interaction term. Otherwise, the simple slopes tests involve the same procedures with which you should now be highly familiar.

In case you'd like to check your work, here are the results of the two simple slopes tests. First, remembering that we are focusing solely on lovers of *The Three Stooges,* the simple slopes tests showed that among men who were below average in BMI, there was no association between how much they'd like to lose (more) weight and how much they liked Dr. Pepper Ten, $\beta = -.201$, $p = .095$. (If anything, there might be a weak tendency for these thin men to dislike Dr. Pepper Ten more the more they'd like to lose weight. We'd need a replication, preferably with a much larger sample size, to find out.) However, among *Three Stooges* lovers who were above average in BMI, there was a very strong positive association between how much they said they'd like to lose weight and how much they liked Dr. Pepper Ten, $\beta = .685$, $p < .001$. If these data were real, you can rest assured that the Dr. Pepper Snapple Group would be offering hefty men who wish to lose weight a lot of free tickets to screenings of *The Three Stooges,* where they would also be plied with lots of free samples of Dr. Pepper Ten, which, by the way, is surprisingly delicious. Nyuck, nyuck.

Appendix 10.2 An Example of How to Report the Results of a Two-Way Interaction in Multiple Regression

If you will forgive your first author for being even more egocentric than usual, we'd like to provide you with an excerpt from an empirical research paper by Shimizu and Pelham (2004). Shimizu and Pelham were interested in the counterintuitive idea, derived from self-verification theory, that people with low self-esteem may actually suffer negative health consequences in the wake of positive life events. The logic of this prediction is that for people low in self-esteem, positive life events may seem nice, but they disrupt such people's chronic negative beliefs about themselves and thus threaten their basic feelings of predictability and control. When Shimizu and Pelham (2004) published their article, Brown and McGill (1989) had already documented this basic finding using a traditional measure of explicit self-esteem. However, Shimizu and Pelham were interested in seeing if this finding also applied to a measure of implicit self-esteem. Furthermore, they wanted to be sure that if they did observe an implicit self-esteem by positive life events interaction in predicting people's self-reported health, this finding for implicit self-esteem was not merely an example of the already established interaction effect for *explicit* self-esteem. Thus, Shimizu and Pelham wanted to see if their predicted two-way interaction involving implicit self-esteem would still hold up after controlling for the same kind of two-way interaction involving explicit self-esteem. Finally, to be sure any effects they observed did not merely reflect negative reporting biases, they also controlled statistically for a frequently used measure of the general tendency to experience negative affect. Thus, their multiple regression analyses were a bit more complex than the ones you've examined in this chapter, mainly because they included additional statistical controls in their test for an implicit Self-Esteem × Positive Life Events interaction. Except for a few necessary complications, though, Shimizu and Pelham followed all of the basic procedures for assessing interactions in multiple regression that we have covered in this chapter. Here is how they summarized their primary results. After reminding readers that they predicted that both implicit and explicit self-esteem would interact (separately) with positive life events to predict illness, Shimizu and Pelham then remind readers that they followed the procedures outlined by Aiken and West for testing and following up on statistical interactions (e.g., centering, doing simple slopes tests). Here are their primary results:

> First, as we expected, the Positive Life Events × Explicit Self-Esteem interaction was significant, $\beta = -.130$, $t(166) = -1.98$, $p = .049$. Simple slopes tests revealed that positive life events were associated with illness only among participants low (i.e., 1 standard deviation below the mean) in explicit self-esteem, $\beta = .276$,

$t(166) = 2.55$, $p = .012$. For participants high (i.e., 1 standard deviation above the mean) in explicit self-esteem, positive life events and illness were unrelated, $\beta = -.018$, $t(166) = -0.19$, $p = .852$. This is a clear replication of Brown and McGill's findings. This same analysis also revealed that the Positive Life Events × Implicit Self-Esteem interaction was significant, $\beta = -.253$, $t(166) = -3.83$, $p = .001$. As we expected, positive life events were associated with illness only for participants low (−1 standard deviation) in implicit self-esteem, $\beta = .433$, $t(166) = 4.07$, $p < .001$. For participants high (+1 standard deviation) in implicit self-esteem, positive life events were marginally associated with lower levels of illness, $\beta = -.166$, $t(166) = -1.70$, $p = .091$. Except for the fact that our findings were somewhat stronger for implicit as opposed to explicit self-esteem, the findings for implicit and explicit self-esteem were strikingly similar. Building on the original conclusions of Brown and McGill, we suggest that positive life events may disrupt the identities of people low in implicit as well as explicit self-esteem.

Notes

1. If you are wondering why the correlation is $r = .577$ rather than $r = .500$ (9 matches minus 3 mismatches/12 = .50), notice that for the criterion variable (missing cookies), cookies were missing only 3 out of 12 days, meaning that the criterion does not meet the precise requirements ($p = .50$ for all variables) that allow the super simple correlation formula to work correctly.

2. We add the qualifier *considered* because some extremely lean but muscular people can have very high BMI scores. The Incredible Hulk, for example, would be technically obese, as are some NFL running backs. However, these exceptions are rare enough that BMI does an excellent job of predicting many important health outcomes.

3. It is also possible to deal with categorical variables that have more than two levels by using **unweighted effects coding,** which is sometimes more appropriate than dummy coding. See Aiken and West (1991, pp. 127–130) for a discussion of the pros and cons of this alternative way of dealing with categorical variables in multiple regression.

4. If these data were real, you might also choose to focus on specific BMI scores that medical researchers and health psychologists consider to be representative of specific weight groups, or to create and use multiple BMI categories such as "underweight," "normal weight," "overweight," and "obese." The specific details of how you would do the analysis might change greatly based on this kind of decision, but the basic logic of testing the three-way interaction would not change. It would always boil down to selecting a meaningful two-way interaction that was qualified by the third predictor of interest. For example, it might be meaningful to run the tests differently to see how weight loss desire moderates the two-way BMI × Liking for *The Three Stooges* interaction. There are always multiple ways to conceptualize and break down three-way interactions.

11
ANCOVA, Covariate-Adjusted Means, and Predicted Scores

Introduction: Ends to a Mean

When you're presenting research findings to an interested lay audience (e.g., journal editors, great aunts, newspaper reporters), it is often useful to present audiences with a set of two or more **means** that summarize a research finding. Means are a bit like pictures in that they convey a lot of information very efficiently. For example, suppose that you wanted to give your great uncle Karl (a smart but poorly educated 81-year-old retired potato farmer) a good sense of whether there is gender inequity in pay among American men and women. Which of the following statements do you suppose he would find more informative?

"My study of about 400 representative Americans showed that the correlation between gender and income was $r = -.36$, where being female was arbitrarily coded as 2 and being male was arbitrarily coded as 1."

"My study of about 400 representative Americans showed that, on average, women made $26,032 per year whereas men made $41,442."

We hope you agree that the second way of summarizing the data is a lot easier on your audience. One simple piece of advice about the best way to present data is to report means whenever you possibly can. Thus, if you have a design such as the one above (i.e., if your only independent variable is a categorical variable such as gender), you should almost always conduct a *t* test or one-way analysis of variance (ANOVA) and present your results in the form of two or more means. In the preceding example, we hope you can see that there is nothing wrong with conducting a point-biserial correlation and reporting the correlation. In fact, the *p* value associated with this point-biserial correlation will always be *the same* as the *p* value associated with the logically identical *t* test. However, the *t* test yields means, and means make it very easy for your audience to understand what you've found.

Unfortunately, once you graduate to more complex research designs, it's not always so easy to present means. For example, suppose you wanted to know if the gender difference in income that we just discussed still holds true once you control for the fact that (in samples of middle-aged or older Americans), men tend to be better educated than are women. Do men *still* make more money than women if you statistically control for the higher level of education traditionally enjoyed by most men? We hope you realize that you will need to use some kind of covariance technique (e.g., multiple regression analysis) to address this question. However, it may have dawned on you that, if you were to conduct a multiple regression analysis, you would no longer be able to look at means to get a good picture of what was going on in your data. After all, multiple regression analyses do not readily yield means. We will return later to the issue of how you can get around this problem in multiple regression analyses. Before we discuss that issue, however, let's consider an alternate statistical technique that has many of the same features as multiple regression analysis but *does* readily yield means. This technique is the analysis of covariance (ANCOVA), a slightly more sophisticated cousin of ANOVA.

The Analysis of Covariance (ANCOVA)

Many statisticians have described ANCOVA as a blend of multiple regression and ANOVA. One way to think about ANCOVA is that it is *very much like conducting an ANOVA on a set of data once you have statistically removed the effects of a third variable* (referred to as a covariate) on the dependent variable. For example, suppose you conducted a true experiment in which you randomly assigned people to sit in hot versus cool rooms, with the expectation that the people in the hot rooms would stick more pins in a voodoo doll that resembled their statistics instructor. That is, suppose you expected that heat would influence aggressive behavior. Because you randomly assigned people to conditions, you began the study with the assumption that the average level of disliking for one's statistics instructors would be more or less identical in the two experimental conditions. However, due to simple bad luck, imagine that the people who were randomly assigned to the *hot* room happened to dislike their statistics instructor more (prior to taking part in the study) than did the people who were randomly assigned to the *cool* room. This is an unfortunate, de facto confound that crept into your design—despite your valiant efforts to be a good methodologist. Nonetheless, as long as you measured how much people liked their statistics instructor, you could "repair" this confound by using ANCOVA. In essence, your approach would be to adjust people's scores on the dependent measure of aggressiveness (i.e., how many pins people chose to stick in the voodoo doll). You could do this, for instance, by running a multiple regression analysis in which you predicted people's aggressiveness scores from their naturally occurring level of dislike for the instructor. You could then use the **residuals** based on these predicted scores (the variation each pin-sticking score that is *left over* after you statistically remove the portion of each pin-sticking score that is predictable from each person's score on the disliking measure) as the replacement scores for your original dependent measure. The residuals, then, would reflect the variation in people's levels of aggressiveness that *still existed* between the two experimental groups above and beyond the variation attributable to differences in their preexisting distaste for the instructor.

Of course, if you could predict people's aggressiveness scores perfectly from their level of dislike for their instructor, there would be no variation in aggressiveness left over to predict from room temperature. But as you know, it is practically never the case that a covariate perfectly predicts a dependent measure. So the question in ANCOVA is whether any independent variables of interest still predict people's scores on the dependent variable once we have statistically adjusted for the effects of one or more covariates. In this hypothetical study of temperature and aggression, the covariate was a coincidental confound that turned up based on simple bad luck. Much more often, covariates correspond to natural confounds that exist not because of bad luck but because an independent variable that the researcher could not manipulate (e.g., gender) happens to be confounded with a rival independent variable. Thus, quasi-experiments and passive observational studies (i.e., "correlational" studies) are often prime candidates for analysis by means of ANCOVA. To give you a better sense of how ANCOVA works and how it compares with multiple regression analysis, this chapter introduces you to three new data sets.

Data Set 1: Gender Differences in Income

The first data set has to do with gender differences in income. The simple question you will address (to start with) is whether men have higher incomes than women, but your question will eventually become more sophisticated than this. Using the data from an SPSS practice file, [**Employee data.sav**], you will want to march systematically through several different analyses to learn about ANCOVA and how it compares with ANOVA and with multiple regression analysis. (Notice that the SPSS instructions for these analyses come *after* the first two data analysis questions.)

QUESTION 11.1. First, see if men do make more money than women in this sample. To facilitate a comparison between ANCOVA and ANOVA, conduct a simple ANOVA and interpret the means. By the way, you could simply conduct an independent samples *t* test, but if you run an ANOVA, you can eventually work from the same general set of ANOVA commands to convert your ANOVA into an ANCOVA.

QUESTION 11.2. Second, conduct an ANCOVA and generate and interpret your *covariate-adjusted means.* Your independent variable should be gender, your covariate should be education, and your dependent variable should be income. By adding education to the design, you can address whether the men in your sample simply make more money because they are, on average, more highly educated than the women.

Image 1 contains some clues about how to get started on these analyses. Once you open your data file and go to "Analyze," you'll want to go to the "General Linear Model" and then click the "Univariate…" command.

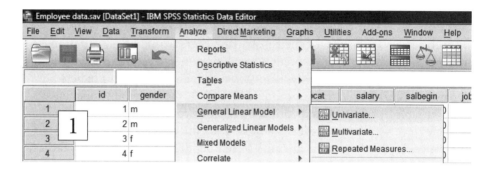

This will open the "Univariate" box you see in Image 2. From this box, send gender to the "Fixed Factor(s):" (independent variable) list. Then send salary to the "Dependent Variable:" list. Next, click "Options…" and you'll open a dialog window much like the one in the right half of Image 2. After you send gender to the "Display Means for:" box, click on all three of the options you see clicked under "Display." Collectively, these three display options will generate a lot of useful statistics.

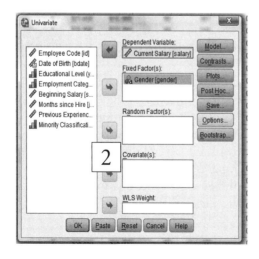
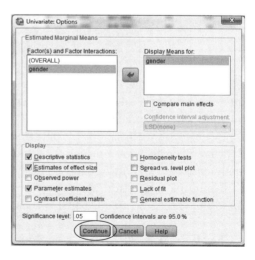

When you click "Continue" in the "Univariate: Options" dialog box (circled in the right half of Image 2) and then click "Paste" when you return to the main analysis box, you'll create a syntax file that includes your ANOVA command. If you run this syntax command, you should

Chapter 11 ANCOVA, Covariate-Adjusted Means, and Predicted Scores

produce an SPSS output file, the most important part of which looks very much like Image 3.

Descriptive Statistics

Dependent Variable: Current Salary

Gender	Mean	Std. Deviation	N
Female	$26,031.92	$7,558.021	216
Male	$41,441.78	$19,499.214	258
Total	$34,419.57	$17,075.661	474

Tests of Between-Subjects Effects

Dependent Variable: Current Salary

Source	Type III Sum of Squares	df	Mean Square	F	Sig.	Partial Eta Squared
Corrected Model	2.792E+10[a]	1	27918533029	119.798	.000	.202
Intercept	5.353E+11	1	5.353E+11	2296.792	.000	.830
gender	27918533029	1	27918533029	119.798	.000	.202
Error	1.100E+11	472	233046530.5			
Total	6.995E+11	474				
Corrected Total	1.379E+11	473				

a. R Squared = .202 (Adjusted R Squared = .201)

3

We won't offer you any help interpreting these findings. However, we *will* offer some help finding out if these findings merely reflect the fact that men are more educated than women. To do so, you need to repeat the analysis you just ran and *control for gender differences in education.* This means you will make education a **covariate** in an ANCOVA. This is pretty easy. It amounts to doing exactly what you just did to run the ANOVA with the exception that you make an extra click or two (to add the covariate). The left-hand portion of Image 4 shows you what the appropriate part of the "Univariate" command box should look like once you've added your covariate to the equation. That's right; you just click on education ("educ") in the main variable list and then send it to the "<u>C</u>ovariate(s):" box.

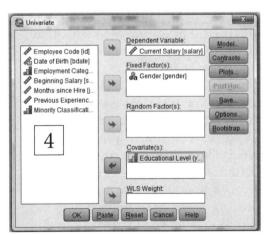

4

Gender

Dependent Variable: Current Salary

Gender	Mean	Std. Error	95% Confidence Interval	
			Lower Bound	Upper Bound
Female	29834.645[a]	864.413	28136.061	31533.229
Male	38258.107[a]	786.204	36713.206	39803.009

a. Covariates appearing in the model are evaluated at the following values: Educational Level (years) = 13.49.

Once you run this ANCOVA, you'll get a more complex version of the SPSS output file you got for the basic ANOVA. By the way, it's important to note that *if you were starting this ANCOVA from scratch (without having just run an ANOVA), SPSS would not know which options you just asked for.* You'd thus have to march through the same options we marched through for the original ANOVA. However, given that we have just done this, *without closing and reopening the SPSS data file,* SPSS assumes that we'd like to keep the same options we just asked for in the ANOVA. That's very convenient because people don't usually want to get rid of any options when they move from one analysis to a slightly different one. When you click on "Paste" this time, and run the ANCOVA commands from your syntax file, the more complex version of your results will include the **covariate-adjusted means** that we pasted into the right-hand portion of Image 4. Covariate-adjusted means are a lot like regular means except that they are adjusted statistically for the effects of any covariates you included in the model. In this case, the covariate-adjusted means you see correspond to the separate income levels for women and men after statistically equating women and men for education. If your covariates (or, in this case, your one covariate) are pretty strongly correlated with both your dependent variable and your independent variable, your covariate-adjusted means could be different than your raw means (sometimes very much so).

Now, for purposes of comparison, let's see what happens when we address the same basic statistical questions by conducting a simultaneous multiple regression analysis rather than an ANCOVA.

QUESTION 11.3. Conduct a simultaneous multiple regression analysis on the same data. Compare (a) the regression coefficients and associated *p* values from this regression analysis with (b) the main effects and covariate effects, as well as the associated *p* values, from the ANCOVA you ran to answer Question 11.2.

Because you should already know how to conduct a multiple regression analysis, we won't review all of the details about how to run a regression analysis. Remember, though, that you'd like to compare the effects of your two predictor variables in a regression analysis with the effects you observed for your independent variable (gender) and your covariate (education) in the ANCOVA. We should further tell you that, to run the regression analysis, you'll need to address a problem. The problem is that, in this data file, gender is not coded as a number but rather as *m* or *f.* Unlike ANOVA, which doesn't mind categorical variables, multiple regression analyses require that all variables in the analysis take on numerical values (e.g., 1 and 2 rather than M or F). This will require some simple recoding, which is trickier than you might think because of how SPSS treats string variables in IF statements. With this in mind, here is an example of how you could make the appropriate conversion in your syntax file. Notice that the letters *m* and *f* in the IF statement as well as the words *male* and *female* in the VALUE LABELS statements are enclosed in quotation marks. It is crucial in SPSS 20 to use standard (double) quotation marks in the VALUE LABELS statement (as we did here). If you try using single quotation marks in SPSS 20, SPSS will not create your value labels correctly. However, it is your primary author's memory that in older versions of SPSS, one had to use single quotation marks to get the value labels correct. So be ready to experiment a bit depending on what version of SPSS you are using.

IF (gender eq "m") new_gender = 1.

IF (gender eq "f") new_gender = 2.

exe.

VALUE LABELS new_gender 1 "male" 2 "female".

exe.

The Ghosts in the Machine: Generating Predicted Scores in a Multiple Regression Analysis

This should get you through Question 11.3, but if you really wish to compare the pros and cons of multiple regression analysis versus ANCOVA, you need to get a multiple regression analysis to generate something resembling means rather than the *B*s (unstandardized regression coefficients) or beta weights (standardized regression coefficients) that you traditionally get from a regression analysis. As it turns out, this isn't all that difficult. However, it requires you to do two new things. The easy one is to ask SPSS for an option we have not previously requested from a multiple regression analysis. The second is to generate what we affectionately refer to as "ghost cases." These are false respondents in your data file who have no effect whatsoever on your results (including your sample size) but allow you to generate predicted scores on your criterion that strongly resemble means. This should become a little clearer when you see exactly how we do it.

Let's start by making some ghosts and then move on to asking SPSS for some predicted scores. The screen capture you see in Image 5 shows you the last few cases in the [**Employee data.sav**] data file.

You will notice that beneath the last *real* case in the file, we have added a couple of fictitious (ghost) cases. Importantly, the first ghost is a man, and the second ghost is a woman. However, the two ghosts do *not* differ at all in education. Each ghost has exactly 13.49 years of education. We chose the unusual value 13.49 because this is the mean education level for the 474 respondents who qualified for the crucial multiple regression equation. That is, this is the mean educational level of the 474 respondents who answered (a) the gender question, (b) the education question, and (c) the income question. After all, you *had* to answer all three questions to become a valid case in the multiple regression analysis that involved all three variables. Because we gave these two ghosts the same level of education but varied their gender, we know that if we ask SPSS to generate predicted salary scores for these two ghosts, SPSS will use the complete regression equation to tell us what incomes are most likely for a typical man and a typical woman who each have a precisely average level of education. In other words, we will be able to see (in the form of predicted yearly salaries) whether men *still* make more money than women, on average, once we hold education constant.

By the way, it is *extremely* important that you do *not* include scores for these two ghosts when it comes to the actual salary variable. If you were to do so, you would essentially resurrect these ghosts when you ran the multiple regression analysis, and the scores you decided to give them would influence the results of your regression equation. As you know if you have ever watched those horror movies your parents always warned you about, resurrecting a ghost is one of the last seven or eight things you'd ever want to do. So be careful when dealing with ghosts. For example, if you were to forget that these ghosts were not really part of your data set and simply correlated gender and income with these ghosts included in your data, the ghosts would be treated exactly like real participants and would slightly water down any real correlation between the two variables while falsely inflating your sample size by $n = 2$.

If you have added your ghost scores to your own version of the data file, all you need to do now is to ask for predicted scores when you conduct the simultaneous multiple regression equation predicting salary from both the recoded gender variable ("new_gender") and education. Because of your expertise at running multiple regression analyses in SPSS, we are going to cut right to the part where you ask for predicted scores. The screen capture you see in Image 6 shows what your Linear Regression command should look like when you've chosen your independent and dependent variables.

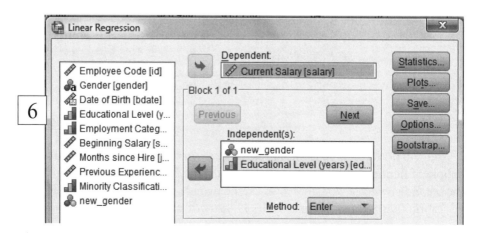

Once you've identified the variables you want, don't forget to click on "S̲tatistics…" so that you can ask for "Descriptives" as well as "Estimates" and "Model fit," as shown in Image 7.

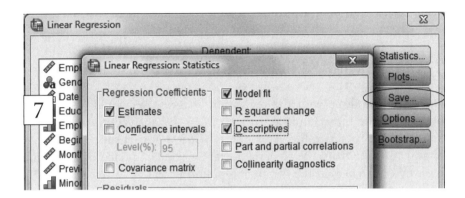

Even more important, though, once you return to the "Linear Regression" box, you'll need to click the "S̲ave…" button (circled in Image 7), which will allow you to click the "U̲nstandardized" option under "Predicted Values," as shown in Image 8.

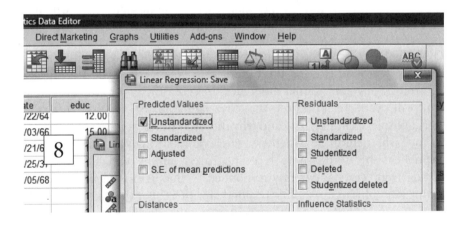

This is the crucial command that will generate predicted salary scores for every respondent in your data file—*including your two ghost respondents*. These predicted scores, by the way, will become a part of your actual data file, and so after running the regression analysis, you will need to return to the data file to see the predicted salary scores of your two ghost respondents.

QUESTION 11.4a. Generate predicted scores in a multiple regression analysis and compare these predicted scores with the covariate-adjusted means that you generated in ANCOVA. If you see any differences, address whether they are meaningful or may just have to do with rounding errors.

QUESTION 11.4b. Use your newly developed ghost skills in these data to see who is likely to have a higher salary: a typical woman who has a master's degree (18 years of education) or a typical man who has a 4-year college degree (16 years of education). Does this suggest to you that regression techniques might be more any more or less flexible than ANCOVA?

Data Set 2: Political Party Affiliation and Attitudes

Once you are done with the regression analysis, we would like you to become a little more familiar with ANCOVA by analyzing some data in another SPSS practice file. This one is [**GSS93 subset.sav**]. These data appear to come from the General Social Survey, conducted on a routine basis by researchers at the National Opinion Research Center since 1972. At any rate, in this second file, you will not need to worry about conducting any multiple regression analyses. Instead, you should merely conduct a couple of ANCOVAs.

QUESTION 11.5a. In one analysis, you should determine whether Democrats and Republicans differ in their attitudes toward evolution.

QUESTION 11.5b. In a second analysis, you should determine whether they differ in their attitudes toward sex education in public schools.

QUESTION 11.5c. If the two groups do differ significantly on either variable, you will want to see if this difference still exists once you statistically control for any educational differences between the two groups. That is, you will want to interpret the covariate-adjusted means from the ANCOVA(s), including how and why they may differ from the raw means.

You will determine whether people are (strongly) Democratic or Republican by doing some recoding based on the variable "partyid." Strong Democrats apparently responded to this item with a 0, whereas strong Republicans apparently responded with a 6. For the sake of simplicity, *you should limit your analyses to participants who are strong Democrats and strong Republicans.* Be sure to provide and interpret both the **raw means** and the **covariate-adjusted means** for the analysis of both dependent measures.

Data Set 3: A Survey of Smoking and Well-Being

The third data file contains a simulation of some real data on smoking status and well-being, specifically how favorably people evaluate their lives. The data file is based closely on an extremely ambitious survey

Chapter 11 ANCOVA, Covariate-Adjusted Means, and Predicted Scores

study of American adults conducted by the polling and consulting firm Gallup. As you may recall from a previous chapter, in January 2008, in collaboration with a health promotion company called Healthways, Gallup began conducting a *daily* telephone survey of 1,000 Americans. Because Gallup uses a random digit dialing procedure and samples people with landlines (traditional phones) and people with cell phones in proportion to their frequency in the U.S. population, this daily survey is highly representative of the U.S. population in the window between January 2008 and October 2009. However, because the full Gallup survey includes hundreds of variables and had a total sample size of more than 650,000 people by the end of October 2009 (and because the data are property of Gallup), I created a simulation of 2,000 cases that strongly resemble the real data. For the purposes of this activity, you may write about the data as if they were real. If you want to see some real results for the complete sample of 650,000 people, go to www.gallup.com/poll/124280/Nonsmokers-Top-Smokers-Across-Incomes.aspx.

Let's assume, for the sake of this analytic assignment, that smokers do evaluate their lives less favorably than nonsmokers. Is this really an effect of smoking per se? Or could smoking just be a marker of poverty? After all, poor people are more likely to smoke than wealthy people. Furthermore, research on well-being has also shown that poor people have lower well-being than rich people. We hope you can see that smoking is *confounded* with income, which makes this research question one very well suited to either an ANCOVA or a multiple regression analysis. The SPSS data file containing these data is called [**smoking and well-being simulation of GALLUP G1K data nov 2009.sav**]. All of the information you need about the variables in this data file is contained in the variable "Labels" and variable "Values" that are part of the data file. However, you will need to do one quick calculation in SPSS, which is to create a single well-being score by averaging wp16 and wp18.

QUESTION 11.6. Analyze these data with an ANOVA, then with an ANCOVA. In the ANCOVA, you should control for income category and see what effect this statistical control has on your covariate-adjusted means. The ANCOVA should give you some covariate-adjusted means that you can compare with the original means. Be sure to explain any differences, however subtle, between your original results and the results of the ANCOVA.

QUESTION 11.7. As an additional way of seeing how ANCOVA (or regression) works, try moving from a one-way ANCOVA with two independent predictors (smoking status as an independent variable and income as a covariate) to a *factorial* ANOVA. A 2 × 4 (smoking status by income category) factorial ANOVA can allow you to hold income constant in a different way—which is to see if there is a relation between smoking and well-being *within* each of the four separate income categories. Are the results of the factorial ANOVA logically consistent with the results of the ANCOVA? Does the factorial ANOVA reveal any useful information not revealed by the ANCOVA? Does it suggest that we may have unknowingly violated one of the assumptions of ANCOVA in the original ANCOVA? (A big clue: The major assumption we have in mind is "homogeneity of covariance.") If so, which set of results should we trust?

Suppressor Variables 12

Introduction: Multiple Regression and Suppression

In the first chapter on multiple regression analysis (Chapter 9), you learned that multiple regression techniques can disentangle a group of competing predictors of some outcome and indicate which, if any, of the predictors is uniquely associated with the outcome. Carefully conducted multiple regression analyses often reveal that a predictor that had a seemingly impressive zero-order correlation (simple correlation) with a criterion was simply masquerading as a true cause—because it happened to be correlated with the real cause (or one of them, at least). Recall that dentistry just happened to co-occur with liquor stores and gun shops. Lisa just happened to co-occur with Bart. In such cases, the beta weight associated with a predictor typically proves to be much smaller than the zero-order correlation involving the same predictor. In the case of cookie thefts, the zero-order correlation between Lisa's presence and missing cookies was a whopping $r = .67$. But in our regression analysis, we adjusted this correlation for the fact that Lisa happened to be confounded with Bart—who proved to be perfectly correlated with missing cookies! This adjustment yielded a beta weight of zero ($\beta = .00$). Of course, these kinds of changes aren't always this dramatic. In fact, sometimes such changes don't occur at all. Bart's beta weight was just as impressive as his simple correlation coefficient. This was true because the extra information about Lisa didn't make Bart look any more or any less guilty. As another example, when two potentially competing predictors are completely uncorrelated with one another ($r = 0$), their beta weights will be exactly as impressive as their correlation coefficients. There is nothing to adjust because neither variable has anything to do with the other. If Bart were no more likely to be in the kitchen when Lisa was present than when she was absent, there would be no need to make any adjustment for his presence when considering her guilt.

In short, multiple regression analyses sometimes show that the true association between a predictor and a criterion is *smaller* than a simple correlation might suggest. At other times, a regression analysis might show that a simple correlation is *perfectly accurate.* Now here's where it may seem to get a little weird. Sometimes multiple regression analyses may show that a variable that was completely uncorrelated with a criterion is, in fact, uniquely associated with the criterion. Stranger still, such an analysis may even show that a variable that was positively correlated with a criterion is actually negatively associated with that criterion (once you control for the presence of other predictors). That is, the sign of an association may actually be reversed. When this happens, researchers refer to the variable whose true unique association with a criterion was hidden as a **suppressor variable.** In our opinion, it would have been much more reasonable to call this kind of variable a "suppressed" variable—because its true effect was being suppressed by some other variable. However this came to be, it is not our job to change the inconvenient label. But it is your job right now to understand suppressor variables, and it is our job to try to help you.

Let's begin with a formal definition of suppression. In their classic statistics text, J. Cohen and Cohen (1975) defined suppression as follows:

> If the [predictor variable] in question has a zero ... correlation with the [criterion variable], the situation is one of classical suppression. If its beta weight is of opposite sign from its [correlation with the criterion], it is serving as a net suppressor. If its beta weight exceeds its correlation with the criterion and is of the same sign, cooperative suppression is indicated. (p. 91)

Without sweating the details too much, notice that this simply reflects three different specific ways in which the correlation between a predictor and some criterion can yield a very misleading picture of exactly how—or even in what direction—that predictor is truly related to a criterion. We won't systematically examine every possible form of suppression. Instead, we'll try to get a general sense of why suppression occurs and how it is possible.

Perhaps we should continue by reminding you that you've already observed a suppressor variable at work—in the marketing data set involving yearly sales at convenience stores. And we're hoping that, with a little help from Wikipedia, or your classmate Ernestine, you understood this example of suppression perfectly. A simple correlation showed that convenience stores had greater annual sales in neighborhoods that had more traffic. Thus, at first blush, it looked like more traffic contributed to greater sales of Slim Jims, Super Gulps, and AA batteries. In fact, the correlation between traffic levels and annual sales was a humungous $r = .77$. However, we hope you recall that traffic levels were strongly positively correlated (i.e., confounded) with both (a) neighborhood income and (b) neighborhood population density. It is easy to see why there is more traffic in wealthier neighborhoods that have more people living in them. It should be even easier to see that a lot of rich people probably spend more money than a few poor people, especially if the poor people are stuck in traffic. *All else being equal,* traffic is bad for sales. But both people (population density) and money (per capita income) are quite good for sales. Traffic happened to go hand in hand with two other things that both seem to facilitate sales.

Uncovering Causes: Attribution Theory and Suppression

Over the years, there has been a lot of debate about suppression and suppressor variables. Some early thinkers argued that suppression was merely a misleading statistical artifact. Although we know of no contemporary statisticians who believe this, some people are still highly suspicious of evidence for suppression. There are also plenty of thoughtful statisticians who have noted that some apparent evidence of suppression probably reflects the work of overzealous regression runners—who have merely capitalized on chance (typically by running numerous regression analyses and then selecting the one, after the fact, that yielded the most interesting or supportive results). If the presence of a suppressor variable is highly consistent with theory and/or common sense, the researcher who documented this is unlikely to be raked over the statistical coals. However, she may need to replicate her findings at least once before some of her more persnickety critics accept them. Notice, as in the case of other statistical analyses, independent replication is the best insurance against capitalization on chance.

Our view of suppression and suppressor variables is heavily influenced by research on how people understand the causes of other people's behavior. In the early days of research on attribution theory, Harold Kelly (1971) argued that thoughtful people who wish to understand the behavior of other people often make use of two highly useful attributional principles—namely, *discounting* and *augmenting*. When people engage in discounting, they adjust a trait judgment they might make about a person based on something about the *environment* (something outside the person that logically facilitates the same kind of behavior). If Stella broke a school record in the 100-meter dash, you might be tempted to conclude that she is an incredibly gifted sprinter. But if you learned that Stella was using steroids, you should give her a lot less credit for her stellar performance. Some of her speed may have come from a gift for running, but some of it may have also come from a pill or a syringe. This example of discounting is like the classic case when two variables both predict the criterion in a multiple regression but turn out to share some variance. Once we learn that high blood pressure tends to co-occur with high cholesterol levels (which happen to be very strongly correlated with susceptibility to heart attacks), we duly decide that high blood pressure doesn't deserve *all* the credit for causing heart attacks.

Suppression involves exactly the same kind of careful logic. However, in the case of suppression, you *increase* a person's (or a variable's) causal credit rather than decreasing it. To go back to the previous example, let's assume that Stella is not a cheater after all. If she broke a school record in the 100-meter dash despite running into a headwind, on a very cold day, while carrying a bag of concrete on her back, we should be ready to lobby for her position on the next Olympic team. In this case, the variables that happened to accompany Stella were the kind that would normally *interfere* with a good performance. Giving Stella the extra attributional credit that she deserves for a good performance is known as *augmenting*. The logic of suppression in multiple regression is exactly like the logic of augmenting in causal attribution. That's a fancy way of saying that there is nothing inherently odd or curious about suppression, just as there is nothing inherently odd about augmenting. It occurs, for example, when a variable that may promote some outcome happens to be positively correlated with another variable that is *negatively* associated with the same outcome.

A Practice Example of Suppression: Running and Squatting

Let's visit the world of sprinting a third time and consider a concrete statistical example of suppression. First, imagine that we measured people's physical strength by seeing how much weight they could squat. In the SPSS file of hypothetical data in Image 1, called [**running speed data for suppression july 2008.sav**], there is no meaningful correlation between how much a person can squat and his or her average running speed (in miles per hour) in five 100-meter races ($r = .001$). You can easily confirm this by running a correlation in SPSS. In these same data, however, there is a strong positive correlation ($r = .563$) between how much people can squat and how much they weigh. Heavier people can squat more. Duhhh! This is why they have weight classes in weightlifting competitions! However, weight is strongly and *negatively* correlated with running speed ($r = -.712$). This is why you'll probably never see a Sumo wrestler win any Olympic 100-meter race. However, if we were to predict running speed from strength (squat performance), *controlling for body weight,* we might see that stronger people can, in fact, run faster. It's just that greater strength normally comes with the liability of being *greater* (i.e., being heavier).

	firstname	squat	avgspeed	weight
1	tim howell	385.00	22.90	165.00
2	ben	290.00	19.90	160.00
	jesse o	485.00	23.10	190.00
	carl	300.00	19.40	175.00
5	donovan	440.00	15.70	225.00
6	jim	405.00	17.50	235.00
7	justin	345.00	21.00	180.00
8	keith m	450.00	20.20	205.00
9	leroy	370.00	20.80	160.00
10	mauricio	215.00	22.90	145.00
11	lincoln	410.00	23.00	160.00
12	mo	457.50	22.70	180.00
13	eric p	280.00	19.40	180.00

1

Chapter 12 Suppressor Variables

To get a better sense of how suppression works, you should use the above data to find out the association between strength and running speed *after controlling for body weight*. If you conduct the multiple regression analysis correctly, you'll see that the zero-order correlation of .001 between strength and speed increases to a pretty impressive beta weight when you control for body weight. To do this, you'll want to conduct a simultaneous multiple regression analysis in which your criterion variable is "avgspeed" and your two predictors are "squat" and "weight." When you've done this, the important part of your output file should look a lot like Image 2.

Coefficients[a]

Model		Unstandardized Coefficients B	Std. Error	Standardized Coefficients Beta	t	Sig.
1	(Constant)	30.983	2.594		11.942	.000
	squat	.017	.006	.586	3.014	.013
	weight	-.091	.017	-1.042	-5.360	.000

a. Dependent Variable: avgspeed

As you can see in Image 2, there is a positive and significant unique association between "squat" and running speed, $\beta = .586$, $p = .013$. *Controlling for weight*, stronger people *can* run faster. To give you an even better feel for what is going on here and to practice sorting data files a bit, take a look at Images 3 and 4. These images give you some clues about how you could sort these data first by body weight (from lightest to heaviest runner) and *then* by strength (based on how much weight people could squat). It's important to sort on *both* variables to make it easy to see the patterns in the data.

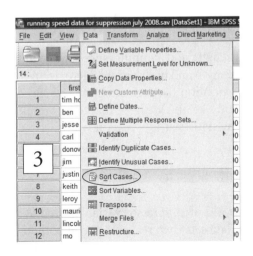

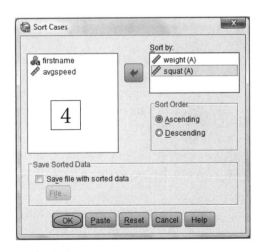

If you sort these data, you'll make your SPSS data file look very much like the one depicted in Image 5. This file gives you a very useful view of the running speed data. Now you can begin to assess the correlation between running speed and strength *for individual runners whose weights are held constant*. Notice the three runners who all happen to weigh 160 pounds. Lincoln is stronger than Leroy, and he can outrun Leroy (not that he should ever need to—given his superior strength). Furthermore, Leroy is stronger than Ben, and he is faster than Ben. Now look at the three runners who all happen to weigh 180 pounds. Mo is the strongest and also the fastest. Eric is the weakest and the slowest. Clearly, strength goes hand in hand with running speed, but only once we adjust for the natural confound between strength and body weight. (By the way, if you're wondering why your friend Usain is not in this data set, it is because he demanded a $2 million appearance fee.)

5

	firstname	squat	avgspeed	weight
1	mauricio	215.00	22.90	145.00
2	ben	290.00	19.90	160.00
3	leroy	370.00	20.80	160.00
4	lincoln	410.00	23.00	160.00
5	tim howell	385.00	22.90	165.00
6	carl	300.00	19.40	175.00
7	eric p	280.00	19.40	180.00
8	justin	345.00	21.00	180.00
9	mo	457.50	22.70	180.00
10	jesse o	485.00	23.10	190.00
11	keith m	450.00	20.20	205.00
12	donovan	440.00	15.70	225.00
13	jim	405.00	17.50	235.00

The true association between strength and running speed was suppressed by the correlation between strength and body weight. Here strength is a suppressor variable. The true association between strength and speed emerges only when you control for body weight. We hope that this example gives you a good feel for how and why statistical suppression works.

Practice With Suppression: Three Data Sets to Analyze

To immerse you more fully in the world of suppressor variables, we have provided you with three additional data sets (two real and one hypothetical) that all document a highly plausible case of statistical suppression. You should analyze each data set, identify any suppressor variable(s) at work, and come up with a reasonable explanation for the suppression. To make this more concrete, you'll want to do the same thing with each data set. Specifically, for *each* data set, do the following two things and report your answers:

A. First, run and interpret a set of correlations between the predictor variables and the criterion variable. You don't have to be very detailed here. Just make it clear that you know what each correlation indicates.

B. Second, run and interpret a simultaneous multiple regression analysis that includes all of the important predictors. Be sure to explain, for example, why a predictor that wasn't correlated with the criterion turns out to predict it well (classical suppression) when you control for the other important predictor(s). It might even be the case in one or more of these data sets that the sign of a predictor will be reversed (net suppression) as you move from the correlation to the regression analysis. Your job will be to explain, based on logic, theory, or both, whatever kind of suppression you observe in the data.

Data Set 1: Anagram Difficulty and Self-Pay

The first data set is called [**anagram difficulty & self-pay.sav**]. The data are real. They come from a pilot experiment in which John Hetts and the first author examined the effect of self-perceived performance on self-pay. In this pilot study, students had a few minutes to solve as many of 24 anagrams as they possibly could. Unbeknown to the students, we manipulated the difficulty of the anagrams. Thus, one group of students got easy scrambled words such as "owyell" and "riseusn." The other group got difficult scrambled words such as "litlecam" and "clodfasf." This manipulation worked. Students who got the easy anagrams solved about four times as many as those who got the hard ones. Students realized this. The difficulty manipulation was also very highly correlated with the number of anagrams students later *thought* they solved, $r(25) = -.91$, $p < .001$. (The harder the anagrams actually were, that is, the fewer anagrams people thought they solved when asked to report this a few minutes after they had finished.) We know what you're thinking. This less-than-fascinating correlation just means that we know how to create hard and easy anagrams. But remember that our goal was to examine predictors of how much people would *pay themselves* for their work. Perhaps most people pay themselves for their effort. If so, those who got the hard anagrams should have charged us more for all their trouble. Alternately, maybe people pay themselves strictly for their output. In this case, people should have demanded more pay for working on easier anagrams (because they solved more of them). You might also wonder if people's perceptions of how other people might have done on the anagrams would have any effect on their self-pay. If Anna thinks she only solved five of the impossibly difficult anagrams but thinks other people would have solved only one or two, maybe she'll demand a high premium for her premium performance. In these data, your criterion variable is called "selfpaid." This is the amount of money, in dollars and cents, people asked for (and got). The two predictor variables are (a) "howidid," the number of anagrams that people thought they personally solved, and (b) "howmostdid," the number of anagrams people thought most of their classmates solved.

QUESTION 12.1a and 12.1b. Answer the two questions (A and B) for these data. Make sure to pay special attention in answer B to which predictor looks very different in the correlation and in the regression analysis. If you need a good theory to help you along, check out what an introductory social psychology text tells you about social comparison theory (Festinger, 1954; Morse & Gergen, 1970).

Data Set 2: Predicting Voting Behavior

The second data set is called [**hypothetical election data spring 2005.sav**]. These hypothetical data focus on the 2004 presidential election. For the sake of this activity, you should assume that the data come from a representative sample of real voters. Because this is an election study, the criterion variable is "presvote," which refers to whether people voted for Bush (coded as 1) or Kerry (coded as 2). Thus, the lower of the two possible scores on this dummy-coded measure means a vote for Bush. That'll be important to remember when it comes time to interpret your correlations. You have three predictors this time. The first is *political party affiliation* ("partyaffil"), coded such that Republican = 0 and Democrat = 1. Notice that if things go as we would normally expect, this coding scheme would promote a *positive* correlation between party affiliation and voting behavior. The second predictor is *personal income* ("zincome"). This variable has been converted to z scores, so higher scores simply mean greater income (in standardized units). The third predictor is "fundamentalism," which refers to *religious fundamentalism*. As you could confirm by looking at "Values" when in the "Variable View" mode of the SPSS data file, this variable is scored so that higher scores mean greater levels of fundamentalism.

QUESTION 12.2a and 12.2b. Answer the two questions (A and B) for these data involving voting behavior.

Data Set 3: Predicting Homicide Rates From Country-Level Statistics

The third data set is a collection of country-by-country statistics tabulated by organizations concerned with human well-being and human rights. More specifically, the first author merged some data available from the United Nations Development Project (UNDP), Freedom House, and the CIA Factbook. The UNDP routinely reports a wide range of indicators of health and well-being in almost every country in the world. Freedom House pursues a similar mission, except that it focuses on human rights and freedoms rather than health and well-being. One indicator Freedom House publishes every few years is worldwide country comparisons of basic human rights. In addition to categorizing every country in the world as "free," "partly free," or "not free," Freedom House also rates each country on a 7-point scale based on whether its citizens enjoy basic human rights and civil liberties. It is important to remember that higher ratings on this scale reflect *less* freedom (i.e., more oppression). Based on these sources, this data file includes five important variables for more than 100 countries: (a) Freedom House 2005 human rights ratings; (b) per capita GDP or gross domestic product, a measure of national wealth; (c) the percentage of a country's residents who are younger than 15 years (a measure of a country's youth); (d) the number of people in each country (out of 100,000) who were in prison in 2007; and (e) average homicide rates (per 100,000 residents) between 2000 and 2004. Incidentally, it would be ideal if all of the records came from the very same year, but not all of these records are collected or tabulated every year, and so we would like you to forgive this problem and assume that it merely introduces a little random error into the data. It is worth noting, however, that all of these properties of nations are pretty stable over time. It is exceedingly unlikely, for example, that the GDP of Syria or Madagascar will exceed that of Luxembourg (the world's richest country in these data) any time soon. The SPSS data file you will need for these analyses is called [**worldwide freedom homicide GDP.sav**].

QUESTION 12.3a and 12.3b. Answer the two questions (A and B) for these data involving homicide rates.

A Cautionary Note Regarding Multicollinearity

When suppression is at work in a data set, it is more important than usual to pay attention to the potential problem of multicollinearity (predictor variables that are very highly correlated with one another). Multicollinearity is worrisome because when two or more predictors are very highly correlated with one another, changing one or two scores for one or two predictors can sometimes have a huge effect on the results of a multiple regression analysis.

To return to a very old example, if Bart and Lisa are almost always together, changing Bart or Lisa's cookie theft data on a single day (or worse yet, having one day's data entered erroneously, perhaps by an absent-minded professor) could radically change the results of a multiple regression analysis.

As a more recent example, you may wish to revisit the data on traffic and sales at convenience stores. To be more specific, you might wish to check the tolerance or **variance inflation factor** (VIF) scores for each predictor variable in the final regression equation (you can choose this from the "Statistics" options by clicking on the "Collinearity diagnostics" button). Incidentally, because tolerance is simply 1/VIF, a high VIF score mathematically guarantees a low tolerance score. It is thus arbitrary which of these indicators you worry about, but we prefer to focus on the VIF because multicollinearity is a *problem* and higher scores on the VIF indicate *more* of the problem. The VIF for at least two of these predictor variables is very high (greater than 10.0) and thus is worrisome. This would suggest, for instance, that we should be reluctant to trust these results until we see a replication. One frequent sign of a very high VIF is a very high beta weight for one or more predictor variables. In fact, one sign that suppression is at work in a data set is observed beta values near or greater than 1.0.

Coda: Why Suppression?

Given that some people are naturally suspicious about suppression, and given that testing for suppression often comes with extra headaches (i.e., extra worries about multicollinearity, extra-skeptical reviewers), you might wonder why we devoted an entire chapter in this text to suppression. The answer is that the raison d'être ("reason for being") of inferential statistics is to identify meaningful patterns in data—patterns that might often be invisible to the naked eye.[1] When patterns are *visible* to the naked eye, of course, all you need is folk wisdom, not statistical analysis, to know what is going on the world. From this perspective, there is no greater statistical feat than to identify a valid suppressor variable in data when either folk wisdom or a statistically inferior analysis of the same data would lead one to conclusions that are the opposite of what proves to be the case. Thus, from our perspective, understanding statistical suppression is at the very heart of inferential statistical analysis.

Note

1. We hate it when people try to sound smart by throwing around French or Latin terms, and we apologize (*lo siento mucho*, as we say in Spanish) for imposing on you in this way. However, there are some foreign terms that have a certain *je ne sais quoi* that, *ceteris paribus,* we just find irresistible.

Mediation and Path Analysis

13

Introduction: Disentangling Competing Causes

One of the most important developments in the history of statistics is the creation of techniques for what one might call "disentangling causal spaghetti." You have already learned quite a bit about one specific technique in this large family of techniques. In particular, you have learned that **multiple regression** analyses can be used to see (a) which of several potential predictors of an outcome are uniquely associated with the outcome (controlling for everything else in the picture) and (b) which potential predictors only happen to be spuriously (i.e., coincidentally, not-so-meaningfully) related to the outcome—because they happen to be correlated with the variables that truly matter. Revisiting the observational study of cookies thefts yet again, remember that whereas Lisa initially *seemed* to be responsible for at least some of the cookie thefts, her connection to stolen cookies proved to be spurious (i.e., fake, coincidental). She simply had the misfortune of usually visiting the kitchen on the same days when Bart did, because she is his sister. A careful analysis of the days on which only one child was present placed all of the blame on Bart.

If you completed the statistical activity using the "World95" data set, you used this same technique (multiple regression) to try to disentangle several potential predictors of female longevity across different nations. You were probably able to show that factors such as (a) female literacy rates and (b) the number of children the typical woman bore in a given country were associated with female longevity—even when these two predictors were disentangled not only from one each other but also from powerful but less interesting predictors such as national wealth or level of technological development. In short, multiple regression helped you sort out multiple predictors of female longevity—to help you understand which variables prove to be most important. For instance, if female literacy had been correlated with female longevity only because better heath care exists in countries where more women can read, then the association between female literacy and female longevity probably would have disappeared once you controlled for indicators of the quality of health care in the various countries in this worldwide sample. This hypothetical finding would have suggested that female literacy is *not* directly connected to female longevity. Instead, literacy would just be a marker for the more important variable of quality of health care. Recall, however, that female literacy rates *did* prove to be an independent predictor of female longevity in the actual data, suggesting that female literacy has a unique connection to female longevity. In other words, holding variables such as national wealth constant, women who live in countries where a higher percentage of women can read have longer life expectancies.

Third Variables Versus Causal Starting Points

So as it turns out, multiple regression analysis can often tell us which variables are **directly** connected to an outcome and which are only **indirectly** connected to the outcome (by means of their connection to some other predictor variable). In short, regression can help us rule out confounds (i.e., the third variable problem). This familiar fact is illustrated in Figure 13.1. Suppose we identified two reliable predictors of smoking among adolescents (hereafter referred to as "kids"). First, assume that kids who take up smoking tend to be low in self-esteem. Second, assume that kids who take up smoking tend to have highly permissive parents (parents who fail to create and/or enforce predictable rules). Figure 13.1 illustrates that these two predictors of adolescent smoking are confounded. If you argue that a permissive parenting style promotes smoking, I can point out that parenting style is confounded with kids' self-esteem levels. And if I argue that low self-esteem among kids promotes smoking, you can point out that kids' self-esteem levels are confounded with their parents' levels of permissiveness. (We are assuming, by the way, that highly permissive parents have kids who are lower than average in self-esteem; see DeHart, Pelham & Tennen, 2006.)

By now, you should know that if we had the right data at our disposal, a multiple regression analysis could help us resolve this debate. (We said *help*, by the way, because you'll notice that—for the time being—we are ignoring thorny problems such as causal order.) Let's imagine that we assessed both kids' self-esteem levels and their parents' parenting styles in a large, representative survey study of adolescent smoking behavior. Now let's consider two different sets of results we might observe if we analyzed these data in a multiple regression analysis.

First, remembering that we are continuing to indulge ourselves in causal language, imagine the situation depicted in the left-hand portion of

Figure 13.1 The Third Variable Problem as It Relates to Two Correlated Predictors of Smoking

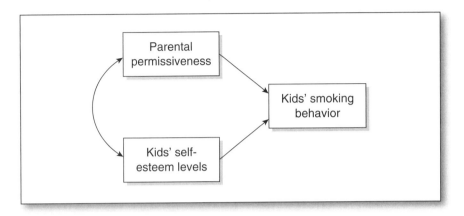

Figure 13.2. That is, imagine that (a) parental permissiveness still predicted kids' appetite for cigarettes once we controlled for kids' self-esteem levels, whereas (b) kids' self-esteem levels *no longer* predicted their smoking behavior once we controlled for parental permissiveness. The most reasonable conclusion from this finding is that kids' self-esteem levels are not the reason why parental permissiveness is associated with smoking among kids. Parental permissiveness per se seems to promote smoking among kids directly (even if a kid happens to be high in self-esteem). And parental permissiveness just happens to be correlated with kids' self-esteem levels (e.g., because parental permissiveness directly or indirectly contributes to low self-esteem among kids). So, in short, through whatever route parental permissiveness may promote smoking, it does *not* do so via the route of reducing kids' self-esteem levels. The left-hand portion of Figure 13.2, then, summarizes one reasonable causal account of the original data—based on one specific set of hypothetical findings from a multiple regression analysis.

Now imagine another reasonable pattern of results. Imagine that (a) parental permissiveness *no longer* predicted smoking once we controlled for kids' self-esteem levels, whereas (b) kids' self-esteem levels still predicted smoking once we controlled for parental permissiveness. This set of findings is depicted in the right-hand portion of Figure 13.2. Now it appears that the only reason that parental permissiveness was correlated with kids' susceptibility to smoking was because of its association with kids' self-esteem levels. So regardless of how kids come to be low in self-esteem (e.g., because of a genetic predisposition, poverty, or excessive exposure to music by Justin Bieber), kids who prove to have low self-esteem are at increased risk for becoming smokers—even if their parents happen to do an excellent job of establishing and enforcing rules for their

Figure 13.2 Interpretations of Two Different Sets of Multiple Regression Findings When We Predict Adolescent Smoking Behavior From Both Parental Permissiveness and Kids' Self-Esteem Levels

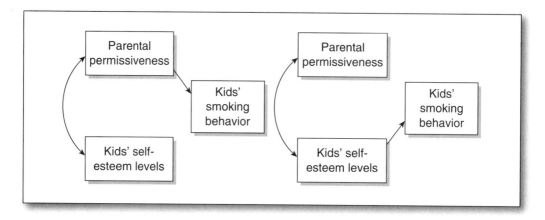

kids' behavior. Parental permissiveness is *not* the *direct* route through which low self-esteem promotes cigarette smoking. Instead, low self-esteem seems to have a direct effect on smoking.

Causal Plausibility

So we're assuming for now that kids' self-esteem levels won the multiple regression footrace and uniquely predicted smoking, whereas parental permissiveness did not. Now here comes the tricky part. From a purely statistical perspective, it would be completely reasonable to label parenting style an annoying confound and dismiss it as causally unimportant. In the case of these particular data, however, imagine that we were pretty sure that (a) permissive parenting *is* an important cause of kids' self-esteem levels, whereas (b) kids' self-esteem levels are *not* a cause of parents' levels of permissiveness. Would we really want to dismiss parental permissiveness as an unimportant (spurious) part of this interesting story? Or would it be more appropriate to say that kids' self-esteem levels turn out to be one of the important *routes through which* parental permissiveness promotes smoking? In case you're not sure, we'll tell you. In this specific case, it would be more appropriate to say that kids' self-esteem levels mediate the previously established connection between permissive parenting style and kids' smoking behavior. We used to know that permissive parenting is associated with teen smoking, and now, because of the results of our regression analysis, we have a better idea of *why* this is the case. It doesn't matter much that there is more than one way for kids to become low in self-esteem. What matters is that kids' self-esteem levels appear to be the reason why permissive parenting is associated with smoking among kids. Permissive parenting is not merely an annoying confound. Instead, it is an important starting point in a meaningful causal chain.

This tricky part can be even trickier than it seems at first blush. For example, if we continue to assume that permissive parenting influences kids' self-esteem levels (rather than the reverse), the findings summarized in the left-hand portion of Figure 13.2 do *not* suggest that parenting mediates the associate between kids' self-esteem levels and smoking. The reason why these findings do not suggest this is that we know (or have decided to assume) that kids' low self-esteem levels do not *cause* parents to become more permissive. In other words, when you are trying to figure out whether one variable mediates the association between another variable and your criterion (i.e., outcome) variable in a multiple regression, you need to begin with some reasonable assumptions about what causes what. Variable M can only mediate the association between X and Y if there is good reason to believe that X causes M, which then causes Y.

Sometimes you may know something about the direction of causality (e.g., you may know that X causes Y and not the reverse) because of previous research. In other cases, you may actually have collected experimental or quasi-experimental data yourself—which will usually allow you to make strong assumptions about the direction of causality. (That's right; it's OK to analyze experimental data using multiple regression techniques; see Judd & Kenny, 1981; MacKinnon, 2008, pp. 55–61, 69.) In other cases, sheer logic may do the trick. If Erica is disproportionately likely to marry Eric, we should worry about whether Eric and Erica happen to be equally wealthy or equally White, but we do not have to worry very much about reverse causality (because few people change their first names to resemble their partners' names after getting married). Certain kinds of causal assumptions are sometimes possible even when dealing with passive observational data. A psycholinguist studying the evolution of language would not have to worry very much about whether the animal name *hog-nosed snake* existed before the word *hog*.

To summarize, the same kind of logical and statistical analyses that can help us resolve the third variable problem can also help us resolve whether variable M mediates the association between variable X and variable Y (our criterion or outcome). Before we delve any deeper into mediation, though, it would be useful to discuss some of the traditional terminology of mediation. Figure 13.3 illustrates an abstract model in which X causes Y (with no assumptions about mediation).

The familiar terms X and Y refer to the independent and dependent variables of interest, respectively. Furthermore, the path from X to Y (a simple correlation) is referred to as c. Path c is often referred to as the *total effect*.

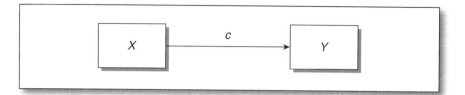

Figure 13.3 The Simple Association Between X and Y

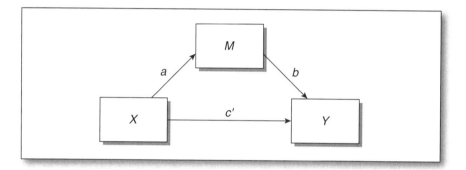

Figure 13.4 The Abstract Model in Which M Mediates the Association Between X and Y

However, the connection between X and Y often occurs because of a mediating variable that, conveniently enough, is usually referred to as M. Notice that the various paths also have labels. The original path from X to Y is now referred to as c' to indicate that it has been statistically corrected for any effects on Y of the hypothesized mediator (M). If there is any significant effect remaining after controlling for the mediator, this path (again, c') is referred to as the direct (nonmediated) effect of X on Y. Furthermore, the simple path from the independent variable (X) to the mediator is labeled a, and the unique path between the mediator and the outcome (Y) is labeled b. It is good to be highly familiar with this terminology. For example, statisticians often refer to the mediated path as ab to indicate that the mediated path consists of *both* the association between X and M and the unique association between M and Y. It is called ab because it is literally the product of path a and path b.

Using the language of this abstract example, three separate things all need to be true to document that M is a likely mediator of the association between X and Y. First and foremost, as suggested in the previous discussion, you need **causal plausibility.** That is, if you lay out your three variables in a temporal "timeline," like the ones you see in Figures 13.4 and 13.5, the causal ordering of events needs to makes sense. For example, it makes sense in Figure 13.5 that parents' income levels (X) could influence the quality of education their kids get (M), which could then influence kids' adult salary levels (Y). It makes a lot *less* sense to reverse this causal sequence (i.e., to suggest that Kym's salary as an adult influenced the quality of education she got as a child, which then influenced her family's income level when she was an even younger child). Unfortunately, there is no one simple rule for determining causal plausibility. Moreover, there are many cases in which there is more than one very reasonable way to arrange the causal sequence of three (or more) variables. As long as you do not specify a causal sequence that (a) is illogical (e.g., because it requires the present to influence the past) or (b) is inconsistent with established experimental or longitudinal research, you have some degree of wiggle room when it comes to causal plausibility. On the other hand, the more causal plausibility you have going for you, the better. For instance, we'd rather have to defend the specific causal model depicted in Figure 13.5 than either of the models depicted in Figure 13.2. This is because it is theoretically possible to rearrange the causal order of the variables in Figure 13.2 (kids' self-esteem levels *might* conceivably cause parental permissiveness). In contrast, rearranging the sequence in Figure 13.5 would require the future to influence the past.

Empirical Plausibility

So first and foremost, if you want to argue that M mediates the association between X and Y, it needs to be logically and/or empirically reasonable that X causes M and that M causes Y. However, two additional things also need to be true. These two additional things are not based on logic or on past research. Instead, they are based on the specific results

Figure 13.5 A Likely Causal Sequence for Two Variables Associated With Adult Salary Levels

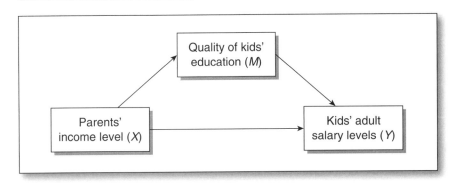

of your own empirical study. The second requirement for mediation is that, in your own data, both variable X and variable M have to be (significantly) correlated with variable Y. That is, both of your predictor variables (the hypothesized distal cause and the hypothesized mediator) must be correlated with whatever outcome you have chosen to study. If parental income levels during childhood have nothing whatsoever to do with kids' adult salary levels, it is hard to argue that parental income contributes (directly or indirectly) to kids' adult salary levels. Similarly, if the quality of the education that kids receive has nothing to do with kids' adult salary levels, then quality of education cannot be the reason why parental income is associated with kids' adult salary levels (cf., however, MacKinnon, 2008, who identifies some exceptions to this general rule). Third, and perhaps most important, to establish that M mediates the association between X and Y, it should be the case that the association between X and Y is reduced or eliminated when we control statistically for the association between M and Y (Baron & Kenny, 1986; and see MacKinnon, 2008, for some sophisticated refinements of this general position).

A concrete example of all this might be a mediational model of running speed in a long-distance footrace. It is well established that a superior ability to take in and use oxygen (VO_2 max) is one of the immediate (direct, proximal) causes of fast running times in distance races. It is also well established that regular training increases a runner's personal VO_2 max level. Because VO_2 max level is a direct, physiological determinant of immediate running performance, a mediational analysis of training history (e.g., being a regular runner versus. being a couch potato) and VO_2 max levels (assessed shortly before a race) would probably show that the association between training history and performance would be greatly reduced once we take into account the association between VO_2 max levels and race performance. In other words, one very good reason why regular aerobic training is associated with superior distance running times is that regular aerobic training produces increases in VO_2 max. In contrast, the habit of training frequently is probably *not* the main reason why having a high VO_2 max is associated with superior running times. Tests for mediation are tests for reasons or likely causal pathways. You can probably think of other physical or psychological examples. One reason why weight training is associated with being able to bench press more weight might be because of differences in people's muscle tone or muscle mass. One reason why attending religious services regularly might be associated with superior physical health is because of the social support people often receive from those with whom they worship. Complicating things a bit, though, some of the reasons why social support is associated with health are probably physiological, whereas others are clearly behavioral (e.g., see S. Cohen, Doyle, Skoner, Rabin, & Gwaltney, 1997; Uchino, 2008).

Moderation in All Things—Except for Mediation

Unfortunately, people frequently confuse the terms **mediation** and **moderation**, and there are several possible reasons for this confusion. Memory researchers would probably argue that both terms are close semantic associates of words such as *highfalutin'*, *multiple*, and *regression*. In contrast, Freudian theorists might assume that it is because both terms vaguely resemble the word *masturbation*. Whatever the reason for the confusion, there is no good excuse for it. As you know by now, the term *moderation* indicates that there is a statistical interaction between two or more predictors of an outcome. When this is the case, it is appropriate to refer to either predictor variable as a moderator variable. For example, we might say that gender moderates the association between stressful life events and depression. This means that stress and gender interact to predict depression. For example, it might mean that the positive correlation between stress and depression is stronger for women than for men. However, it would be equally correct to say that the association between gender and depression is stronger when people are experiencing a lot of stressful life events than when they are not. Either independent variable can be conceptualized as a moderator. When you hear *moderation*, then, think interaction. However, mediation is different. Mediational analyses tell you whether, or to what degree, a third variable intervenes to explain a correlation or association in which you are interested. Mediators don't change the association between two variables. Instead, they explain it. A mediator is the specific reason why two variables are associated.[1]

Just to be sure you're intimately familiar with the concept of mediation, let's briefly consider two additional examples. First, based on recent biomedical research, it would be appropriate to say that cumulative damage to a tumor-suppressing gene (known as p53) mediates the association between cigarette smoking and lung cancer. That is, at least one reason why smoking causes lung cancer is because smoking cigarettes deactivates a natural biological process that normally suppresses the growth of tumors. To choose a psychological example, recent evidence suggests that negative emotions such as anger mediate the well-known association between frustration and aggression. In other words, frustration does not lead *directly* to aggression. Rather, frustration produces negative emotions such as anger, which then fuel aggression.

In a little while, we're going to examine mediation in a hypothetical data set to see how you might test the meditational model that frustration leads to anger, which then leads to aggression. By working with this model, we hope you'll get a good sense of how mediation works. In addition, we'll explore this model in SPSS to address the sometimes tricky question of exactly how you establish the statistical significance of any evidence you observe for mediation.

Before we begin this discussion, however, we feel compelled to note that there are many different ways to assess whether mediation is significant. In this text, we are going to discuss only two of them: (1) the simplest possible way followed by (2) one of the popular ways that seems to hold up pretty well to expert scrutiny regarding Type I and Type II errors. We'll examine both of these ways by examining how anger mediates the association between frustration and aggression.

A Mediational Model of How Frustration Leads to Aggression

In the late 1980s, Leonard Berkowitz (1989) summarized a great deal of research on the frustration-aggression hypothesis. Berkowitz's insight, which helped clear up some confusion regarding the original frustration-aggression hypothesis, was that frustration probably does not lead directly to aggression, as many researchers had believed for several decades. Instead, Berkowitz argued, frustration usually leads to hostile negative emotions (such as anger), which are usually the more immediate cause of aggressive behavior. For the sake of understanding mediation, let's assume that Berkowitz's reformulated frustration-aggression hypothesis is correct. Then let's see how, if it *is* correct, we can analyze a supportive hypothetical data set to document that anger mediates the association between frustration and aggression. To do so, we'll begin by reviewing the three logical requirements Baron and Kenny spelled out in their classic (1986) paper on mediation. However, we'll also follow up on this logical analysis with a couple of statistical tests of whether the apparent mediation observed is statistically significant.

As we noted earlier, it can sometimes be very difficult to know for sure whether an apparent case of mediation is statistically significant. In fact, it can be so difficult that statisticians have written entire books on the best ways to test for mediation (e.g., see MacKinnon, 2008). Many statistical tests for mediation, for example, are extremely sensitive to whether the meditational paths being analyzed follow most of the basic assumptions of a traditional multiple regression analysis (e.g., normally distributed predictor variables, homoscedasticity, but see Dave Kenny's website listed at the end of the chapter, which notes that some degree of multicollinearity is to be expected). To give you a sense of why testing for mediation can be so tricky, consider the following: One very good way to think about mediation is that there is evidence for mediation in a data set precisely to the degree that the *product* of two different regression coefficients (*ab*) is substantial. One drawback of this is that the sampling distribution of the *product* of two normally distributed regression coefficients often is *not* normally distributed. The best way to deal with this problem involves bootstrapping procedures that are beyond the scope of this intermediate text (see MacKinnon, 2008; Preacher & Hayes, 2004).[2] With this limitation in mind, we focus in this chapter on the popular Sobel significance test for mediation. Although the Sobel test can be

Chapter 13 Mediation and Path Analysis

problematic with small samples, Monte Carlo studies suggest that as long as you have a *large* sample (i.e., $n = 100$ or greater), the Sobel test is well calibrated—and superior to most other tests for the significance of mediation. Although we will require you to conduct a few hand calculations in this chapter to familiarize yourself with the Sobel test, we also provide a couple of websites at the end of the chapter that will do all of the hand calculations for you, so long as you have already run the appropriate multiple regression analyses.

But as Wayne would say to Garth, we digress. Let's return to the conceptual question of whether anger mediates the association between frustration and aggression, and let's apply Baron and Kenny's logic (and then a Sobel test) to this question. Recall that three things are logically necessary to document clear-cut mediation. First, the independent variable (X) must be correlated with the dependent variable (Y). If anger mediates the association between frustration and aggression, there better be an association between frustration and aggression to begin with. Second, the hypothesized mediator (M) must be correlated with the dependent variable. If anger is the intervening reason why frustration leads to aggression, anger better be correlated with aggression. After all, how can anger take us from frustration to aggression if anger doesn't go there? Third, the association between the independent variable (X) and the dependent variable (Y) must either disappear or get smaller one you control statistically for the association between the hypothesized mediator (M) and Y.

It is this third requirement that is the trickiest. In obvious cases of complete mediation, a very large correlation between X and Y disappears completely when you control for the unique and statistically significant effect of the tested mediator M. In fact, one of the oldest and simplest ways of assessing mediation is based on this logic. Let's check out this very simple way of documenting mediation and then a somewhat more complex way, using a data set called [**hypothetical frustration anger aggression for mediation.sav**]. All of the variables are coded intuitively. Thus, higher scores on the variables frustration and anger mean, respectively, that participants said they felt more frustrated and got angrier. Furthermore, higher scores on the variable aggression mean that people engaged in more aggressive behavior (let's say, slapping someone with a fish; see Visher & Nawrocki, 2002). Here's what we'll need to test for mediation in the simplest possible way:

1. Run a multiple regression analysis in which we predict Y (aggression) from X (frustration),

2. run a multiple regression analysis in which we predict M (anger) from X (frustration), and

3. run a multiple regression analysis in which we predict Y (aggression) from both X (frustration) and M (anger).

In case you haven't run a multiple regression analysis today, the screen capture in Image 1 should remind you of how. We'll begin by just seeing if frustration predicts (i.e., is correlated with) aggression. (This means your first regression analysis will have only one predictor.)

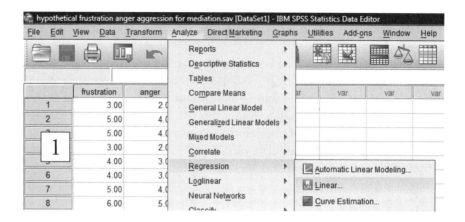

So that you can check your work, here is what you should find in these three regression analyses (numbered to match the three steps listed earlier):

1. Frustration does predict aggression ($\beta = .472$, $p < .001$).

2. Frustration does also predict anger ($\beta = .638$, $p < .001$).

3. When frustration and anger are both entered as predictors, anger still predicts aggression ($\beta = .622$, $p < .001$), but frustration no longer does ($\beta = .075$, ns).

The simplest possible way to show that mediation is significant is to show that there is both (a) a significant path between X and M (this is path a, covered in point 2) and (b) a significant (unique) path between M and Y (path b, covered in point 3). In other words, the predictor is significantly correlated with the mediator, and the mediator is uniquely and significantly associated with the criterion (in the logically and/or theoretically correct way). The icing on the cake in this traditional approach occurs when the once-significant association between X and Y disappears (or is reduced to nonsignificance) when the mediator is included in the regression model. That is exactly what happened here. So from a very simple perspective, we seem to have observed strong evidence of mediation. Although we will continue to ask you to think about this simple approach in some of the data analysis exercises that follow, it is important to note that there are some problems with this approach. For example, imagine we had observed the following results in our hypothetical study of frustration and aggression.

1. Frustration does predict aggression ($\beta = .593$, $p < .0001$).

2. Frustration also predicts anger ($\beta = .227$, $p = .006$).

3. When frustration and anger are both treated as predictors, anger predicts aggression ($\beta = .145$, $p = .049$), but frustration also continues to predict aggression ($\beta = .367$, $p < .001$).

It is not quite so obvious now whether there is mediation. It is pretty clear, though, that if mediation is occurring, it is far from complete, mainly because there is still a very strong association between frustration and aggression *after controlling for anger*. On the other hand, frustration is associated with anger, and anger is uniquely associated with aggression, even after controlling for frustration. It would be nice to do some kind of significance test to see whether the *partial* mediation we seem to have observed is statistically significant.

Formal Testing for the Significance of Mediation Requires Knowledge of Standard Errors

Tests (such as the Sobel test) that are sensitive to partial as well as complete mediation are a bit more demanding than the traditional two-step process of merely seeing whether the paths *a* and *b* are both significant. The main way in which they're demanding is that they require you to use information about **standard errors** to create a single statistic that represents the entire meditational path (i.e., *ab*, which includes both the *X* to *M* path and the *M* to *Y* path). Luckily, SPSS regression always gives you the standard errors associated with all of the standardized regression coefficients (beta weights) in a regression analysis. Before we discuss exactly what to do with standard errors to test for the significance of mediation, however, let's briefly discuss exactly why we need those standard errors in the first place.

Recall that the gist of any inferential statistical test involves comparing a measure of a result (e.g., a difference between two means) with a standard error that takes into account (a) some measure of observed error variance and (b) the sample size. In the familiar case of analysis of variance (ANOVA), for example, recall that an *F* ratio is a ratio of between-groups variance to within-groups variance. The fact that your experimental group delivered 10 more volts of shock than your control group would surely be statistically significant if the average standard deviation within each of the two experimental conditions were 5.0 and your sample size were $n = 200$. However, if the standard deviation were 50 and the total sample size were $n = 12$, that same 10-point difference between your experimental and control groups would not even approach statistical significance. Conceptually, then, we need standard errors (not just regression coefficients) to decide whether a mediated path of a certain magnitude is statistically significant.

Fortunately, Sobel worked out a pretty simple statistic (a version of the z statistic) that can often be used to test for the significance of the complete meditational path ab. The formula for the Sobel statistic is

$$\text{Sobel} = \frac{ab}{\sqrt{a^2 s_b^2 + b^2 s_a^2}}.$$

You should not be surprised to learn that this formula is a ratio consisting of the product of the two crucial regression coefficients already discussed (ab = the numerator) divided by the standard error of the product of these two regression coefficients (often referred to as s_{ab} = the entire denominator). The a^2 and the b^2 in the denominator are simply the squared values of the separate a and b coefficients that make up ab. The value s_b^2 in the denominator is the squared standard error for regression coefficient b, and the value s_a^2 is the squared standard error for regression coefficient a.

If you refer back to your SPSS output file, you should be able to see that the standard errors of the two regression coefficients a and b were .081625 and .172199, respectively. Thus, the Sobel statistic for mediation in this case is as follows:

$$\text{Sobel} = \frac{.638 \times .622}{\sqrt{(.638^2 \times .172199^2) + (.622^2 \times .081625^2)}},$$

$$\frac{.396836}{\sqrt{(.0120697) + (.0025774)}},$$

which is $\frac{.396836}{\sqrt{.014647}}$,

which is $\frac{.396836}{.120125}$,

which is 3.279.

If you recall that the Sobel statistic is a form of the z statistic and that the critical value for z is 1.96 (at $\alpha = .05$, two-tailed), it should be obvious that the mediation observed in this hypothetical example is statistically significant. To summarize, then, two traditional ways to test for mediation are (1) to see if path a and path b are both statistically significant (if they both are, this is the simplest possible evidence for mediation, and this evidence is strengthened if path c is no longer significant once you control for path b) and (2) to conduct a Sobel test to see if path ab is significant. If the absolute value of your Sobel statistic exceeds 1.96, your result is significant at $p < .05$. If the absolute value exceeds 2.58, your result is significant at $p < .01$. Now that you are armed with both of these approaches to testing the significance of mediation, let's get some further practice testing for mediation in a couple of new data sets.

What Mediates the Association Between Self-Esteem and Relationship Satisfaction?

The first data set you'll analyze on your own in this chapter is a hypothetical data file based very closely on a series of real studies of Murray et al.'s theory of *dependency regulation* in close relationships. The essence of Murray et al.'s theory (e.g., see Murray, Holmes, & Collins, 2006) is that people cannot allow themselves to view their relationship partners too favorably unless people feel secure in their partner's regard for them. Rick cannot afford to think that his partner Rhonda is too wonderful if he has lingering doubts about whether she truly loves him. According to the dependency regulation model, this explains a well-established finding from the literature on self-esteem and close relationships. Sadly, people who are low in global self-esteem (i.e., people who score low on measures such as the Rosenberg Self-Esteem Scale) tend to be less satisfied with their relationships than are people high in self-esteem. In the not too distant past, researchers knew that this was true (i.e., that low self-esteem is associated with relationship dissatisfaction) without really knowing much about *why* it is true. In combination with theories about how our expectancies about the world shape our perceptions of the world, dependency regulation theory suggests one clear reason for the link between low self-esteem and dissatisfaction. If people low in self-esteem tend to view the world through thorn-colored lenses, then perhaps people low in self-esteem believe (wrongly) that their partners view them negatively (even if this is not truly the case). That is, maybe one reason why low self-esteem is associated with relationship dissatisfaction is that people low in self-esteem wrongly assume that their partners perceive them negatively. Maybe perceptions of our partners' regard for us (e.g., how Rick thinks Rhonda views him) mediate the association between self-esteem and relationship satisfaction. This is the hypothesis you should test in the first data set you'll examine on your own in this chapter.

You can find these data in a file called [**self-esteem study for mediation april 2005.sav**]. You should begin by familiarizing yourself with the data. As you do so, be wary that data from couples can get a bit confusing. Specifically, when dealing with data from couples, it's easy to forget who reported what about whom (e.g., is a variable such as "satisf" Rick's satisfaction with Rhonda, Rhonda's satisfaction with Rick, or some mixture of the two?). In these data, we've simplified things by focusing mainly on the perspectives of husbands. Thus, when you see the word *target* (person of focus), this will always refer to husbands, and when you see the word *partner*, this will always refer to wives. Furthermore, a variable that makes no reference to a target or a partner will always refer to the target. For example, the variable "satisfy" means how satisfied these husbands were with their marriages. So what variables are in your file? First, you have an arbitrary identification number or participant number and a variable for gender (notice gender is always 1 because we have data only from the husbands in this file). Next the file contains the following variables: (1) "selfestm" or self-esteem (Rick's global feelings of self-worth), (2) "percregard" (Rick's perception of how favorably his wife regards him), (3) "truregard" (Rhonda's actual regard for Rick—as reported privately by Rhonda herself when she filled out her own, anonymous survey), and (4) "satisf" (Rick's level of satisfaction with his marriage). In a little while, you'll see why it's important to know not only what Rick thinks Rhonda thinks about him but also what she herself really does think about him.

So where do we go from here? First of all, it'd probably be useful to get a picture of the exact mediational process that is predicted by the dependency regulation model. So before you go any further, take out a separate sheet of paper and sketch out the simplest possible mediational model. You may wish to refer back to Figure 13.4 as a reminder of the general format for representing a mediational model. Recall that, at an abstract level, distal variable X directly leads to hypothesized mediator M, which directly leads to the dependent variable (variable Y). You just need to decide which variables will serve as your X, M, and Y variables and then conduct the three multiple regression analyses that will allow you to test for mediation. Abstractly, recall that you want to see if

1. X predicts Y,
2. X predicts M, and
3. M predicts Y controlling for X.

Although you could test Steps 1 and 2 by running a simple correlation, recall that you'll save yourself some work later if you run regression analyses to answer all three questions.

QUESTION 13.1. By conducting the three crucial regression analyses, report the step-by-step results that will allow you to see if perceived regard mediates the association between husbands' self-esteem levels and their marital satisfaction. Make sure that you report both (a) the simplest possible approach to whether there is mediation (Are a and b both significant? Is c' no longer significant?) and (b) the results of a Sobel test of mediation (note that the Sobel test will require some hand calculations, which you should include in your answer).

Mediation Analysis as a Specific Case of Path Analysis

If you remember the title of this chapter, you may be wondering when you are going to learn about path analysis. The answer is that you just did. As it turns out, mediation analysis *is* path analysis. To be more specific, mediation analysis is merely the simplest possible form of path analysis. This is because path analysis is a technique for testing the plausibility of a hypothesized causal sequence involving at least two predictors and a criterion. In other words, the main difference between mediation analysis and path analysis is that path analysis usually involves more than two predictors of a criterion (whereas mediation analysis typically, but not always, involves only two). Of course, adding additional predictors to the picture means, by necessity, complicating the picture of what leads to what. That is, path analysis is all about figuring out a potentially complex set of causal chains or pathways between the various predictors and the outcome. For example, in a typical path analysis, two or three family variables that were assessed when children were born (e.g., parents' socioeconomic status, the number of children's books in people's homes) might be used to predict both childhood IQ and academic self-concept in Grade 6, and each of these childhood variables might be used, along with all of the original family variables, to predict an adult outcome such as educational attainment or adult incarceration status (e.g., Did kids grow up to do time in prison?).

But before we deal with any complex cases such as this one, let's begin with a much simpler case. Specifically, let's revisit the study of marriage, for which you just analyzed the data. First, let's assume that you saw some pretty good evidence of mediation. Can you think of any good criticisms of your findings? For instance, based on this mediation analysis, can you be sure that low self-esteem targets who thought their partners viewed them negatively were *wrong*? What if the husbands who were low in self-esteem were simply reporting *correctly* that they had married women who viewed them unfavorably? If Rhonda thinks Rick is a loser, and if Rick *realizes* that she thinks this, would it be all that surprising if this made Rick feel dissatisfied? By now, you may recognize that we have just identified a potential confound in this nonexperimental design. Is it *perceived* regard or *actual* regard that accounts for (i.e., mediates) the association between self-esteem and marital satisfaction? By now, you may also recognize that, like any other potential confound, actual regard could be pitted against perceived regard in a multiple regression to see which of these two variables (perceived regard or actual regard) is the true mediator of the association between self-esteem and marital satisfaction.

Figure 13.6 represents this slightly more complex set of affairs. To be more specific, Figure 13.6 represents the state of the world according to dependency regulation theory. Notice, for example, that the direct path from perceived regard to marital satisfaction is still expected (as represented by a solid causal arrow) even after we control for any possible direct

Figure 13.6 A Predicted Path Model That Considers the Potential Confound Between *Perceived* Regard (Among Husbands) and *Actual* Regard (How Wives Actually View Their Husbands)

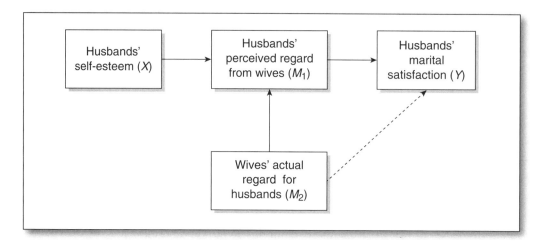

path between actual regard and marital satisfaction. In other words, *according to the dependency regulation model*, it is perceived regard, not actual regard, that is the important mediator of marital satisfaction. Incidentally, you might be wondering why there is no arrow between husbands' self-esteem and wives' actual regard for husbands. Although this kind of path is highly plausible, past research (including past research on dependency regulation) has suggested that there is not a very strong correlation between a person's level of global self-esteem and how favorably that person is viewed by his or her partner. Nonetheless, it would not change the logic of this model to include this additional causal arrow (which is probably weakly positive rather than nonexistent). In fact, if one really takes seriously the criticism that *actual regard* is the true mediator of the association between husbands' self-esteem levels and their levels of marital satisfaction, it might be best to include an extra arrow in Figure 13.6 to reflect the cynical counterclaim that actual rather than perceived regard mediates the association between husbands' self-esteem and their marital satisfaction. Without sweating this detail, your introduction to path analysis is to test the more complex version of the dependency regulation predictions shown in Figure 13.6—by conducting a more thorough multiple regression analysis.

By the way, because you have already conducted the meditational analysis for perceived regard alone and because you are addressing a critique of these original results, it's not really necessary (as it might be otherwise) to conduct a separate meditational analysis using variable M_2 (without including M_1 in the model) as a way of deciding whether to include M_2 in the model. If you were starting from scratch, that is, and you had no reason to expect actual regard to behave any differently than perceived regard, you might simply repeat all of the analyses you just completed for M (now called M_1), using M_2 instead of M_1. And if both mediators proved to be significant, you would have an empirical reason to pit them against one another and run the more complex meditational analysis that includes both mediators at the same time. In this case, however, we are trying to address a critic's point that actual and perceived regard are confounded. Thus, for theoretical and logical (rather than purely empirical) reasons, we would include actual regard in the model even if there wasn't very strong initial evidence that actual regard behaved like perceived regard (e.g., even if actual regard weren't highly correlated with husbands' satisfaction scores). To make this more concrete, you do not need to retrace all of your steps from your original mediational analysis. Instead, you simply need to add actual regard to your Step 3 regression analysis so you can see if the ab_1 path (involving husbands' self-esteem and perceived regard) is still significant when you use the adjusted ab_1 value and the adjusted standard errors that all take into account the association between M_2 and satisfaction. On the other hand, this is not everything. The most thorough test of this critique should also include a separate test for whether actual regard (the ab_2 path) mediates the connection between self-esteem and satisfaction. After all, it is possible, in principle, that both path ab_1 and path ab_2 are significant.

QUESTION 13.2. Run two new sets of analyses to see if (1) the mediational effect of perceived regard holds up when controlling for actual regard and (2) actual regard has a mediational effect once you control for perceived regard. In other words, which of the two variables, *perceived* regard or *actual* regard, seems to do a better job of mediating the association between husbands' self-esteem levels and their marital satisfaction? Make sure that you explicitly say whether support for the dependency regulation model is any stronger or weaker than it seemed to be after the original mediation analysis. In the interest of thoroughness, test each of these ab paths both the easy (are both a and b significant?) and the slightly less easy (Sobel) way. Be sure to provide hand calculations for the Sobel tests and to report your main results in plain English, as you would in a journal article.

The Logic of Path Analysis

Recall that path analysis is usually used to test causal models that are at least a bit more complex than these two models of dependency regulation. In addition, path analysis is ideally used in prospective studies, that is, studies in which researchers have assessed people's beliefs and/or behaviors over time. You might also recall that, all else being equal, causal plausibility is increased when you assess hypothesized causes prior to hypothesized outcomes—as you might be able to do, for instance, in a longitudinal study. All of this means that there are two tricky things you need to do to conduct any path analysis. The first tricky thing is to decide on the proposed causal ordering of *all* the variables in your model—that is, to decide what is likely to cause what. The second tricky thing is to test the model in a step-by-step process that allows you to separate *direct* and *indirect* effects of all the variables you are studying. The main rule to remember when conducting a path analysis is that, once you have specified your exact model, *you have to assess the association between every variable in your model and every variable that exists "upstream" (i.e., earlier in the proposed causal sequence) from that variable.* Let's take a closer look at that rule.

Consider a pretty simple path model and how you would test it. Imagine that you were a health psychologist who wanted to assess whether a healthy diet contributes to the physical health of older adults. Imagine, further, that past research had identified two good predictors of eating a healthy diet. First, on average, women have healthier diets than men do (e.g., women eat more tilapia and drink less tequila). Second, on average, wealthier people have healthier diets than do poor people (poor people eat food, whereas rich people eat "cuisine"). Perhaps you'd like to know, for example, if one of the reasons why older women tend to be healthier than older men is because, on average, women eat healthier diets than men do. That is, you might want to know if diet mediates the association between gender and physical health. Completely independent of this idea, you might also propose that diet mediates the association between income and physical health.

Figure 13.7 summarizes the way you might represent this model if you had access to some longitudinal data. The first thing to notice is that gender and income, the two variables that begin your hypothesized causal model, are both treated as *source* variables in the model. They are at the beginning (Step 1) of your hypothesized causal chain. Another thing to notice is that a person's health status at the beginning of the study is also treated as a source variable. Remember that you'd like to account for *changes* in people's health over time. To do so, you need to control statistically for people's initial health status when predicting their health status at Time 2 (e.g., their health status 5 or 10 years after your initial assessments in this longitudinal study). Initial health status is a background variable, an existing quality of the person at the beginning of the study. So it makes sense to place this variable at the beginning of the causal chain, just as we did for gender and income. You should also notice that diet is placed downstream from these three source variables and that physical health at Time 2 is placed at the very end of your model (it is downstream from everything, because it is your criterion or outcome measure). It is also worth noting that this model specifies only indirect effects of income and gender on physical health. That is, as things are proposed here, gender and income only contribute to physical health because they influence diet. If this model is correct, these two variables will not have a direct effect on physical health once we remove the effects they have on diet. In contrast, the model specifies that Time 1 physical health has both a direct and an indirect effect on health at Time 2. In other words, even if we control for any effects of diet on Time 2 health, people who are physically healthier at the beginning of the study will also tend to be healthier at Time 2. This model is based on the reasonable assumption that diet is not the *only* reason why healthier people today tend to be healthier people down the road.

So how would you go about testing this model in a path analysis? Once you have clearly stated your model by identifying each variable's exact position in your proposed causal stream, your work is practically done. To conduct a path analysis, you simply move "downstream" step by step—by predicting every downstream variable in the model from all of the upstream variables. You do this by first predicting whatever variable or variables are one step downstream from your source variables. You then predict whatever variable or variables are two steps downstream from your source variables. Your predictors *now* include not only the source variables but also the Step 2 variables that are presumably influenced by these source variables. *In other words, at every step in a path analysis, you predict the variable on which you are currently focusing from every variable that either is at the same causal step in the model or is upstream in the model.*

This means that a path analysis testing the model summarized in Figure 13.7 would involve two distinct steps (and thus two separate simultaneous

Figure 13.7 A Proposed Path Model Predicting Changes in Health From Background Factors and Quality of Diet

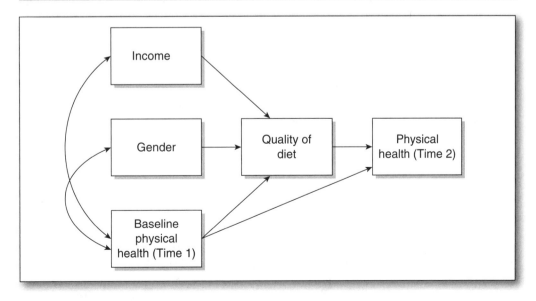

multiple regression analyses). First, you would predict quality of diet from all three of your source variables (income, gender, and baseline health). Preliminary support for the model would consist of significant paths (significant beta weights) between each of these three source variables and quality of diet—because we specified that each of these three predictors would make a separate contribution to diet. Second, in a *separate* simultaneous multiple regression analysis, you would predict Time 2 health from all four of the variables that are upstream from Time 2 health (including quality of diet). The strongest possible support for the model would consist of (a) a direct path between diet and health (i.e., diet directly predicts health at Time 2 even after we control for gender, income, and Time 1 health) and (b) a direct path between Time 1 and Time 2 health (indicating that Time 1 health predicts Time 2 health in ways that are independent of diet—because of variation in variables such as cholesterol levels, exercise, social support, and stress).

By the way, when planning a path analysis, it is crucially important to specify carefully the exact model you hope to test (by sketching it out or creating a formal figure) because you have to keep careful track of where each variable exists in your proposed causal stream. Thus, regardless of what theory you are testing or what kind of data you have at your disposal, you always begin by predicting all of the variables that appear as the second causal step in the model from all of the source variables that are further upstream (i.e., from all of the variables that could reasonably be expected to be causal starting points). In Step 2, you always predict all of the variables that appear in the third causal step from all of the variables that are upstream from (or at the same point in the causal stream as) this next variable or set of variables. You continue in this fashion until you have made your way to your dependent variable, and when you get there, you predict scores on the dependent variable from everything upstream. Any variable that predicts the dependent variable significantly at this point is said to have a direct effect on the dependent variable. After all, if a variable only influenced the dependent variable by means of another variable in the model (i.e., indirectly), there would be no direct effect of the variable left over.

We should note that although this description covers the essence of path analysis, it is not the whole picture. For example, Figure 13.7 is simplified somewhat because it does not include errors (little arrows showing, for example, that other things that we did not measure in our study are also good predictors of the quality of people's diets). In addition, researchers who perform path analyses usually begin by drawing a hypothesized model such as Figure 13.7 but end by drawing a separate, final model that sometimes differs substantially from the original model (e.g., because the results weren't in line with predictions). In addition, the final model usually includes the exact beta weights that were obtained in the results of the multiple regression analysis. Figure 13.8 illustrates a potential set of results for a path analysis based on the predictions of Figure 13.7 (although it leaves out a few details, such as error terms). The

Figure 13.8 Final Path Model for a Hypothetical Study of Diet and Physical Health

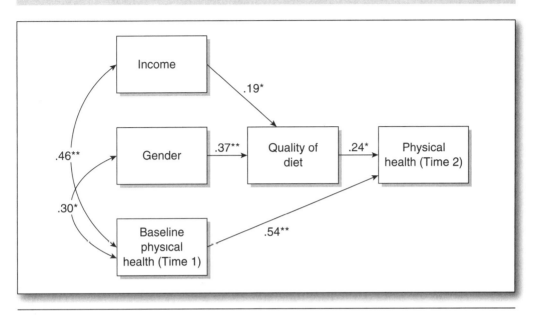

Note: *p < .05. **p < .01.

only meaningful difference between Figure 13.8 and the predictions of Figure 13.7 is that, for some reason, baseline heath at Time 1 did *not* prove to be a significant predictor of quality of diet. This is why the causal arrow from Time 1 health to quality of diet was removed in Figure 13.8. The hypothetical data simply did not support this piece of the model. A final important detail regarding Figure 13.8 is that the arrows on the far left (connecting the source variables) are double arrows. The use of double-sided arrows rather than traditional arrows reflects the fact that we are not making any assumptions about direction of causality among these source variables. In addition, because the source variables are starting points in the model, the coefficients connecting these variables to one another are simply zero-order correlation coefficients. Because correlation coefficients are always standardized, it's best to report path coefficients in standardized form, too—that is, to report beta weights rather than unstandardized *B*s for all of the remaining paths in the model.

A Hypothetical Path Model Involving Positive Beliefs and Health

Now that you have a sense of how path analysis works, let's stick with the topic of health psychology and conduct a path analysis for a hypothetical longitudinal study of positive beliefs and health. Imagine that a health psychologist recruited a representative sample of older Americans and assessed their baseline level of physical health at Time 1. At Time 1, the health psychologist also assessed two distinct measures known to be correlated with health. First, she assessed optimism, a personality trait that has to do with whether people generally expect good things or bad things to befall them in life (e.g., "I believe that every cloud has a silver lining."). Although optimism is, in fact, correlated with many health outcomes, the questions in dispositional measures of optimism never refer to health or health behavior. That is, the measure of optimism used most frequently by health psychologists does not assess people's optimism about their health. Instead, it assesses people's level of optimism about *life in general* (e.g., see Scheier & Carver, 1985; Scheier, Carver, & Bridges, 1994).

Second, imagine that at Time 1, the researcher also assessed specific health-related expectancies. This specific measure includes questions about whether a person believes that health-relevant behaviors are, in fact, connected to physical health (e.g., Is exercise good for you? Does taking multivitamins really contribute to health? Does smoking really kill you?). Next, assume that, at Time 2 of this study, the researcher was able to obtain good records of (a) whether people adhered to medical advice for taking prescription medications (e.g., Did a diabetic patient take his insulin regularly?) and (b) whether people ate well—not only by eating healthy foods but also by continuing to take in sufficient calories on a daily basis later in life. Finally, assume that this healthy young researcher was able to wait around long enough for everyone in the sample to die. Thus, her final (Time 3) measure was simply longevity (At what age did each person in the study die?). These longitudinal data, which you can find in [**health beliefs longevity data for path analyss.sav**] would provide the researcher with a unique opportunity to control for people's initial health status and then assess whether, and if so through what route, the two different psychological measures (optimism and specific health expectancies) influenced longevity.

QUESTION 13.3. Summarize the complete results for this hypothetical longitudinal study of optimism, health-related expectancies, and longevity. Your summary of the results should include not only a verbal description of all of the paths in the final model (including any you expected to see but did not) but also a figure that includes all of the appropriate correlations and/or standardized regression coefficients (beta weights). Assuming that optimists do live longer than pessimists, your analysis should let you say whether it is because (a) they are more likely than pessimists to take their medicine, (b) they make sure they get enough to eat (even when eating isn't as pleasurable as it used to be), or (c) they do other things (not assessed in this study) that lead directly or indirectly to a longer life. One could ask similar questions about specific health-related expectancies. Finally, one could ask whether the direct or indirect effects of these two beliefs on mortality were independent of one another. Do optimists live longer than pessimists, for example, *only* because they happen to have specific positive expectancies about health (as assessed in the health expectancies measure)?

Because we want you to give a lot of thought to the question of how to conduct a complete path analysis on these data, we offer you some general reminders to help you plan your analyses. First and foremost, you should always begin by sketching out the exact causal picture that you think is most reasonable. If two variables have equal causal significance (and if they were also measured at the same time), you should place them at the same point in the "causal stream" of the model. In the case of this model, it makes a lot of sense that you would have (a) three predictors that would serve either as distal causes (or as control variables), (b) two predictors that seem likely to be influenced by one or more of these Step 1 predictors, and (c) an outcome (your dependent variable) that could be influenced, in principle, by any of the five upstream variables that go into the model.

Remember that because you need to determine unique paths for *every* variable in a path model, you will need to conduct a separate multiple regression for each of the middle-level variables in the model. For example, in the present model, you will need to conduct one simultaneous multiple regression to predict diet and a different, albeit very similar, multiple regression analysis to predict adherence to medical treatments instructions.

In case this all seems a bit overwhelming, remember that it is only by remaining optimistic, and working through it step by step, that you can take the plunge into path analysis. No one said that disentangling causal spaghetti is always easy, but if you can begin to master this skill, you will be well on the road to being a true expert at data analysis. In case your optimism requires a little booster shot, you might want to consult the appendix at the end of this chapter, which includes the results of a simple and clearly described path analysis (a meditational analysis) from a published research paper by Kruger, Gordon, and Kuban (2006). Familiarizing yourself with this example may make it a little easier to report the results of your own path analysis.

Finally, let us remind you again that path analysis is simply a way of formally assessing the paths through which two or more variables presumably influence a dependent variable. Remember that the only major difference between path analysis and testing for mediation is that testing for mediation, in its simplest form, only involves two predictors. In contrast, path analysis usually involves multiple predictors, some of which may occupy different temporal points in a hypothetical causal chain.

For Further Reading

Baron, R. M., & Kenny, D. A. (1986). The moderator-mediator variable distinction in social psychological research: Conceptual, strategic, and statistical considerations. *Journal of Personality and Social Psychology, 51*,173–182.

Berkowitz, L. (1989). Frustration-aggression hypothesis: Examination and reformulation. *Psychological Bulletin, 106,* 59–73.

MacKinnon, D. P. (2008). *Introduction to statistical mediation analysis.* Philadelphia: Psychology Press.

Pedhazur, E. J. (1997). *Multiple regression in behavioral research: Explanation and prediction* (3rd ed.). Fort Worth, TX: Harcourt Brace. (See Chapter 18, "Structural Equation Models With Observed Variables: Path Analysis".)

Scheier, M. F., & Carver, C.S. (1985). Optimism, coping, and health: Assessment and implications of generalized outcome expectancies. *Health Psychology, 4,* 219–247.

Useful Web Pages

UNDERSTANDING AND CALCULATING THE SOBEL TEST

http://www.danielsoper.com/statcalc/calc31.aspx

http://www.people.ku.edu/~preacher/sobel/sobel.htm

MEDIATION

http://www.davidakenny.net/cm/mediate.htm#ST

http://www.public.asu.edu/~davidpm/ripl/mediate.htm

PATH ANALYSIS

http://userwww.sfsu.edu/~efc/classes/biol710/path/SEMwebpage.htm

Appendix 13.1: An Analysis of Teasing From Kruger, Gordon, and Kuban (2006)

Would you be impressed if your primary author told you that he's psychic? Probably not. But what if I told you that my psychic powers are telling me right now that you have a chronic problem with body odor? I can't quite tell—from this tremendous distance—if it's halitosis or excessive perspiration, but there's definitely a problem. Oh wait, it's coming in clearer now. It's both.

If you were ever intrigued by my claims of psychic prowess, the intrigue is probably wearing pretty thin. So before you send me any additional hate mail, I'd like to note that there is a statistical point here. I promise! First, you must have known that I'm not *really* psychic. Second, I was just teasing. Third, and most important, if you are like most people, *your* perception of this incidence of teasing is probably a lot less favorable than mine. From *my* perspective, I was just playfully teasing one of my thoughtful and diligent readers, whom I have come to respect and admire after spending all this quality stats time together. From *your* perspective, I didn't make enough dumbass jokes in the body of the text so I had to add the worst one of all in an appendix—and make fun of a smart, decent, and honest person in so doing. If this rings true to you at all, you would enjoy Kruger et al.'s (2006) research paper because this was precisely their point. *Teasers* and the *targets* (aka victims) of teasing often have very different perceptions of a teasing incident. As Kruger et al. put it, teasing "is often seen as innocent and playful by the teaser" but "it tends to be construed as considerably more malicious by the target" (p. 412).

To document this effect, Kruger and colleagues (2006) conducted five different studies involving different kinds of teasing and different kinds of relationship partners (e.g., roommates, romantic partners). In Study 3, they examined teasing among friends and family members. In this particular study, they also looked at teaser-target differences in *perceived intentions* (Did the teaser mean to be playful or hurtful?) as one of the *reasons* why teasers and targets view teasing differently. Specifically, their participants either thought of (a) a time when they *teased* a friend or family member or (b) a time when they *were teased by* a friend or family member. They then answered two questions about the construal (interpretation) of the tease (involving how *humorous* it was and how *lighthearted* it was) and three questions about the perceived intention of the teaser (e.g., To what extent, if at all, did the teaser *intend* to hurt the target's feelings?).

Kruger et al.'s (2006) predictions were (a) that teasers would perceive the tease more favorably than targets and (b) that perceived intentions would *mediate* this association. To set the stage for their meditational analysis, we will tell you (as they told their readers) that teasers did, in fact, construe the tease more favorably than did targets. The respective means (on an 11-point scale) were 8.49 for teasers and 6.53 for targets (where higher scores mean more favorable perceptions). But did perceived intention mediate this association? Here is our slightly edited version of their results (edited because the authors were describing two sets of results at once, and we are focusing here on only *one* set of their results).

Meditational Analysis. As in the previous study, we next conducted a path analysis to explore whether awareness of good intentions mediated the link between participants' role in the tease (teaser or target) and their construal of it. Because the independent variable was between subjects rather than within subject, we used the familiar Baron and Kenny (1986) procedure to test for mediation.

The results are illustrated in Figure 1. As can be seen, the IV (independent variable; participants' role in the tease) was a significant predictor of the perceived intentions behind the tease. As well, the significant relationship between role and tease construal was reduced when perceived intent was held constant. Sobel (1982) tests revealed that this reduction was significant ($Z = 2.67, p = .008$). Perceived intent also continued to significantly predict the DV (dependent variable) when the effect of the predictor variable was controlled ($b = 60, p < .01$). These results suggest that perceived intention partially mediated the link between teasers and targets in their construal of the tease.

Figure 1 Betas from a path analysis of teaser role (teaser or target) and tease construal, Study 3. The beta in parentheses shows the relationship between teaser role and construal after controlling for the proposed mediator (perceived intention)

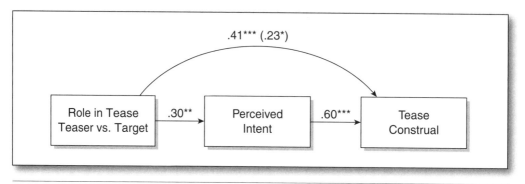

Source: Adapted from Kruger, Gordon, and Kuban (2006).
*$p < .05$, **$p < .01$, ***$p < .001$.

Of course, your path analysis assignment in this chapter is somewhat more complex than this simple test of mediation, but we hope this simple and elegant example gives you a feel for how to graph and report the results of a path analysis. Good luck!

Author's note: Did you notice how first author *Justin* used the word *just* not once but twice in this article title, and how he (Kruger) collaborated with Kuban? This is only an anecdote, of course, but it seems to support a theory called implicit egotism.

Notes

1. This discussion oversimplifies a complex story. For example, as MacKinnon (2008) notes, there are some statistical situations in which moderation strongly implies mediation. Further mediation and moderation can both occur in the same data set (a situation referred to as moderated mediation).

2. Both Preacher and Hayes (2004) and MacKinnon (2008) provide step-by-step instructions, using both SAS and SPSS, on how to use bootstrapping as a superior way to test the statistical significance of mediation.

Data Cleaning 14

Introduction: Data Cleaning

It is difficult to overstate the number of ways in which personal computers have made life easier for people lucky enough to be able to use them on a frequent basis. In our opinion, one of the most wonderful ways in which this is true has to do with differences in how people collect data today and how they did so as recently as 15 or 20 years ago. About 20 years ago, the first author usually collected both experimental and survey data by asking participants to mark their answers to his questions right on their paper survey pages. He and his unlucky research assistants then spent hundreds of hours entering these data into SPSS data files. This meant that a crucial stage of the research process involved translating the little marks people had put on sheets of paper into the words and numbers they presumably had intended to mark. Like Al Gore (who reputedly still has nightmares about hanging chads), the first author lost a lot of sleep. He grew angry, bitter, and confused. To see why, check out Image 1, which contains the responses of a hypothetical participant in one of his old pencil-and-paper surveys.

```
        1         2         3         4         5         6         7
     Disagree                    neither agree                    Agree
     Very Much                   nor disagree                     Very Much

      7  1. I feel that I am a person of worth, at least on an equal basis with others.
      1  2. I feel that I have a number of good qualities.
      1  3. All in all, I am inclined to feel that I am a failure.
      7  4. I am able to do things as well as most other people.
      1  5. I feel I do not have much to be proud of.
      7  6. I take a positive attitude toward myself.
      7  7. On the whole, I am satisfied with myself.
      2  8. I wish I could have more respect for myself.
      2  9. At times, I feel that I am useless.
```

1

This person seems to *agree very much* that he is "a person of worth," who has a number of good qualities (ignoring penmanship, of course). However, based on the marks he made next to Questions 3 and 5, he *also* seems to *agree very much* that he is a "failure" who "does not have much to be proud of." Although this is possible, in principle, it is very rare to see a person who strongly agrees with both the positive and the negative items in the Rosenberg Self-Esteem Scale. With the benefit of a systematic handwriting sample, we might have been able to disambiguate these highly ambiguous responses.

123456781   113456789

If the *first* guy filled out this survey, he was probably indicating that he *dis*agrees very much with the negative self-esteem items. However, if these same responses came from the second guy, we'd have to take his unusual pattern of responses at face value (unless the rest of his survey responses suggested that he wasn't paying attention to the survey questions). For the present, we are going to skip the details of how you might make data "decoding" decisions in the absence of a systematic handwriting sample (e.g., Do you always leave ambiguous individual scores blank? Do you get different research assistants to blindly code the individual scores?). Suffice it to say, though, that one principal consideration in disambiguating individual data points should be avoiding experimenter bias (e.g., "He *must* be high in self-esteem because he reported in the section on childhood experiences that his principal caregiver was highly nurturing."). See Rosenthal and Fode (1963), Rosenthal and Jacobson (1966), or Pelham and Blanton (2013) for details.

Now for the good news. One of the most wonderful research developments in the past couple of decades is the development and refinement of computer software programs (such as MediaLab and E-Prime) that greatly reduce these kinds of data coding problems. When researchers use data collection software such as MediaLab, this makes a survey or an experiment more like an interview conducted by a careful and polite computer. Thus, researchers who collect data on the computer don't usually have to worry about sloppy handwriting. They can also stop worrying about whether their participants used the appropriate 7-point scales to answer the self-esteem questions and then dutifully switched to a true-false format for the self-monitoring survey (designed by a different research team). This is because data collection software packages such as MediaLab typically remind participants, if necessary, of what kinds of answers are acceptable to specific survey questions. Suppose, for example, you had taken part in Mitsuru Shimizu's dissertation study of self-esteem and health. One thing Shimizu wanted to know was how many of about

20 illnesses or symptoms people had experienced in the week preceding their participation in his study. If you tried to answer "3" to a yes-no question involving whether you had suffered from a headache, the computer wouldn't accept this response. Instead, it might politely remind you to answer "Y" for yes (I had that symptom or illness), "N" for no (I didn't have it), or "S" (skip) for "I'd like to skip this question."

Having said all this, we must also note that sophisticated data collection software can't solve every possible coding problem. For example, if you allow people to report their weight in pounds or kilograms, you may want to leave this question open ended. Thus, someone who is 5′9″ may sometimes enter a weight of 1,555 instead of 155 because of a sticky computer key or a careless entry error. Similar problems could apply to self-reported income, age, or daily spending. For ethical reasons, researchers are also obliged to give participants the option of skipping any specific question that the participants don't feel comfortable answering. Finally, if you do not have access to a computer for data collection, or if you ever analyze data sets that were collected *before* data collection became such a delight, you'll often have to deal with missing or out-of-range data. This means that you should have at least a passing familiarity with data cleaning. The goal of this chapter is to introduce you to some of the most common data-cleaning problems you are likely to face and suggest some simple solutions to each of them. Although dozens of specific problems may require data cleaning, the bulk of data-cleaning problems can be boiled down to just two. The first problem occurs when data are *missing*, and the second problem occurs when continuous data are *not normally distributed*. If you happen to work with potential voters or middle school students, you are probably familiar with this "rule of two": Either (1) people don't show up or (2) they *show* up but don't *line* up the way they are supposed to. Let's begin with statistical absenteeism, otherwise known as missing data.

Missing Data

If you recall that the goal of most research is to draw firm conclusions about a population from a sample, you can probably see that many research headaches involve some kind of missing data. If only 10% of the people you contacted respond to your survey, how can you be sure that the mundane 90% of people who refused to respond resemble the noble 10% who responded? If everyone agrees to be surveyed, but half of these bozos refuse to answer the three highly sensitive questions about their sexual fantasies involving circus clowns, how can you be sure that the 50% who answered have the same kind of sexual fantasies as those who decide to skip these painful questions? What about the people who answer every single question but whose responses suggest that they may have considered the survey an opportunity to clown around? Many years ago, your primary author conducted a study involving the perception of highly ambiguous (low-pass filtered) speech. Typical guesses about what the speaker was saying included things like "I wanted to get good grades," "I asked them to close the drapes," and even "I was haunted by her grave." A much more atypical guess was "I wanted to paint myself like a giraffe and become a go-go dancer on an army base." Your primary author no longer remembers whether he used any of this participant's responses to the *other* questions, but he does remember that he coded this particular response as missing.

Because there are many reasons why data can be missing, as well as many things researchers wish to do with data, there are many different solutions to the problem of missing data. Interestingly, two of the most serious problems with missing data occur at two ends of the continuum of the complexity of a research question. Sometimes, for example, the answer to a research question is quite simple. In fact, sometimes it's a single number. Examples are the percentage of voters who plan to vote for a specific former bodybuilder or professional wrestler in a gubernatorial race, how much money the average American earned last year, and the percentage of cola drinkers who prefer Coke over Pepsi in a blind taste test. When questions are this simple, important things such as votes, tax policies, or Super Bowl commercials often hinge on them, and so pollsters and marketers are interested in making sure that the opinions they measure truly describe everyone in the population of interest. So failing to get answers from a substantial percentage of those surveyed is problematic. At the other extreme, sometimes the answer to a research question is quite complex. For example, researchers who study well-being would probably love to know the relative contribution to well-being of (a) absolute wealth (how much do you make?), (b) relative wealth (do you make more than your neighbors?), (c) education, (d) occupational status, (e) career satisfaction, (f) gender, (g) age, (h) ethnicity, (i) objective physical health, (j) subjective physical health, (k) diet, (l) exercise, (m) negative mood, (n) positive mood, (o) neuroticism, (p) extraversion, (q) optimism, (r) quality of close relationships, (s) social support, and (t) work stress. Let's assume that you are one of those researchers and that you conducted a longitudinal study of 1,000 people to assess the relative contribution of each of these potential predictors to well-being over a 12-month period.

Let's begin by very generously assuming that anyone who ever took part in any phase of your longitudinal study always answered your well-being questions. Let's further assume, more realistically, that you worked hard to keep your attrition rate down to only 15% over the course of the year. That is, let's assume that 850 of the 1,000 participants who took part in Wave 1 of your longitudinal study also took part in Wave 2 of the same study. Finally, let's assume that for each of the 20 predictor variables listed above (income, ethnicity, extraversion, etc.), there is exactly a 90% chance that each individual participant answered that question. (The truth is that it would usually be much higher for questions such as gender but much lower for questions such as income, but 90% is a reasonable value for a demonstration.) This means that we'd only expect about 12% of your original sample to have complete data on all 20 predictor variables at Time 1 (that's based on $.9^{20} = .1216$)! This is important because the default approach to a multiple regression analysis in SPSS is to *exclude cases "listwise,"* which means that if a participant is missing a single score on even one of the 5 (or 10, or 21) predictors in the multiple regression analysis, that participant is removed completely from the analysis. So if you accepted this default setting and ran a regression analysis, your Wave 1 sample of 1,000 participants would be reduced to a useable sample of only about 122 people.

Assuming that you'd like to capitalize on your longitudinal design and run a multiple regression analysis in which you predicted well-being at Time 2 from the 20 Time 1 predictors (controlling also for Time 1 well-being), you'd likely have a final sample of about 100 participants! (Remember that 15% of the sample dropped out of the study completely at Time 2, reducing $n = 122$ to about $n = 100$.) Notice that you now have only about 5 participants per predictor, which is far from ideal. Moreover, even if this tremendous loss of statistical power (and source of unreliability) didn't worry you, you'd have to be worried that the 10% of your participants who were diligent enough to answer every single question you ever posed might be a bit unusual. If nothing else, they'd probably be much higher than average in the personality trait of conscientiousness. In this case, a researcher would have to worry a great deal that the results of any regression analysis would apply only to that subset of the population who are very much like this highly conscientious 10% of the original sample.

This is a long and winding way of saying that the more predictors you include in a multivariate statistical model, the more problematic it becomes to have missing data. In some ways, the preceding example understates the problem. If you had worked with any multi-item scales in this survey (e.g., if your extraversion measure were based on the sum of the answers to eight different questions), anyone who failed to answer even one of the extraversion questions might fail to receive a score on the extraversion scale. This is because the default approach to calculating a scale score is to fail to give someone a score on the scale if that person omitted even one scale item. Finally, whether you are conducting a simple multiple regression analysis or trying to pull off a much more sophisticated confirmatory factor analysis, there are other ways in which you can lose participants. Most notably, in some multivariate statistical analyses, you are sometimes obliged to delete a few of the participants who answered every single question you asked because these participants proved to be "multivariate outliers" (i.e., statistical oddballs). We address this issue in some detail later in this chapter.

So whether you want to answer a simple question about all of your participants, a complex question about all of your variables, or, worse yet, a complex question about all of your participants and all of your variables, missing data can be a very serious problem. This means that you may often have to wrestle with questions about missing data. Although we can't promise you solutions to every possible problem, we're happy to say that there are some simple, and sometimes clever, solutions to some of the missing data problems researchers face most routinely. In the next two exercises, we'll examine a couple of hypothetical data sets in which we work through some decisions about missing data.

Let's begin with a hypothetical data set based pretty closely on a real data set collected by Tracy DeHart (see DeHart, Pelham, & Tennen, 2006). DeHart was interested in implicit self-esteem. Moreover, she strongly suspected that part of the reason past researchers had previously observed only weak associations between implicit self-esteem and some of the variables that ought to be related to it was because implicit self-esteem is pretty volatile (i.e., temporally unstable). If you want to know about people's *trait* level of implicit self-esteem, she argued, you should ideally assess implicit self-esteem on multiple occasions. The average of a person's implicit self-esteem scores over time should be a much better indicator of chronic implicit self-esteem than that person's score on any one occasion. DeHart actually measured implicit self-esteem on six separate occasions. We're going to simplify things a bit and pretend that she measured implicit self-esteem on only three occasions. In addition, we're going to greatly simplify her study and assume that all she wanted to do was see if people's implicit self-esteem scores correlated with their mother's reports of how nurturing (i.e., sensitive, loving, caring) these mothers were to the participants when the participants were growing up (see Baumrind, 1971, 1983, or Bowlby, 1982, for theoretical reasons to expect this association).

Our measure of implicit self-esteem, by the way, will be based on people's liking for three things: (a) their first initial, (b) their last initial, and (c) their birthday number. For the purposes of this exercise, we are going to consider each of these three scores an individual item in an overall implicit self-esteem score. The SPSS data file [**hypothetical name letter ISE data based loosely on DeHart.sav**] contains all three of these implicit self-esteem scores at each of three time periods (e.g., *frstlik*, *lastlik*, and *bdaylik* are the three scores at Time 1; see the SPSS variable labels for the Time 2 and Time 3 variable names). Crucially, the same data file also contains a variable called *cg_nurture*, which is each participant's mom's own independent report of how nurturing she was when the participant was growing up. Impressively, in her actual study, DeHart convinced about 90% of the mothers of her participants to take part in her survey study by mail. Nonetheless, 90% participation converts to 10% data loss, which is roughly the loss rate you'll see for the moms' independent responses in these hypothetical data.

Once you've opened the data file, you'll want to get a general idea of what kind of missing data problems you're facing. In this case, the easiest way to do this is to run a simple frequency analysis on (a) all of the individual

Chapter 14 Data Cleaning

variables that will go into your implicit self-esteem (ISE) score and (b) the cg_nurture variable that you'd like to correlate with ISE. A quick reminder of how to do this follows in Images 2 and 3.

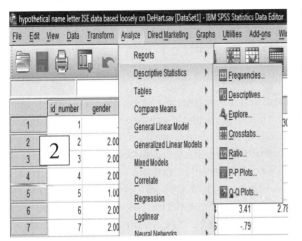
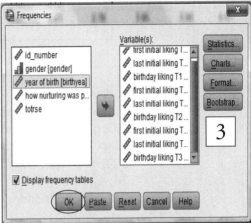

For the moment, by the way, we are only interested in the percentage of participants who have missing data, and so we are going to forego asking for any charts or statistics and simply click "OK" to run the frequency analysis that will tell us how many participants are missing data on each selected variable. The first part of your SPSS output file should resemble the partial screen capture in Image 4.

Statistics

		first initial liking T1	last initial liking T1	birthday liking T1	first initial liking T2	last initial liking T2
N	Valid	176	175	178	146	145
	Missing	55	56	53	85	86

Frequency Table

You can see, for example, that between 53 and 56 of 231 total participants have missing data for the Time 1 (T1) individual letter or number liking questions that are part of the ISE score. It gets worse at Time 2. As you can see in the small circled area, more than a third of participants (85 of 231) have missing data for first initial liking at Time 2. By the way, to get the exact percentages of missing data, variable by variable, you simply need to scroll down in the output file for the detailed frequencies that are provided for each specific variable. Image 5 shows you the bottom of the output table for first initial liking at Time 1, and you can see that the 55 missing cases you saw in the left-hand portion of Image 4 make up 23.8% of the 231 total cases.

		4.43	3	1.3	1.7	97.7
		4.49	3	1.3	1.7	99.4
		4.97	1	.4	.6	100.0
		Total	176	76.2	100.0	
Missing	System		55	23.8		
Total			231	100.0		

Recall that DeHart was interested in improving upon previous measures of implicit self-esteem by assessing implicit self-esteem measures on multiple occasions. This means that, in an ideal world, she would get complete data from all of her participants on multiple occasions and add them up to get an overall implicit self-esteem score. In the not so ideal world in which we actually live, what would happen if you just averaged the nine implicit self-esteem items (three items per time period) and correlated this ISE score with the crucial cg_nurture score? Let's find out, using a syntax file that you can create yourself. You should begin by averaging the nine implicit self-esteem items to create an overall implicit self-esteem index. You should do so by very carefully entering the following compute statement directly into a blank SPSS syntax file.

COMPUTE TOT_ISE = (frstlik+lastlik+bdaylik+frstlik2+lastlik2+bdaylik2+frstlik3+lastlik3+bdaylik3)/9.

EXE.

Once you successfully enter and run this compute command (including the "EXE" part), you'll find a new variable in your SPSS data file (TOT_ISE). Now we *could* ask you to correlate this ISE score with cg_nurture, but to save us all a little valuable time, we'd like you to hold off on running that correlation until you calculate ISE in two other ways. Please do so by very carefully entering each of the following two alternate compute commands:

COMPUTE TOT_ISEm = mean (frstlik, lastlik, bdaylik, frstlik2, lastlik2, bdaylik2, frstlik3, lastlik3, bdaylik3).

EXE.

COMPUTE TOT_ISEm3 = mean.3 (frstlik, lastlik, bdaylik, frstlik2, lastlik2, bdaylik2, frstlik3, lastlik3, bdaylik3).

EXE.

If you want to save a little typing, you might notice that the second of these two alternate commands is just like the first except that (a) the second variable name ("TOT_ISEm3") has a 3 added at the end and (b) the second variable is calculated using the "mean.3" command rather than the "mean" command. You may recall from previous exercises that the mean.x command calculates a scale mean but *only* for those participants who have a minimum of *x* scores from the list of variables provided in parentheses. Thus, the "mean.3" command will only calculate a mean ISE score for participants who filled out at least three of these nine individual scores (regardless of *which* three they answered). If you had used "mean.9" rather than "mean.3," you would have gotten scores only from participants who answered all nine ISE questions. The "mean" command without a suffix, by the way, means the same thing as "mean.1." This version of the command will give a person a scale score even if that person answered only a single question in a multi-item scale.

Once you have corrected all of your typos (we're pretty sure both Dwight and Kevin typed "firstlik3" rather than "frstlik3"), make sure to highlight and run these two additional compute commands. Now you should have three different versions of the ISE variable in your original data file. You can now correlate each of these three versions of the ISE score with the moms' reports of how nurturing they used to be to these lucky young adults. A quick reminder of how to do this follows.

Once you click "Bivariate..." as shown in Image 6, make sure to select (a) all three versions of the ISE variable and (b) the cg_nurture variable (as shown in the "Variables:" box in Image 7).

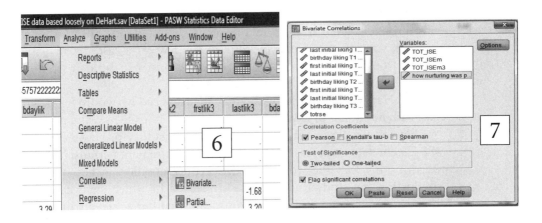

QUESTION 14.1a. Briefly explain why you had so few cases for the tot_ise variable and why you had a few more cases for tot_isem than you did for tot_isem3.

QUESTION 14.1b. In addition to offering a simple conceptual interpretation of the correlations observed in this hypothetical study, provide a brief methodological guess about why the correlation between nurturance and tot_ise is stronger than the correlation between either nurturance and tot_isem3 or nurturance and tot_isem.

QUESTION 14.1c. Although this would be unlikely to happen, assume that a researcher insisted on using the ISE score that made use of data from all nine ISE items (from all three waves of the study). An obvious criticism of this finding is that it might apply only to the 43% of participants (99 of 231) who (with a little help from their moms) were diligent enough to provide complete data on all of the individual ISE items. How might the researcher use the data from the alternate analyses to address this criticism?

You may be wondering which of the three ways of calculating implicit self-esteem is best. Although there is no simple and sovereign rule about how to handle missing data within a specific scale or index, we hope it is obvious that the present case is a bit unusual. Most studies are cross-sectional, for example, which means that you won't usually be missing data from large groups of participants. Thus, for example, if you lose only 3 of 200 participants because 3 of them left a few items blank in one particular scale, you might not bother to use the mean.x command (or a variation thereof) to pick up the extra 3 cases. Presumably, it is going to be extremely rare to get results with $n = 197$ of your participants that differ meaningfully from the results for the total sample of $n = 200$ (and if you do, something odd is probably going on). Our philosophy on this is that it is a shame to ever lose data, and the only reason to ignore data from a participant is that you have good evidence that the person wasn't really paying attention (e.g., because the person left many items blank in more than one spot in a survey or experiment). On the other hand, if you assign a scale score to a person who answered only 1 item in a 20-item scale, you have probably gone too far in the other direction. In cross-sectional research, you can sometimes pick up quite a few extra cases by allowing people who answered the majority of questions in a scale (e.g., 7 of 10 questions) to be included in your analyses.

In DeHart's actual study, by the way, she balanced her desire for inclusion with her desire to have the most temporally reliable measure possible by allowing people in the final sample as long as they provided implicit self-esteem data on at least two of the four measurement periods. It is also worth noting that she worked hard to reduce attrition in her study by conducting the study during class—in a class in which she knew that attendance rates were very high. This meant that her actual attrition was somewhat lower than what you saw here. If you're starting a study from scratch, there are many things you can do to minimize attrition (e.g., sending participants polite reminder emails about the study the day before they are scheduled to participate, offering participants a bonus of some sort if they participate in most or all sessions of a multisession study).

Unfortunately, the example we just covered (missing data on some items in a scale based mainly on simple attrition) is not the most common form of missing data. As we suggested earlier, researchers often measure a crucial independent or dependent variable with a single question—a question that some highly diligent and thoughtful participants simply chose not to answer. For example, for a wide variety of reasons (e.g., paranoia, modesty, cultural factors), many people do not like to tell researchers exactly how much money they make. This is a big problem for economists, sociologists, and health psychologists, for example, because how much money people make turns out to be tied to a lot of interesting physical and psychological outcomes. For example, no one yet knows for sure *exactly* why rich people seem to live much longer than poor people. Some of the reasons are obvious (e.g., lower rates of smoking among the wealthy; e.g., see Erikson & Torssander, 2008). However, other reasons may be much more subtle (e.g., greater feelings of predictability and control among the wealthy).

Notice that we are *not* arguing that it is impossible to know how much money Americans (or Swedes or Lithuanians) make as a group. A few Swiss bank accounts and some under-the-table jobs notwithstanding, the IRS has a very good idea how much money the average American makes. However, the IRS has good access to useful things such as people's pay stubs. That is, they tabulate something that people's employers legally have to report, whether people like it or not. Furthermore, the IRS doesn't care much about psychological wealth. This means they never have to survey people about the richness of their social networks, how invested they are in being parents, or whether they feel taxed and depleted by chronic work stress. To understand the role of income in people's physical and psychological well-being, you have to rely on the willingness of most people to *tell* you how much money they make—in surveys of physical and psychological well-being. It is thus a big problem for many researchers that many people don't like to do so.

For example, there is a large team of psychologists, behavioral economists, and other social scientists at Gallup who study the connection between income and well-being, both in the United States and worldwide (e.g., see Deaton, 2008; Kahneman & Deaton, 2010). In the Gallup-Healthways Daily U.S. poll, for instance, researchers survey a different representative sample of 1,000 Americans about 360 days a year! Almost every day, then, 1,000 typical Americans agree to do an interview with a total stranger. Furthermore, on any given day, at least 990 of these 1,000 interviewees willingly answer a lot of personal questions—including whether they experienced anger a lot of the day yesterday, whether they are satisfied with their lives, whether they are divorced, and whether they have ever been diagnosed with cancer. This near 100% response rate for personal questions involving divorce and cancer drops to about 60% when you just ask people how much money they make (although a considerably higher percentage of people will usually provide a ballpark income figure).

If you have ever worked with suspected criminals or middle school students, you are probably familiar with this "interviewer's dilemma": The more interested you are in getting people to answer a question, the more likely people are to refuse to answer it. Few questions fit this principle better than the question of income. With this dilemma in mind, let's examine some hypothetical data that are based loosely on (a) the previously mentioned

Gallup-Healthways U.S. Daily poll and (b) the General Social Survey (GSS), which, as you may recall, is an in-depth, face-to-face survey conducted every 2 years by the National Opinion Research Center.

Believe it or not, all we want to know from these hypothetical data, using [**hypothetical US data resembling Gallup G1K & GSS jan 2011. sav**]**,** is the correlation between income and education (on a scale of 1–6 that reflects total years of formal schooling; see the values for the variable "educ"). Let's jump right in and correlate education with three different measures of income. You'll see in a moment exactly *why* there are three different measures of income. But first, let's explain each measure.

The first income measure in our hypothetical survey is "income_dollars." This variable is people's responses to an open-ended question about income: "What was your average *monthly* household income last year in dollars (gross income, before taxes)?" The second income measure is "income_cat." This measure is similar to the one used in both the GSS and the Gallup daily poll. Assume that the actual question was the following: "I'm going to show you [or read you] 11 labeled income categories. Which income category best describes your total monthly family income, from all sources, last year, before taxes?" The third measure of income is "income_needed_cat." This third question, which was always asked after the first two, was this: "Now we'd like you to think about the same category scale we just showed [or read] you for monthly household income. Using this scale, how much monthly income would you say a household like yours needs these days, just to get by?"

You may be wondering whether it is really appropriate to call this third measure a measure of "income." Rest assured that many people would not trust this measure unless (a) it correlated very highly with the two straightforward income measures and (b) it proved to have some meaningful advantage over the first two measures. Part of your job in this missing data exercise is to figure out whether either or both of these two things is true. To start doing this, simply correlate your measure of education with each of the three income measures.

QUESTION 14.2a. Report and interpret each of the three relevant correlations (between education and each of the three alternate income measures). Make sure to report the number and percentage of your sample who provided data for each correlation, and say something about why you think the sample sizes differed the way they did for the three different income-education correlations.

QUESTION 14.2b. First, based on the correlations between the third measure (income_needed_cat) and the two more straightforward income measures, how reasonable would it be to treat the third measure as an income measure? Second, imagine that you are working as a statistical consultant for a client who wishes to use (a) the first measure of income (and that measure only) when it is available, (b) the second measure of income (and that measure only) when the first measure is not available, and (c) the third measure of income only when neither the first nor the second is available. Write and run the SPSS syntax code that will do this and then report and interpret the correlation between this alternate income measure and education. Make sure to report which simple but extremely important transformation you should make on each of the three income measures (before you can safely mix and match them) to maximize your sample size in this way.

If you'd like an embarrassingly huge hint about what kind of transformation to conduct before treating these three different income measures as interchangeable, consider the screen captures that follow in Images 8 and 9.

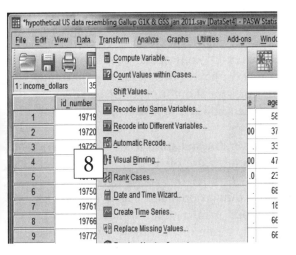
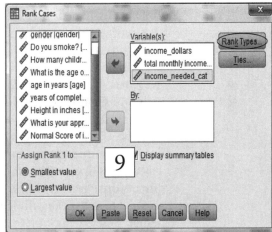

Image 8 shows that one kind of transformation you can conduct on a numerical variable is to "Rank Cases...." Of course, we don't really want to rank the cases, but if you click the "Rank Cases..." option highlighted in Image 8, you'll be allowed to select all three of the income measures as variables using the dialog box that you see in Image 9. If you then click "Rank Types" (circled in Image 9), you'll open yet another dialog box (not pictured) that allows you to (a) deselect "Rank," (b) select "Normal Scores," and then (c) click "Continue." This will return you to the dialog box you see in Image 9, where you can simply click OK to create z scores for all three of your income measures. If you do this successfully, your data file will contain three new variables, each of which has a funky name that will only barely allow you to tell which original variable the new z-scored variable is based on. However, SPSS is considerate enough to have created a label for each of these variables that should help you keep them straight. You'll want to use these three new variables in some creative IF statements to create the specific measure of income that your client requested.

That's Not Normal: Outliers

As we noted near the beginning of this chapter, missing data are not the only reason why researchers need to clean their data. As we noted very briefly in Chapter 3, a more common reason for data cleaning, especially when dealing with continuous interval or ratio scores (e.g., income, extraversion, the number of hours in one's life spent playing the Nintendo DS), is that a set of scores that one would *expect* to be normally distributed deviates in some serious way from normality. In many cases, distributions that deviate from normality deviate because of the influence of one or more **outliers.** A statistical outlier is simply an extreme score—that is, a score that differs dramatically from most other scores in a set. In many realms of life, people aspire to be outliers. Olympic gold medalists, billionaires, and world champion chess players are all outliers on the respective dimensions that distinguish them from causal athletes, the middle class, and people (like your primary author) who have trouble remembering what you call those horse-shaped pieces. Of course, being an outlier can sometimes be a serious problem. Consider having been struck by lightning five times, having been arrested 10 times for shoplifting, or being afflicted with Lou Gehrig's disease (which affects about 1 person in 10,000 in the United States).

To be a little more technical, however, the statistical term *outlier* refers not to an entire person but to a person's score on a particular variable. Thus, the same person could possess positive outlying scores on some dimensions, average scores on other dimensions, and negative outlying scores on still other dimensions. Your primary author has a brother named Stacy, who, pound for pound, is one of the physically strongest people the author has ever met (including people in national weightlifting competitions). Stacy's score on a specific measure of strength, for example, might make him an outlier on that particular variable if we measured it in a study of physical performance. If we remember that first names have frequencies in the population, there is also a sense in which Stacy's name makes him an outlier. For every man named Stacy in the United States, there are at least 500 men named James! However, Stacy is roughly average in height, and he seems to be very near the population mean in the personality traits of agreeableness and openness. In contrast, he is probably a negative outlier in his culinary knowledge. Although he makes a truly excellent lemon pie, he cooks very little else, and he would gladly eat steak and potatoes at every meal, including breakfast. So we can only answer the question of whether he is an outlier on a variable-by-variable basis.

As you may recall from Chapter 2, outliers can be a very serious problem because one or two outliers can sometimes have a huge effect on whether a particular statistical test proves to be significant. Although no one is in favor of spurious research results, we should note that there is a wide range of statistical opinion about the best way to deal with outliers. For example, researchers who make frequent use of structural equation modeling consider it an absolute necessity to screen carefully for multivariate outliers before they do any sophisticated model fitting. In fact, failing to eliminate multivariate outliers (to be defined below) can sometimes make sophisticated model fitting mathematically impossible (e.g., see Kline, 2005). In contrast, experimenters who rely on simple t tests or analyses of variance (ANOVAs) are often much more reluctant to remove research participants who happen to have extreme scores on a dependent variable. In fact, experimenters often engineer situations in which it is *impossible* for participants to have extreme scores on any *independent* variables—because they design the simple, categorical conditions (e.g., hot room or cool room, aggressive primes or control primes) to which they randomly assign research participants. The ubiquitous 5- or 7-point Likert scales that experimenters find so appealing also reduce—but do not eliminate—the risk that experimenters will have to deal with any outliers on their *dependent* variables (especially if experimenters pilot test their measures for problems such as restriction of range). When you combine all this with the fact that tests such as the t test are pretty robust to some of the more common violations of normality, it should be easy to see why some conscientious experimenters are extremely reluctant to *ever* delete *any* participant's data from their experiments. In fact, when laboratory experimenters do delete participants' data, it is not usually because the participant's data are outliers. Instead, it is usually because the participants (a) were suspicious of a cover story or (b) failed to follow instructions.

It is reasonable to be concerned when a large percentage of participants in an experiment either failed to believe a cover story or failed to follow instructions (e.g., Does this mean that the experimental results only generalize to people who are easily duped? Can we be positive that the criterion for "failing to follow instructions" was set without any knowledge whatsoever of which participants tended to confirm the experimental predictions?). Without trying to resolve any thorny ethical or philosophical debates, let us merely affirm (a) that (unconscious) experimenter bias is always waiting to rear its ugly head and (b) that if others cannot replicate your experimental findings, the findings will ultimately have little scientific impact. This means, for example, that good researchers should be just as interested in deleting outliers who happen to *confirm* their predictions as they are in deleting outliers whose scores happen to be highly inconvenient. In other words, to paraphrase the primary author's former UCLA colleague Thomas Wickens, you should be reluctant to trust the results of any supportive inferential statistical test if you could render it nonsignificant by deleting a couple of highly influential cases. This point is applicable to all kinds of research, not merely experimental research. To get back to the main point, outliers are usually a bigger, and trickier, worry in multivariate research than in experimental (often univariate) research. For this reason, we pay only brief attention in this chapter to univariate outliers while paying much more attention to multivariate outliers.

Before we discuss this distinction, though, it is worth noting that some kinds of outliers are much easier to deal with than others. For example, Tabachinck and Fidell (2007, p. 73) identify four different reasons why a score could be an outlier. The first two reasons are (a) the score was entered or coded erroneously or (b) the person who created a data file forgot to specify missing value codes—for example, an unsuspecting user confused a missing value score (such as 99 or 999) for a real score. No one, not even "The Rain," would be reluctant to drop or correct such erroneous scores in the first case. The second case is even easier to fix (by simply entering the appropriate missing value codes). The third reason why a score may be an outlier is that the participant who has the outlying score never should have been sampled in your study (e.g., you accidentally sampled a man named Stacy in your Internet survey of women's responses to work stress). This, too, is easy to fix by simply dropping the erroneously sampled participant.

The fourth reason for outliers is where things get tricky. This fourth reason is that a participant was truly a member of the population you wished to sample but "the distribution for the variable in the population has more extreme values than a normal population" (Tabachnick & Fidell, 2007, p. 73). We hasten to add that even if you know that the population is normally distributed on the variable of interest but you just happened to sample an extreme outlier or two (because of dumb luck), this does not solve the problem. In other words, we think things are even worse than Tabachnick and Fidell (2007) suggested based on this perspective. When this fourth problem occurs, the best approach is often to retain the offending scores but to alter the scores to make them less extreme—so that no single score has an undue influence on whatever statistic(s) you are calculating. However, as Tabachnick and Fidell astutely note, "Deciding between alternatives three and four, between deletion and retention with alteration, is difficult." In other words, because you often do not know for sure whether an outlier came from the wrong population, was the result of a coding error, or is a valid but überextreme score, it is not easy to know what to do with some outliers. Perhaps the most important rule of thumb we can offer is transparency. Whatever you do, be explicit about it so that the consumers of your statistical work can come to their own conclusions. In the best of all possible worlds, you might be able to report that your results change very little regardless of whether you delete one or more outliers, truncate them, or retain them all without alteration. In the more realistic world of small samples and sometimes inconvenient truths, you might have to report that some analyses are highly supportive of your predictions, whereas others yield only highly qualified support or no support at all.

Chapter 14 Data Cleaning

Identifying and Dealing With Univariate Outliers

If your research question is very simple (i.e., if you are reporting a simple mean or the results of an independent samples t test), it is very easy to identify a simple (univariate) outlier. A univariate outlier is a case in your sample that is extreme on only one variable (or at least on the one and only variable on which you happen to be focusing). Many statisticians have agreed that if your sample size is small ($n = 80$ or lower), a score is a univariate outlier if the score lies more than 2.5 standard deviations from the sample mean. Note that this means that if you convert your raw scores to z scores, any score whose z score–converted absolute value is 2.5 or greater ($|z| \geq 2.5$) is an outlier. For samples larger than 80, statisticians often raise the bar to a z score equivalent of ± 3.0 or even ± 4.0. If you're merely interested in central tendency, one easy solution to the problem of univariate outliers is simply to note that a distribution is skewed (as discussed in detail in Chapter 2) and then to report the median score as well as the mean. As you probably recall, the median is relatively impervious to outliers. If your research question is one step more complicated and you'd like to make a statement about group differences in means (e.g., by conducting an independent samples t test), it is useful to see if you have any univariate outliers on your *dependent measure*. If you do, one simple solution to the problem is to **winsorize** your scores on the dependent variable so that all scores whose z score equivalent is 3.0 or greater are trimmed to whatever score corresponds to a z score of 3.0. Finally, if you are correlating two continuous scores (e.g., height and weight, income and longevity in years) and are worried about outliers, winsorizing both the independent variable and the dependent variable is a good way to protect yourself from the undue influence of extreme scores (see Erceg-Hurn & Mirosevich, 2008).

Although it is beyond the scope of this intermediate text to cover every possible worry regarding outliers, it is worth noting that outliers come in more than one form. In fact, as we suggested earlier, the more sophisticated your statistical analysis is, the more likely it is that you might have concerns about outliers. Unfortunately, simple (univariate) outliers are not the only kind of outliers. As soon as you graduate from univariate statistics (such as t tests) to multivariate statistics (such as multiple regression analysis, path analysis, and confirmatory factor analysis), you have to worry about **multivariate outliers.** Multivariate outliers turn out to be a bit harder to identify, and thus harder to correct, than univariate outliers.

Identifying and Dealing With Multivariate Outliers

The first author used to be on a track team with Scott Jones, who frequently referred to himself as the "world's fastest redhead." Scott admitted that he was, at that time, about 0.8 seconds shy of being the world's fastest *person* (in the 100 meters). However, based on Scott's personal theory that redheads are notoriously slow, he took some pride in his unusual *combination* of speed and redheadedness (we will for now conveniently skip the question of Scott's red*neck*-ness—because those who live in glass houses shouldn't throw stones). If we remember that, as an athlete, Scott's goal was to be an extreme outlier, this example provides a nice illustration of the meaning of multivariate outliers. In a sense, Scott admitted that he was not an *extreme* univariate outlier. If you simply focus on running speed alone, that is, Scott did not quite make it to the highly rarefied athletic air to which he aspired. However, if you consider the highly unusual combination of being both quite fast and quite redheaded, Scott *was* an extreme outlier (especially if you accept his implication that the mean sprinting speed for redheads is well below the human average). In statistical terms, Scott was arguing that it is possible for someone to be a multivariate outlier without being a univariate outlier on any of the specific traits that go into making someone a multivariate outlier. Tabachnick and Fidell (2007) would have agreed. As they put it, "Multivariate outliers are cases with an unusual combination of scores on two or more variables" (p. 73). It is implicit in such a definition that the two or more variables are often correlated in a specific way that makes the multivariate outlier an outlier. For example, Tabachnick and Fidell note that (a) being 15 years old does not make one an outlier in age and (b) making $45,000 per year does not make one an outlier in income. However, they add that a 15-year-old who makes $45,000 per year probably *would* be a multivariate outlier—because it is so unusual for a person so young to make a substantial income. At age 15, even the surprisingly speedy (for a redhead) Scott Jones could not have delivered enough newspapers to generate that kind of yearly income.

Having noted this interesting possibility, it is important to add that if a case in your sample is a univariate outlier on one or more of the individual variables that you are grouping together to decide whether a case is a multivariate outlier, this does, in fact, make it much more likely than usual that the case in question will be a multivariate outlier. For example, if you were predicting life satisfaction from (a) height and (b) income, both Shaquille O'Neal and LeBron James would almost certainly be multivariate outliers, because both Shaq and LeBron are univariate outliers on both of the two variables that you decided to include as predictors in your multivariate statistical analysis. Furthermore, they would be multivariate outliers *regardless of which dependent variable (aka criterion) you decided to predict.* In other words, whether a case represents a multivariate outlier is based on the extremity of and intercorrelations among each of the individual predictor variables under consideration, without regard for what variable you are predicting or how well you can predict it. This means, for example, that you can run diagnostic tests to see if a case in your data file is a multivariate outlier by predicting anything

Chapter 14 Data Cleaning

whatsoever in a multiple regression model from the specific predictors you care about.

Another important thing to know about identifying multivariate outliers is that doing so is an *iterative* process. You can't just identify all the multivariate outliers in a data set, eliminate them, and be done with it. Assuming that your initial approach is to delete multivariate outliers, let's further assume you are able to identify three extreme multivariate outliers in your data set. Once you have deleted these three outliers, you would need to run the same diagnostic test for outliers *again* to see if any scores that did not qualify as multivariate outliers before you deleted the three problematic cases now *do* qualify as multivariate outliers. You'd need to repeat this iterative process until you reach a point at which there are no more multivariate outliers. Only then are you done.

To see exactly how this works, let's identify some multivariate outliers in a data set inspired by Malcom Gladwell's (2008) best-selling book *Outliers*. In this intriguing book (which has much more to do with cultural and social psychology than with statistics), Gladwell tackles many interesting questions about high performers. One of the most interesting is why Asians perform so well in school, especially at math. In addition to noting, for example, that many Asian languages make it really easy to think about, talk about, and remember math-related words, Gladwell also argues that in many Asian countries, a lot of farmers happen to be rice farmers. Farming rice turns out to be extremely labor intensive. A rice farm can sometimes be as small as a large room, but if it is painstakingly nurtured, it has the potential to support an entire family. Gladwell's argument is that the powerful cultural norms about discipline, mastery, and self-sufficiency that have evolved over the centuries in rice farming cultures virtually guarantee that people in such cultures internalize the values of hard work, learning, and achievement. He thus argues, for example, that the main reason why Asians are good at math is cultural rather than biological. Furthermore, he identifies the cultural trappings of rice farming as one of the key historical drivers of Asians' superior intellectual performance and focus on learning. We will have to forego some of the details of this provocative cultural argument. Suffice it to say, though, that we found this argument interesting enough that we decided to put it to a rough-and-ready test. The country-level SPSS data file [**UNDP Gallup learn & grow and USDA world rice production data feb 25 2011.sav**] contains several real variables that will allow us to do so.

In the interest of full disclosure, we must tell you that we were a little sloppy with one of the important variables in this data set. But let's review all of the variables. First, because countries rather than people are the unit of analysis, *country* identifies each country in the sample. Second, *GDP* is 2005 per capita gross domestic product, adjusted for true purchasing power (purchasing power parity [PPP] corrected) and converted to U.S. dollar equivalents. We harvested the variable *learn* from the Gallup Worldview website, where you can learn all kinds of things about more than 150 countries worldwide. The learn variable is the percentage of those surveyed in each country, circa 2008, who reported that "most children in _____ [respondent's country] have the opportunity to learn and grow every day." The advantage of this variable over a more objective variable (such as national averages on standardized test scores such as the PISA [Programme for International Student Assessment]) is simply that it was available for more than 100 countries worldwide when we harvested the data in 2008 and 2009. Finally, using national rice production records available from the U.S. Department of Agriculture (USDA; http://usda.mannlib.cornell.edu/usda/fas/grain-market//2000s/2007/grain-market-12-01-2007.pdf), we were able to identify the 16 countries worldwide that produced the greatest amounts of rice in 2007. The raw rice production variable (expressed in thousands of metric tons) is rice_production2007. We then divided this total rice production variable by the population of each country to correct for the obvious fact that more populous countries should be expected to produce more of anything. The statistic we care about most, then, is "rice_per_cap." A potentially serious limitation of this measure is that the USDA provided rice production figures only for the world's top 16 rice producers (one of which, Burma, was not surveyed by Gallup). However, because these top 16 countries produced 93% of the world's rice in 2007, we can *estimate* the rice production of the other 185 or so countries in the world to be zero (for every country). Needless to say, this estimate of zero for all of these other countries will be off the mark in a few rice-producing countries. However, for the purpose of the present exercise, this imperfect estimation approach yields a set of $n = 101$ worldwide rice production scores that are decidedly nonnormal (i.e., the distribution is extremely positively skewed). How better to learn to deal with multivariate outliers than to work with a slightly contrived data set that is (a) inspired loosely by the book *Outliers* and (b) chocked full of outliers?

Image 10 shows you the first few cases in your SPSS data file, and it is worth noting that the countries have already been ranked for you based on rice_per_cap. You now know, for example, that in 2007, Thailand produced more rice per capita than any other country in the world. (Remembering

	country	GDP	learn	rice_production2006to7	rice_per_cap	gotrice	va
1	Thailand	$8,677	.87	18250	.00028196		
2	Cambodia	$2,727	.85	3946	.00027497		
3	Vietnam	$3,071	.86	22922	.00027253		
4	Bangladesh	$2,053	.69	29000	.00020091		
5	Indonesia	$3,843	.82	35300	.00015827		
6	Philippines	$5,137	.85	9775	.00011556		
7	China	$6,757	.93	127200	.00009697		

Chapter 14 Data Cleaning

that the original figures were in thousands of metric tons, that was 1,000 × .00028196 metric tons, or about 600 pounds of rice per person in Thailand in 2007.) You might also note that none of the countries at the top of this list are very wealthy. This is important because, not surprisingly, the public perception that one's country is a place where most children have the chance to learn and grow every day is pretty highly correlated with national wealth (e.g., most people living in poor countries do not generally say that they live in a place where kids have the chance to learn and grow every day).

You now know why we're going to include both GDP and rice_per_cap as predictors in our simultaneous multiple regression analysis in which the criterion variable is "learn." We want to know if, controlling for national wealth, people living in countries that produce a lot of rice are more likely to report that their countries are places where most kids can learn and grow on a daily basis. However, because rice_per_cap is pretty badly skewed (+3.67), we should be prepared to deal with the problem of multivariate outliers. By the way, you may recall that you can sometimes reduce skew, for example, by taking the square root, the base 10 logarithm, or the inverse of a skewed variable. You should feel free to try these transformations, but we can tell you, for example, that the square root transformation will only help a little bit—and the log transformation won't help at all. This means you have little excuse but to learn the sometimes painstaking approach of identifying and deleting multivariate outliers. So let the painstakingness begin.

One popular way to find out if a specific case is a multivariate outlier on the variables in question is to calculate the **Mahalanobis D^2** statistic for that case, based on whichever set of variables you are treating as predictors. Fortunately, SPSS regression will calculate this D^2 statistic for every case in a data file. Even more fortunately, the D^2 statistic is really just a χ^2 (chi-square) statistic with degrees of freedom (df) equal to the number of variables under consideration. With this in mind, let's see how many multivariate outliers there are in these data, and then let's take the statistically aggressive approach of deleting them. Before we do so, however, remember that outliers are, by definition, extreme. Thus, statisticians usually set the bar pretty high for deleting multivariate outliers. Only cases whose D^2 score has a p value < .001 are considered multivariate outliers. It is worth noting that the critical value for chi-square for p = .001 with 2 df (two predictors) is 13.82. This is good to remember so that we can keep an eye on whether SPSS is doing its job properly when it converts our individual D^2 statistics to p values.

Because you have had plenty of practice running a linear multiple regression analysis, Image 11 cuts right to the point where you have already chosen to run a regression analysis. You can see that we selected both rice_per_cap and GDP as the predictors. Although we also selected learn as the dependent variable, recall that we could have chosen *anything* (so long as it is numeric) as the dependent variable. To get the crucial D^2

statistic that will tell you whether each case is a multivariate outlier, you need to click "Save..." (circled in Image 11). This will open the Save dialog box (see Image 12) where you can ask for a Mahalanobis distance score, and then click "Continue" from within the Save dialog box (not shown in Image 12). Once you get back to the main dialog window (Image 11), just click OK (circled in Image 11).

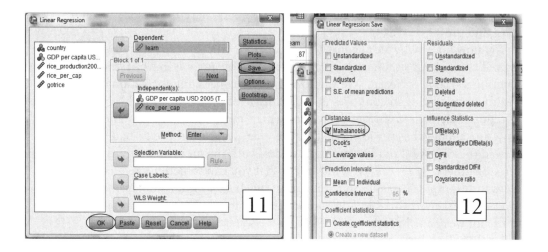

If all goes well, you'll see that your SPSS data file now has a D^2 (MAH_1) score next to each case.

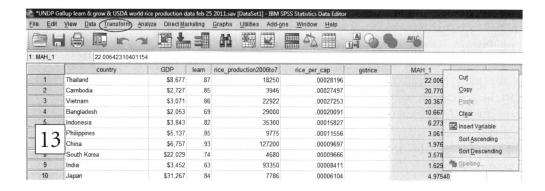

Before you go any further, right click on the MAH_1 (Mahalanobis distance) variable so that you can sort all of these cases (in descending order) on the D^2 statistic. If you remember that our critical value for $p < .001$ is 13.82 with 2 df, you'll be able to see, *after* the sort, that there seem to be three multivariate outliers. To be sure, though, let's get SPSS to provide us with the p values that correspond to each D^2 score. To do this, click on "Transform," which is circled in Image 13, and then "Compute Variable..." (not shown in Image 13). This will open the box shown in Image 14. From here, (a) type in "mah_prob" as the "Target Variable," (b) choose "Significance" from the

Chapter 14 Data Cleaning

Function group box, and (c) choose "Sig.Chisq" from the "Functions and Special Variables" box. Then send it to the "Numeric Expression" box by clicking on the little arrow circled in bold in Image 14.

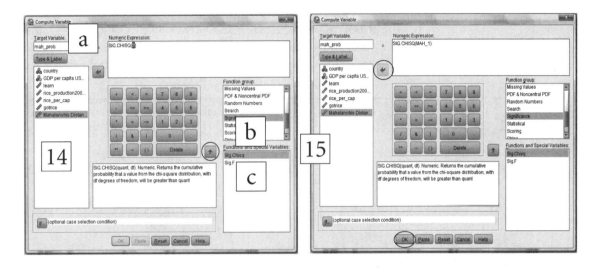

Now be sure to send the newly created variable MAH_1 to the left part of the parentheses under Numeric Expression (by clicking it and then using the arrow circled in Image 15). Then be sure to enter a comma and "2" (for two variables) for the *df* part of the parenthetical expression. In the end, your numeric expression should read "SIG.CHISQ(MAH_1, 2)." Now if you click OK (circled at the bottom of Image 15), SPSS should send the *p* value that corresponds to each D^2 statistic to your SPSS data file. Unfortunately, the default for newly created SPSS variables is only two decimal places. So, as shown in Image 16, you'll need to switch to the Variable View mode and change the variable that you called "mah_prob" so that it has more decimal places (we chose five). Image 17 shows you what scores for this new "mah_prob" variable will look like after you switch back to the Data View mode.

From the Data View mode, you should be able to see that 3 of your 101 cases do, in fact, qualify as multivariate outliers, because their *p* values are all less than .001. As shown in Images 18 and 19, the best way to delete these 3 problematic cases is to use the familiar SPSS "Data... Select Cases..." function. Once you get to the point where you specify your selection criteria, specify "mah_prob gt .001" as your selection rule. Once you've clicked "Continue" and then "OK", you should see (as shown in Image 19) that the three cases that are multivariate outliers have been temporarily deleted from use in the data file.

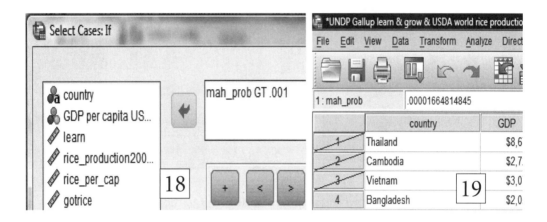

Because deleting multivariate outliers is an iterative process, we are not done yet. Instead, we should start over and do everything we just did on the new, slightly reduced sample—to see if any cases that were not multivariate outliers in the *original* sample have now become multivariate outliers in the *reduced* sample. To do this, you can simply run the same regression analysis you just ran (on the new sample). Furthermore, to minimize confusion, SPSS will automatically give the *new* Mahalanobis distance scores a new variable name: MAH_2. This will require you to make two important changes when you compute the new probabilities that correspond to the new D^2 scores. First, as shown in Image 20, create a new target variable called "mah_prob2." Second, make sure that you base this new variable on MAH_2 rather than MAH_1.

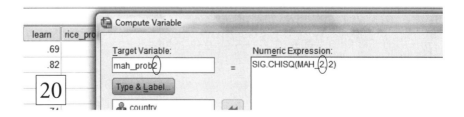

Once you have (a) created this new D^2 statistic for every case in your revised data file and then (b) converted each new D^2 score to a new *p* value, you can sort the data file in descending order based on the new D^2 scores and see if there are any more multivariate outliers. If all went well (e.g., if you remembered to edit "mah_prob2" from the Variable View mode so that it has five decimal places), you should see that two new cases prove to be clear multivariate outliers in the reduced sample (because their "mah_prob2" *p* values are less than .001).

QUESTION 14.3. Identify which two countries now prove to be multivariate outliers in the second iteration of your search, and then report the D^2 scores (for MAH_2) and associated *p* values for each country.

QUESTION 14.4. Now repeat this entire iterative process once more and identify the three countries that prove to be multivariate outliers in the *third* iteration of your search. Report each country's D^2 score (for MAH_3) and its associated *p* value.

Before you get stuck in this endless hall of mirrors, we'd like to point out that things don't usually go *this* badly when you iteratively delete multivariate outliers. Eight percent data loss so far may not sound horrendous, but let us point out that *all eight* of the cases you have already lost from your original sample of about 100 countries were members of the group of only *15* countries that produce a lot of rice! The whole idea of this analysis is to compare rice-producing countries with all other countries, and you are rapidly losing *all* of your rice-producing countries! Fortunately, there is an easy way out of this dilemma. When one independent variable is both badly skewed and resistant to correction (as is rice production), researchers often find a way to create meaningful categories and thus split the sample into one or more groups based on this independent variable. Although it is less than ideal when one of the categories has a lot more members (i.e., cases) than the other(s), creating categories allows you to conduct a *t* test or ANOVA when you have only one independent variable. We hope it has occurred to you, though, that you have two independent variables, and thus, analysis of covariance (ANCOVA) (or multiple regression), rather than ANOVA, is the appropriate analysis. To save you a little trouble, by the way, we should tell you that, unlike rice production, GDP is not so badly skewed. Thus, there should be no problem conducting an ANCOVA in which you (a) make all 15 of these rice-producing countries one group, (b) make the remaining countries the other group, and (c) treat GDP as a covariate in your ANCOVA. Notice that this approach also has the advantage of not requiring you to throw out any cases.

QUESTION 14.5. Consulting Chapter 11, if necessary, analyze the original data using ANCOVA, *without* deleting any outliers, and report the covariate-adjusted means, the *F* statistic, and the *p* value that reflect the unique association between rice production and public opinion about whether one's country is a place where most kids can learn and grow every day. While you are at it, report whether your covariate is significant and make a brief comment about what this means (supported by the appropriate F and p value for the covariate). Do the findings support Gladwell's position?

Putting Your Data-Cleaning Skills to Work

To get one final look at what a multivariate outlier is and how one goes about cleaning multivariate outliers, we would like you to explore a hypothetical longitudinal data set focusing on a military sample whose health was assessed when they entered the military for basic training. Because the study is hypothetical, we were able to follow the recruits while they aged, and ultimately died, without dying ourselves. The outcome measure in this study, then, is longevity in years, and we would like you to predict longevity in a multiple regression analysis from some of the variables that are known (or strongly suspected) to contribute to longevity. These predictors (all measured before age 25, among new recruits) include (a) gender, (b) smoking status, (c) education level, (d) body mass index (loosely speaking, weight relative to height), (e) personal best time in a 5-k time trial, (f) frequency of aerobic exercise (measured in average number of days per week recruits reported that they exercised in the year *prior* to entering the military), and (g) social support (measured dichotomously, meaning that the person either reported having or not having someone they could definitely count on for emotional support during a difficult life event).

The data file for this activity is called [**hypothetical military longevity study for missing data & outliers.sav**]. Before you can find out the relative contribution of these seven predictors to longevity in this hypothetical longitudinal study, you'll want to make sure that there are no multivariate outliers that could be unduly influencing the results. To help you recover from the trauma of having never made it through the multivariate outlier iteration in the last data set, we are going to tell you that you will need to conduct only one or two multivariate outlier search iterations this time. We would also like to remind you that because you now have *seven* rather than *two* predictors in your regression analysis, the df for calculating p values for your D^2 scores will be 7 rather than 2.

QUESTION 14.6. Report the D^2 score and p value for the one (and only one) multivariate outlier you identify.

Now that we've made this so easy for you, we'd like to make things a little harder by asking you to find out for how many of the seven predictor variables under consideration this case proved to be a univariate outlier. Because your sample size is greater than 80, we'll set the bar for any specific univariate outliers at $z = 3.0$. This may sound like a pain to calculate z scores for seven different variables, but now that you have stuck with us for so long in this text, we'd like to reward you by showing you one of the easiest ways

Chapter 14 Data Cleaning

to get SPSS to calculate z scores for you. Furthermore, because gender, smoking status, and social support are all dichotomous (i.e., not interval or ratio) variables, you can leave them off the list. Until the first author's 8-year-old son gets his way and makes all the girls in the world move to Kamchatka, there will be no outliers on gender, for example. So you need to generate z scores for just four continuous variables. Images 21 and 22 show that you can easily do this by (a) asking for "Descriptives..." on the four variables of interest and (b) asking SPSS to "Save standardized values as variables" by clicking the crucial little box that is circled in Image 22.

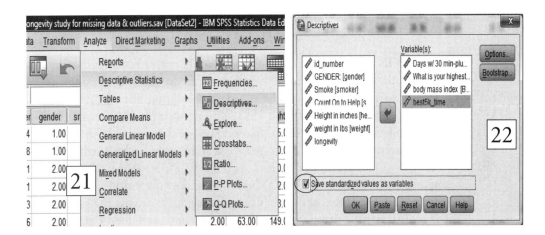

Once you have done this, and clicked OK, you should see that SPSS did, in fact, add the z score equivalents of these four continuous variables to your SPSS data file. The variable formerly known as educ still exists, but it is now accompanied by a z score equivalent known as Zeduc. Similar things should have happened to the other three variables. Now if you simply rank the data file on Zeduc (by right clicking in the Zeduc column), you'll be able to see that there were *no positive or negative outliers on education*. The standardized education scores range from a low of $z = -1.50$ to a high of $z = +1.82$.

QUESTION 14.7. Rank and inspect the other three continuous variables that were a part of the test for multivariate outliers. Then report, for each of the three variables, what the highest and lowest z score was and whether any of these eight z scores was extreme enough to make any case qualify as having a univariate outlier on that specific variable.

QUESTION 14.8. Now that you have documented that no single case in the entire data file possessed a univariate outlier on any one of these four continuous scores that determined whether there was a univariate outlier, explain how it is possible that you did, in fact, have a clear multivariate outlier. The best way to figure this out is to reidentify the multivariate outlier by just reranking the cases on MAH_1. Next, check out the four separate univariate *z* scores of this person, and then look carefully at this participant to see what makes her so unusual. As a pretty big clue, let's just say that this person, if she were real, would have a lot more to brag about than Scott Jones (aka the world's fastest redhead) does.

QUESTION 14.9. Delete this one multivariate outlier (preferably by using the "Select Cases" command). Then check to see if there are any other multivariate outliers, and delete them if there are any. Once you have deleted all of the multivariate outliers, run, report, and interpret the results of a simultaneous multiple regression analysis predicting longevity from the seven predictors. Spend no more than one sentence to interpret each specific regression coefficient.

QUESTION 14.10. Notice that although you had very little missing data in this hypothetical study, you did have a case or two that was missing a score or two. As a result, you should have had a final sample size of $n = 82$ (rather than the original $n = 88$) after (a) disqualifying your one outlier and (b) losing five participants who did not have complete data on all of the variables in the analysis. To reassure any persnickety reviewers who might be concerned about any form of data loss, we'd like you to run the basic regression analysis again while including (a) both the outlier you temporarily excluded and (b) the five cases lost due to incomplete data. To do this analysis, simply (a) reset or turn off the "Select Cases" command, and then (b) rerun the same linear regression analysis while choosing "Replace with mean" rather than the default "Exclude cases listwise" as the option for dealing with missing data (see Image 23, beginning with the "Options" button).

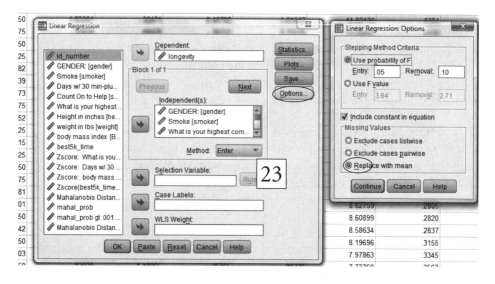

How much did your results change compared with the results from the $n = 82$ sample? In what sense is this troubling or reassuring?

A Final Worry: Multicollinearity

Although it is beyond the scope of this intermediate text to cover every imaginable issue that is relevant to data cleaning, it is worth reminding readers briefly about one other problem that is closely related to the problem of outliers. This problem is **multicollinearity,** which we discussed briefly in Chapter 12. Multicollinearity is similar to the other problems discussed in this chapter (e.g., outliers) in that it represents yet another problematic route by which only one or two scores can have a disproportionate influence on the results of a multivariate statistical analysis. As you may recall, multicollinearity means that two or more variables in a multivariate analysis are so highly correlated with one another that changing just a score or two (whether on just one or more than one variable) could seriously change the results of the multivariate analysis. Another way to put this is that multicollinearity means that two or more predictors are so highly correlated with one another (e.g., $r = |.85|$ or higher) that it is statistically questionable whether it is reasonable to consider them measures of different things. If Larry's new measure of extraversion correlates $r = .96$ with the Rosenberg Self-Esteem Scale, it is going to be extremely hard for Larry to convince people that he is measuring something different than self-esteem. More important even with a very large sample, it is going to be hard to document anything reliable that this new measure predicts that the Rosenberg self-esteem measure doesn't *also* predict. The results of an analysis that included both predictors might shift dramatically, for example, if just one person who scored higher on the "extraversion" measure than on the self-esteem measure had scored a bit higher on the self-esteem measure than on the extraversion measure. If this participant is one of the people whose ambiguous handwriting was featured at the very beginning of this chapter, we hope you can see that Larry's potential insensitivity to multicollinearity might lead to some very serious mistakes.

The following appendix illustrates how dramatically a set of results can change if you merely change one score on one predictor that is very highly correlated with another predictor and how the tolerance or VIF (variance inflation factor) statistics in SPSS can help you flag this potential problem. We hope that, like the rest of this chapter, the appendix will help you develop a good sense of why data cleaning is such an important part of being a good statistician.

For Further Reading

Erceg-Hurn, D. M., & Mirosevich, V. M. (2008). Modern robust statistical methods: An easy way to maximize the accuracy of your research. *American Psychologist, 63,* 591–601.

Appendix 14.1: An Illustration of Multicollinearity

Multicollinearity can wreak havoc on the size, direction, and statistical significance of regression coefficients. As a concrete illustration of this problem, consider the data set [**multicollinearity demo (popularity) feb 27 2011.sav**]. If you carefully examine the data in the first two columns of the SPSS data file shown in Image 24, you'll see that in these hypothetical data, there is an *extremely* high correlation between believing that one is "socially skilled" and believing that one is "outgoing." In fact, if it were not for Participant 12, these two self-report measures would be perfectly correlated.

	socially_skilled	outgoing	popular
1	1.00	1.00	5.00
2	2.00	2.00	4.00
3	3.00	3.00	3.00
4	4.00	4.00	4.00
5	5.00	5.00	6.00
6	5.00	5.00	6.00
7	5.00	5.00	6.00
8	6.00	6.00	7.00
9	7.00	7.00	6.00
10	8.00	8.00	8.00
11	9.00	9.00	9.00
12	10.00	8.00	5.00

Imagine that you were keenly interested in which of these two self-evaluations (socially skilled vs. outgoing) is a better predictor of a person's popularity (as rated independently by his or her peers). Image 25 shows what you'd find if you were to run a multiple regression analysis in which you predicted popularity (aka "popular") from these two very highly correlated predictors:

Coefficients[a]

Model		Unstandardized Coefficients		Standardized Coefficients	t	Sig.	Collinearity Statistics	
		B	Std. Error	Beta			Tolerance	VIF
1	(Constant)	2.735	.705		3.878	.004		
	self-rated "socially skilled"	-1.334	.551	-2.139	-2.420	.039	.039	25.723
	self-rated "outgoing"	1.951	.608	2.838	3.210	.011	.039	25.723

a. Dependent Variable: popularity rated by peers

According to the results of this multiple regression analysis, believing that one is outgoing has a positive unique association with popularity, whereas believing that one is socially skilled has a negative unique association with popularity. Before we accept this conclusion, however, we should consider the extremely large VIF score of 25.72. A VIF greater than 10 is usually a red flag that multicollinearity could be a serious problem. To see just how big that problem is in these data, imagine that Participant 12 had reported that she was an 8 on socially skilled and a 10 on outgoing (rather than the reverse 10, 8 that you see in the data). If it is not obvious to you how the results of the multiple regression analysis would change if you switch these two scores for this one participant on the two predictor variables, then you should edit the SPSS data file to fit this new pattern and then carefully compare the new results with the original results (making sure to ask for "Collinearity diagnostics" when you click the "Statistics..." option from the main analysis screen).

If you are curious to explore multicollinearity a bit further, you might also wish to see what happens if, instead of switching the 10 and the 8, you keep the data from Participant 12 just as you found it but add a 13th participant who has the reversed pattern of predictor scores (as depicted in Image 26). It should be instructive to examine both (a) the change in the regression coefficients and their associated p values, (b) the change in the VIF scores, and (c) the change in the adjusted R-square value compared with the original analysis. Finally, this example might also serve to remind you that tolerance (another popular indicator of multicollinearity) is merely 1/VIF (see the right-hand portion of Image 25 to make the appropriate comparison).

	socially_skilled	outgoing	popular	var	var
1	1.00	1.00	5.00		
2	2.00	2.00	4.00		
3	3.00	3.00	3.00		
4	4.00	4.00	4.00		
5	5.00	5.00	6.00		
6	5.00	5.00	6.00		
7	5.00	5.00	6.00		
8	6.00	6.00	7.00		
	7.00	7.00	6.00		
	8.00	8.00	8.00		
11	9.00	9.00	9.00		
12	10.00	8.00	5.00		
13	8.00	10.00	5.00		

Image 26

15
Data Merging and Data Management

Although most of the skills required to be a good researcher are either statistical or methodological, some are also practical, interpersonal, or even clerical. For example, good experimenters usually make careful notes about the behavior of the individual participants they run in the laboratory (e.g., "This guy seemed to understand the instructions, but he showed up wearing nothing but a feather boa and smelling strongly of gin. Also, he insisted that I call him 'the artist formerly known as Larry'"). Along similar lines, survey researchers who conduct telephone interviews sometimes judge the degree to which each of their specific interviewees seemed to understand their instructions—or seemed suspicious or distrustful during the interview. Interviewer ratings such as these can serve as an important source of data, playing somewhat the same role that manipulation checks play in a laboratory experiment. A detailed discussion of all the practical skills one needs to be a good researcher is beyond the scope of this text (but see Pelham & Blanton, 2013). However, a discussion of basic clerical (i.e., data management) skills is not. In this chapter, you will be introduced to some of the most important clerical skills that you are likely to need as a researcher: (a) how to merge more than one data set collected from the same group of participants and (b) how to merge more than one identical (or nearly identical) data set collected from two different groups of participants. This first situation happens by definition in longitudinal or prospective survey research, but it also happens in many other contexts. For example, laboratory experimenters and quasi-experimenters often make use of pretest data to select or categorize specific participants who are eligible for their research. Thus, a clinical psychologist studying spider phobias might pretest hundreds of PSY 101 participants and then select only those participants who scored above a certain threshold on self-reported fear of spiders. She could then bring all of these potentially spider-phobic participants to her experiment, make sure that they are in fact spider phobic (by means of a structured clinical interview), and then examine the relative efficacy (i.e., effectiveness) of two different treatments for spider phobia. As another example of the first situation, the graduate teaching assistant who needs to keep track of the grades of students enrolled in a large PSY 101 course might need to merge all of the quizzes and/or exam grades for every student in the course. In a class of several hundred students, this is no small feat!

As a concrete example of the second situation, suppose you and a collaborator at a different university ran identical versions of the same laboratory experiment, focus group, or survey study. If you wished to pool your efforts, you would eventually want to merge the two identically formatted data files before conducting your final statistical analyses. Or imagine that you often assess three or four variables in almost all of your survey research. After conducting studies for several years, you might wish to pool the overlapping parts of all of your studies. Doing this would allow you to conduct secondary analyses of your data with a larger and possibly more diverse sample than you ever had in any one specific study. To do this, you would obviously need to merge the different independent data files.

Clerical activities such as these require a good knowledge of how to manage data files. We know of no textbooks in statistics or research methods that teach researchers the basic skills it takes to do this. Thus, we hope this brief chapter will serve at least part of that basic function. Although there are many reasons why you might need to merge different data files, these reasons almost always boil down to some variation on the two kinds of situations we just summarized. Usually, researchers who need to merge data files need to (a) add variables (new scores or measurements) to an existing data file or (b) add cases (new participants) to an existing data file. The developers of SPSS felt this distinction was important enough that they created a different function in SPSS to handle each of these situations. Because the first situation is usually a lot trickier than the second, we will only focus on that first situation in this activity (merging two different data files, each of which contains different variables from the same set of participants). If you can master this first technique, the second technique will usually be a piece of cake. So let's begin with this first situation, the need to add *variables* to an SPSS data file.

Imagine that you are a personality researcher who has collected pretest data on "the Big Five" from a large group of undergraduate students. The Big Five refers to five basic dimensions of human personality. Some researchers believe that almost all personality traits represent subtle variations on one of these five core dimensions of human personality (e.g., see Goldberg, 1990). Although different researchers use slightly different labels for the five traits, an easy way to remember the five traits is to use the acronym OCEAN, which stands for Openness to Experience (sometimes known as "Intellect"), Conscientiousness, Extraversion, Agreeableness, and Neuroticism (also known as "Emotional Instability"). Regardless of whether these five traits truly capture everything that is important about human personality, we hope you can see that the five traits do cover a lot of ground. If your friend Mauricio is highly open-minded, conscientious, extraverted, agreeable, and not at all neurotic, you can imagine that taking a vacation with him would be quite different than taking a vacation with his friend Brett—who is closed-minded, unreliable, introverted, disagreeable (when you can get him to speak up), and highly neurotic. Mauricio agrees to take that limbo class with you, and he shows up on time for it. That evening, he invites you out for a couple of mai tais. While the two of you sit in the cabana, Mauricio freely shares his thoughts and feelings about his total inability to limbo, cracking witty jokes about the part of his anatomy that got in the way the most when he tried to limbo. In contrast, Brett considers the limbo class a waste of money. When you offer to pay for it, he reminds you that he hates all forms of dancing, and you recall that although Brett won't admit it, he is too neurotic to ever dance in public. When you remind Brett that he promised you to try at least one new thing this vacation—and that he has already turned down seven other new things you suggested—he begrudgingly agrees to take the class. Of course, he is 25 minutes late for the 50-minute class, and he berates the instructor for being unclear about her instructions—which were crystal clear to everyone who showed up on time. That evening, he refuses to go out with you because he is still steaming over how you embarrassed him and wasted his time at the limbo class. To get back to the point, the Big Five does cover a lot of ground, including a lot of beach.

OK, so assume that you asked a big group of college students enrolled in a psychology class to report what their personalities are like. Because of your enduring interest in self-esteem, imagine that you conducted a follow-up study several weeks later (on this same group of participants) in which you assessed both explicit (conscious) and implicit (unconscious) self-esteem. We'd be surprised if no one has ever assessed the relation between the Big Five traits and explicit self-esteem. As far as we know, however, no one has ever assessed the relation between the Big Five personality traits and *implicit* self-esteem. Could we learn anything about the difference between implicit and explicit self-esteem by assessing how these two different forms of self-esteem are related to the five core dimensions of personality? To find out, we'd need to merge your Time 1 personality data and your Time 2 implicit and explicit self-esteem data.

How could you blend your two data files together to see how the Big Five traits are related to explicit and implicit self-esteem? There are many ways (e.g., hand sorting, using Excel), but one of the easiest is to use the

"Data" button in SPSS. What follows is a summary of how to merge variables from two related but distinct SPSS data files. However, before you really get going, there are a few things you need to know up front. First, you begin the process of data merging in SPSS by opening *one data file*, to which you attach data from *another data file*. Which file do you open first? If you are merging data from multiple time points, it almost always makes the most sense to begin by opening whatever file you collected *first* (i.e., your "Time 1" data file). Otherwise, there might be some kind of logical reason to treat one file as the main file and a second file as the add-on file. Regardless of where you start, a second very important thing to remember before you begin merging is that you need what might be called a **merging variable.** *A merging variable is a variable that uniquely identifies each individual case in both of the data files that you wish to merge.* Without this key variable, you obviously cannot know which data points from one file match up with which data points from the other file. That is, you cannot be sure that Eric S. Jordan's Time 1 data are correctly matched up with Eric S. Jordan's Time 2 data. Merging variables are often identifying variables such as student ID numbers, Social Security numbers, or arbitrary participant numbers that have been consistently paired with the same individual participants in the two different data files.

Of course, if you wish to merge two data files, it is crucial that you have access to at least one such variable. It is just as crucial that in your two different SPSS data files, you (a) give this merging variable the same name and (b) specify the format of this variable in the same way. Finally, it is crucial that you have sorted this variable in the same way in each file before you ever begin working with either file. If you fail to do any of these three things, you will usually get an error message from SPSS telling you that you have made an unreasonable request of SPSS. (In older versions of SPSS, I seem to recall that the error message read, "I can't do this, you careless dweeb! Get out of my life and stop eating all my steak." In newer versions of SPSS, that message has been replaced by a more polite but equally frustrating message.) Now that you know all this, you are *ready to begin to get ready to begin* to merge your files. Yes, that's right: You are *ready to begin to get ready to begin* to merge your files. You'll soon see what we mean by that.

The file that we will call the original data file is called [**Mitsuru Big Five (Time 1) 2005.sav**]. The file that we will call the second data file (the one that is to be added to the first file) is called [**Mitsuru self-esteem (Time 2) 2005.sav**]. Open the first (Big Five) data file. Now click on the "Data" button and perform a "Sort Cases..." command based on your well-named, properly formatted merging variable ("person_number"). The screen captures you see in Images 1 and 2 should give you an idea of where to point and click to do this. When you are done, be sure to *save* the properly sorted (Time 1) file.

Chapter 15 Data Merging and Data Management

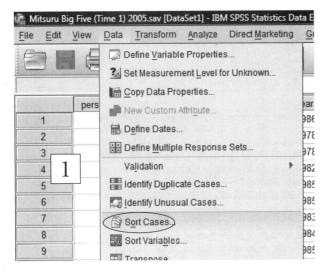 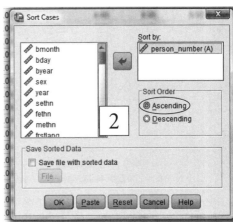

Next, you should repeat this sorting procedure for the second (self-esteem) data file, making sure you accept the default (ascending = small to big) sorting order, just as you did for the original file. You might also want to make sure that the sorting variable has the same name and the same "Numeric 8" variable format in the second file (it should). Once you've sorted the second data file, be sure to save it in its sorted form. By the way, if you are using Version 15.0 of SPSS, or a later version, you will be able to open and view both of these data files at the same time, which makes it a bit easier to merge them. If you have an older version of SPSS, you'll only be able to work with one open data file at a time, which will mean that you'll need to have opened, sorted, and saved the file you wish to merge to your original file before you open (or reopen) the original file to which you wish to add data. The screen captures you see in this chapter are based on SPSS Version 20.0, but the procedures you'll follow if you have an older version are very similar.

Once both of your SPSS data files are sorted and saved, you'll merge the two files by going to your original "Big Five" data file (the file whose variables you want to put at the beginning of your merged file). Once you are here, click on the "Data" button (the same place you clicked to start a sort). This time, however, select "Merge Files" from the drop-down menu. At this point, your menu options are "Add Cases. . ." versus "Add Variables. . . ." You're adding *variables*, so we bet you can guess which of these two commands you want. Once you click "Add Variables" (excellent choice), this will take you to a window, like the one you see in Image 3. This will allow you to select the SPSS data file whose variables you wish to add to the current SPSS data file.

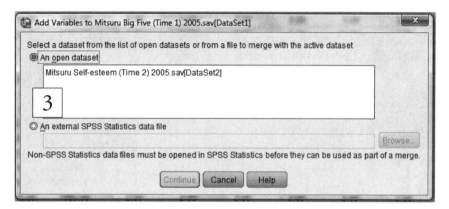

Because we already had the second data file open, SPSS wisely assumed that this was probably the data file we wished to merge with the original data file. This is why you see the "Self-esteem" data file listed in the white box in Image 3. Notice, however, that if the second (to-be-merged) file had been sitting around someplace unopened, we could have clicked on "An external SPSS Statistics data file" to find and merge *that* file instead. You'll probably need to click on the title of the SPSS data file you wish to merge (which will highlight it) before you can click "Continue." Once you click Continue, you'll see the dialog box that appears in Image 4. After clicking on "person_number" in the "Excluded variables:" list, click the "Match cases on key variables..." button (circled in Image 4).

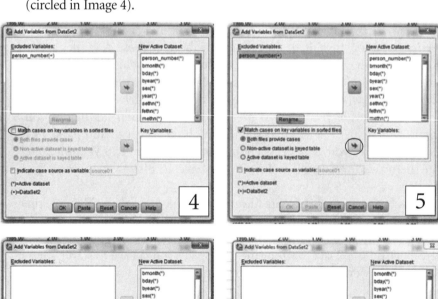

Then simply click the arrow circled in Image 5 to send the highlighted "person_number" variable to the "Key Variables:" list, where it appears in Image 6. Next, click OK and you'll see the final dialog box that appears in Image 7. This box will warn you that you must have sorted your variables in *both* data files for all of this to work out the way you had been hoping and dreaming since you were a child. Because you know that you have, in fact, sorted them both, you can feel confident clicking on the scary "OK" button circled in Image 7.

After clicking the OK button, you'll probably notice that nothing, whatsoever, seems to have happened. In fact, though, if you look at what used to be the end of your list of variables in your data file, you'll see that all of the variables that were once part of the second data file have now been added to your original SPSS data file. In this file, this means that whereas you once had 53 variables in your file (ending with "IS" and then "form"), you now have a long list of additional variables, beginning with se1 and se2. These variables, by the way, were the first variables in the second data file, and if you look at that SPSS data file, you'll see that they're still there.

At this point, it's very important to save this new merged data file under a *new* highly descriptive file name. We recommend a creative label such as **[Merged Big Five Plus Self-Esteem Data 2005.sav]**. Once you've successfully created the merged data file, you'll want to conduct some analyses on the merged data. Before you do this, however, you'll want to conduct a couple of tests to be sure that SPSS merged the two data files properly. Because each of these data files included a couple of basic demographic variables, let's correlate a couple of these variables at Time 1 with the same variables at Time 2. If both of the correlations are a perfect $r = 1.0$, we can feel pretty confident that we merged the data properly. Sex and exact day of the month on which a person was born represent two such useful demographic variables. In these data, the labels for these variables at Time 1 are sex and bday. At Time 2, the same variables are called, appropriately enough, sex2 and bday2. When you conduct a correlation between this set of four variables, you'll notice something a little troubling. Specifically, at least one of these two correlations will not be perfect. SPSS should give you a clue to the nature of the problem with the following warning message:

```
">Warning # 5132 Duplicate key in a file. The BY
variables do not uniquely identify each case on the
indicated file. Please check the results carefully."
```

To resolve this problem, you will need to identify the cases that have the same ID number and look carefully at the other demographic variables for these cases to figure out how to correct the problem. Of course, you could simply delete all of the cases that have the same ID number, but this is not the best possible solution.

QUESTION 15.1. Resolve this duplicate ID problem with a little logic and some detective work, and explain exactly what you did. Once you have done this, recall that the whole reason why we merged these data files was that we wanted to see if the different Big Five traits predict implicit and explicit self-esteem in different ways. You might also want to see whether implicit and explicit self-esteem are related to one another. Although you could probably guess how to accomplish these goals, we'll tell you that you should probably do at least two things. First, conduct a set of correlations between the seven variables (five personality traits and two self-esteem variables) you now have at your disposal. Second, to move beyond simple correlations to assess the unique relations among the various Big Five personality factors and self-esteem, you should conduct a couple of multiple regression analyses. Assume that one of your primary goals is to document differences between the nature of implicit and explicit self-esteem. From this perspective, you might want to know how each of the five personality variables uniquely predicts implicit self-esteem—above and beyond how the other four personality traits predicts it—and above and beyond any overlap between implicit and explicit self-esteem. You can then conduct the same kind of analysis, making explicit self-esteem the criterion (and including implicit self-esteem as a control variable). Do these correlations and regression coefficients tell you anything about the nature of implicit and explicit self-esteem?

16
Avoiding Bias: Characterizing Without Capitalizing

Introduction: Some Common Errors and Biases in Human Thinking

One of the ideas we have emphasized in this text is that statistics are tools that can help us characterize and thus better understand the nature of reality. Whether we wish to know if an experiment supports Piaget's idea of concrete operations or see which of two types of concrete holds up best to heavy foot traffic, statistical analyses provide us with some incredibly powerful tools for getting concrete answers to such questions. However, precisely because statistics (and researchers) are so flexible about how they code, transform, and analyze their data, even the most well-meaning researcher can sometimes observe seemingly beautiful support for a theory not because the theory is correct but because the researcher looked so hard for support for the theory that it would have been very hard *not* to find it. As the primary author's older brother loves to say, "Even a blind hog with a bad cold'll stumble on an acorn every once in a while." And as the author's youngest brother would be quick to add, "I'll see it when I believe it."

Before we discuss exactly what stumbling on acorns has to do with bias in statistical testing, it might be instructive to ask you to play a card game. So if you'll humor us for a few minutes, let's do that. Although not quite as much fun as the amazing card game *Cliffhanger*, this game is quite interesting—and it's very easy to learn. It has some relevance to statistics, by the way, because it requires you to test a hypothesis. Imagine that you are presented with the set of four cards depicted in Figure 16.1 and told that *every card has a letter on one side and a number on the other*. You may

Figure 16.1 A Card Game Involving Hypothesis Testing

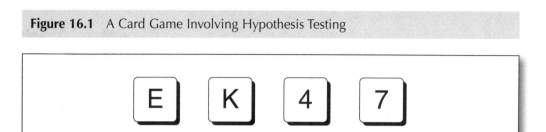

take this as a given (e.g., no cards contain letters on both sides). Having established this much, you would like to test a proposed hypothesis about the cards. That is, you would like to determine for certain whether the hypothesis is correct or incorrect. The hypothesis is that *every card containing a vowel must have an even number on the other side.* If this is true, the hypothesis is correct. If this is not true, then the hypothesis is wrong. Of course, one approach to testing the hypothesis would be to turn over all four of the cards, but this isn't very efficient. Not all cards need to be turned over to test the hypothesis. Your goal is to carry out a complete and accurate test of the hypothesis while turning over *only the cards necessary* to do so. Given this constraint, which cards would you turn over? Please don't read ahead until you have answered this question.

Most people report that they would turn over exactly two cards: the card showing the *E* and the card showing the 4. Let's do this. In this case, assume that the *E* proves to have a 2 on the other side, and the 4 proves to have an *A* on the other side. The obvious conclusion would appear to be that the hypothesis is correct. After turning over the *E* and the 4, this was certainly the first author's conclusion when Bill Swann (the professor who would eventually become his dissertation adviser) asked him to test this hypothesis many years ago. It was also the conclusion of the large majority of participants who were given this same kind of problem by Wason (1971) in his studies of hypothesis testing. The problem with this common approach to hypothesis testing is that it is wrong. First of all, there was no need to turn over the 4 at all. The stated hypothesis was that every vowel *had to have an even number on the other side. This could still be completely true if some or even all of the consonants also had an even number on the other side.* The rule simply didn't say anything about consonants. If you are at all like the primary author of this text, you probably looked at the other side of the card with the 4 because you were somehow hoping or expecting to see a vowel. Strike one. If you are even more like the primary author, you are now saying to yourself, "So big deal, I turned over a card I could have left unturned. I still drew the right conclusion. After all, every vowel *does* have an even number on the other side." Does it? Let's turn over the 7 to make sure. As it turns out, the 7 has a *U* on the other side. *U* is definitely a vowel, and unless you are very, very liberal in rounding numbers upward,

7 is definitely *not* an even number. Strike two. As it turns out, the hypothesis is false. If you are even *more* like the primary author of this text, you are now struggling to find a way to feel rational about your choice of the E and the 4. Perhaps you are complaining that the question was vaguely worded or misleading. Strike three. Don't be so defensive! Unless Bill Swann is standing in front of you right now and smiling because you were so predictably wrong, you still have one up on your primary author.

Wason and his colleagues referred to the systematic bias they observed in this and several other studies as the **positive test bias.** This refers to the tendency for people who are evaluating hypotheses to attempt to *confirm* rather than to *disconfirm* these hypotheses. Of course, this intuitive approach is highly inconsistent with the logic of good science, and that includes good statistics. However, it probably does not surprise you to learn that laypeople do not usually concern themselves with behaving scientifically. More often than not, people who are asked to test a hypothesis look long and hard for evidence that would support the hypothesis while looking past equally important evidence that could disconfirm it. This tendency for people to adopt a confirmatory approach to hypothesis testing is pretty pervasive. For instance, research in social psychology also suggests that when people are testing social hypotheses (e.g., "Is Zoey an extrovert?"), they are inclined to seek out evidence that is consistent rather than inconsistent with their preexisting expectations. For example, when people are asked to find out whether someone like Zoey is an extrovert, most people ask a preponderance of questions that are designed to elicit extroverted responses. For example, people are more likely to ask Zoey what she would do to liven things up at a dull party than they are to ask her when she is most likely to be interested in spending time alone. Almost anything Zoey says about livening up a dull party is likely to make her sound pretty extroverted—regardless of whether she really is the life of the party.

One well-documented social psychological analogue of the positive test bias is **behavioral confirmation,** the tendency for social perceivers to elicit behaviors from a person that are consistent with their initial expectancies of the person (Snyder & Swann, 1978). One of the best-known examples of research on behavioral confirmation is a study by Snyder, Tanke, and Berscheid (1977). These researchers gave some men the hunch that women they were getting to know over a laboratory "telephone" were either sociable or shy (by giving the men fake photos depicting either extremely attractive or unattractive women). The men who *thought* they were talking to highly attractive women later reported that they had expected the women to be highly sociable and entertaining. Moreover, in the course of their conversations with the presumably attractive woman, the men *made these expectations come true*. The men were more animated and entertaining themselves when they thought they were talking with an attractive woman. The most interesting aspect of this study, however, is that the women on the other end of the phone (who knew nothing about the misleading photographs) *confirmed* the men's originally false expectations by behaving in a highly sociable fashion themselves (as confirmed by raters who did not see the photos and only listened to what the women were saying).

Large bodies of research on self-fulfilling prophecies, experimenter bias, and stereotyping tell a very similar story (see Allport, 1954; Darley & Gross, 1983; Hamilton & Sherman, 1994; Rosenthal & Jacobson, 1966). Once we get an idea in our heads, most of us tend to engage in hypothesis-confirming behaviors that may falsely convince us that the idea is correct. Of course, an important consequence of confirmatory judgment biases is that people often believe that they have confirmed hypotheses that are not true. Moreover, once we have been exposed to some tentative evidence in support of our theories or ideas, we also become very reluctant to give them up—even in the face of strong disconfirming evidence that comes along later (e.g., see Ross, Lepper, & Hubbard, 1975; Swann, 1987, 1992). In fact, Dan Gilbert and his colleagues have gathered evidence suggesting that it may be impossible for human beings to *comprehend* a statement without initially encoding the statement as true (see Gilbert, 1991).

If you recall our discussion in Chapter 3 of Rosenthal's classic studies of teacher expectancies, you can imagine that there are many ways in which researchers might unknowingly test their hypotheses in biased ways. Along these lines, it's worth noting that Rosenthal and his colleagues didn't just study teachers. They also studied *researchers*. In one classic study, Rosenthal and Fode (1963) trained a group of highly motivated experimenters and asked them to test the performance of groups of carefully bred "maze-bright" and "maze-dull" rats in a laboratory maze. As Rosenthal and Fode expected, the maze-bright rats learned their way around in the mazes more quickly than did the maze-dull rats. This seemingly trivial finding may seem a little less trivial when you consider that the two groups of rats were identical. Rosenthal and Fode simply got a bunch of run-of-the-mill, maze-mediocre rats and randomly assigned them to the two different groups. The important thing is that Rosenthal and Fode's bright and highly motivated experimenters *thought* that some of the rats were brighter (and perhaps more highly motivated) than the others. Why did the rats that were labeled more favorably perform more favorably? Because the experimenters unwittingly *treated* them more favorably. The experimenters not only petted the presumably bright rats more and handled them with greater care but also encouraged the rats more when the rats were running in the mazes. Rosenthal and Fode also observed evidence of experimenter bias when they looked carefully for coding errors committed by the experimenters. The coding errors favored the ostensibly maze-bright rats quite a bit more frequently than they favored the less highly regarded maze-dull rats. It is important to realize that Rosenthal and Fode's experimenters were not *consciously* engaging in any form of scientific dishonesty. They were simply letting their expectations get the better of them.

Chapter 16 Avoiding Bias: Characterizing Without Capitalizing

Confirmatory Biases + Human Statisticians = Statistical Bias

If you are wondering what Rosenthal and Fode's rats have to do with something as objective and straightforward as statistical analysis, then you really haven't been paying much attention, have you? Remember that, in most studies, exactly which numbers we submit to a given statistical analysis is partly up to us. To be more concrete, deciding which numbers to analyze bears some resemblance to deciding which cards to turn over in Wason's famous card task. Remember, furthermore, that SPSS is just as happy to analyze biased or even fabricated data as it is to analyze data that were collected and coded with excruciating rigor. But of course, you knew *that* much already, even if you weren't paying complete attention. So let's consider a question of bias in statistical testing that is much more subtle, and thus much more interesting, than contrived examples involving fake or obviously biased data.

Consider the case of what to do with a single outlier when reporting a simple correlation. Professor Trew believes that kids who have watched more episodes of the charming, zany, and irreverent TV show *Phineas and Ferb* should perform better on a standard IQ test. Professor Falzss believes there should be no such correlation. Image 1 contains a screen capture of the relevant hypothetical data from 13 participants. Image 2 contains a scatterplot of the two sets of scores (where the *x*-axis is the standardized scores for number of episodes watched and the *y*-axis is the kids' IQ). To facilitate interpretation, we also sorted the data in the SPSS raw data file based on the total number of *Phineas and Ferb* episodes kids said they'd watched. In case you want to explore these data yourself, the SPSS data file is called [**hypothetical Phineas & Ferb data.sav**].

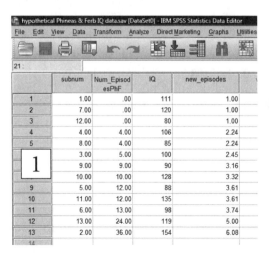
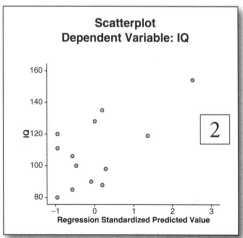

As Dr. Trew would be quick to note, there is a substantial positive correlation between how many episodes of *Phineas and Ferb* these kids watched and their IQ scores. The more episodes of *Phineas and Ferb* kids said they had watched, the higher their IQs, $r(11) = .617$, $p = .025$. (Of course, you could confirm this for yourself using SPSS "Correlate....") However, as Dr. Falzss would be quick to note, the bulk of this correlation seems to be driven by one lone *Phineas and Ferb* fanatic, who also happens to be freakishly smart. In fact, if you eliminate this one participant, the correlation for the remaining 12 participants is $r(10) = .285$, $p = .370$, which does not even approach statistical significance. So who is right? Without collecting a lot more data, it might be impossible to know. However, we might be able to reach a middle ground by (a) assessing exactly how much skewness and kurtosis there is for both the "episodes" and the IQ variables. As it turns out, there isn't *quite* enough skewness and kurtosis for the "episodes" variable to worry most statisticians (the skewness is about 1.5 and the kurtosis is about 2.5). However, when you combine this moderate distributional shape problem with the fact that the participant under debate is more than 2 standard deviations above the mean in IQ (even in this pretty intelligent sample) and is also more than 2 standard deviations above the mean on the "episodes" variable, Dr. Falzss's worries start to seem pretty reasonable. One solution to this problem would be to perform a square root transformation on the "episodes" variable.

Let's try that. Specifically, check out what we did in a syntax file to create the "new_episodes" variable you see in the right-hand column of the SPSS screen capture of these hypothetical data.

COMPUTE new_episodes = SQRT(Num_EpisodesPhF + 1).

EXECUTE.

As it turns out, this transformed variable has only a little skew (+0.456) and virtually no kurtosis (−.020). If there were a significant positive correlation between the transformed "episodes" variable and IQ scores, this would constitute stronger evidence for Dr. Trew's claims. As it turns out, when you put these data to this more stringent test, the correlation is $r(11) = .531$, $p = .062$, which is only marginally significant. Of course, if Dr. Trew truly predicted this correlation in advance, and if he could argue that it would be theoretically and practically uninformative to know that there was a significant *negative* correlation between watching *Phineas and Ferb* and being smart, he might be able to persuade some readers that a one-tailed hypothesis test is in order. Now, of course, $p = .062$ becomes $p = .031$, and Dr. Trew is happy, whereas Dr. Falzss is livid. In this case, Dr. Falzss could make the very reasonable counterargument that there are plenty of practical and theoretical reasons why it would be useful to know if the correlation is negative. Although Drs. Trew and Falzss might never agree about the fairest way to interpret these data, objective observers could draw their own conclusions. One pretty fair conclusion is these data provide some tentative support for the possibility that kids who've watched more episodes of *Phineas and Ferb* are smarter. Of course, it should go without saying that even Dr. Trew probably wouldn't be comfortable arguing that watching this cartoon *causes* intelligence. A more plausible interpretation might be that smarter kids are more likely to prefer this particular show over other shows that, while equally entertaining (think *Gumball* or *Fan Boy and Chum Chum*), are quite a bit less geeky.

Phineas and Ferb Are Just the Tip of the Iceberg

Incidentally, the *Phineas and Ferb* study provides us with fodder for only one of dozens of specific debates we could have about the appropriateness of statistical analyses. Consider this far-from-exhaustive list of some of the other difficult decisions researchers have to make (and justify) about how to analyze their data.

1. Were any participants excluded for failing to follow instructions, for expressing suspicion about a manipulation, or because they were able to figure out the researcher's hypothesis? If these participants are included in the final study anyway (when that is possible), are the results significantly altered?

2. Exactly what criterion did the researcher use to eliminate or trim (e.g., winsorize) outlying scores? Was the same criterion used consistently across different studies? Was it decided a priori or after the fact?

3. For what covariates (aka "confounds") did the researcher control? Is there a theoretical, historical, or practical justification for including each of the covariates in the analyses? How do the results change if one fails to include one or more of the covariates? It is probably much fairer, for example, to control consistently for the same worrisome control variable (in the same way) across multiple studies than to control for a variable in only one study when that variable is not an obvious cause for concern, is not a likely suppressor variable, and bears no theoretical relation to the dependent variable of interest. It might make sense, for example, to control for age and body weight when assessing upper body strength differences between women and men, but it makes a lot less sense to control for participants' skill at juggling or interest in the amazing card game *Cliffhanger*.

4. If a continuous variable was converted to a categorical variable, is there a statistically compelling justification for this? For example, some data are so badly skewed or so full of intransigent outliers that a researcher might have few other options. There might also be well-established clinical cutoff scores that are widely accepted in a particular literature because these scores have clinical significance or facilitate a useful comparison.

Otherwise, if data are continuous, there is usually no reason to create categories, especially dichotomous ones. Furthermore, if there is a compelling reason to create quartiles, quintiles, or the like, most computer software packages (including SPSS) will do this conversion for you. The advantage of putting this decision in SPSS's hands is that SPSS will make the judgment calls for specific break points without any knowledge of which specific cut points lead to the most beautiful pattern of association between the variable of interest and whatever variable you expect to be related to this variable.

5. Were there other independent and/or dependent measures in this study that would have been reasonable to examine? If so, do they show the same pattern showcased in a specific report? This question is particularly important when a researcher engaged in "data mining," which refers to looking in data sets to uncover hidden patterns that may have eluded others who have analyzed the same data. Like prospecting for gold, data mining can be incredibly rewarding. However, the risk of reckless data mining without making any correction for conducting numerous statistical tests is that it is very easy to capitalize on chance (also known as sampling error), especially in small data sets. One possible solution to this problem, at least when large data sets are available, is to split your data set randomly into two or more subsamples. A researcher can then data mine in one part of the sample and then see how well the resulting model performs in the random portion (or portions) of the sample that was set aside during data mining. A more sophisticated solution to this problem involves various kinds of bootstrapping procedures, which involve sampling repeatedly from one's sample (preferably a large one) to see how often a particular result holds up.

6. Did the researcher have an a priori plan about how many participants to run? Was the plan based on a careful power analysis? This question becomes most relevant when a researcher (hopefully without realizing that this is highly problematic) runs some participants, analyzes the data to see if a particular effect is significant, and—if the effect is *not* significant—runs some more participants and analyzes the data again. In the extreme case of stopping to analyze one's data on a frequent basis, it would be quite easy to capitalize on chance when there was no real effect whatsoever to be found.

7. Did the researcher really make an a priori prediction that would justify a powerful and generous approach to data analysis such as planned contrasts in analysis of variance (ANOVA)? If not, did the researcher provide any evidence of replication in a separate sample?

In the past couple of decades, researchers have begun to take difficult questions such as these very seriously. In one way or another, all of these questions boil down to whether a researcher allowed his or her confirmatory biases to influence a specific statistical analysis by knowingly or unknowingly capitalizing on chance. When this happens, the statistics presented in support of a hypothesis do not really provide support for the hypothesis because the researcher has unknowingly (at least we *hope* unknowingly) violated the assumptions on which the statistical test is based. In the case of data mining, for example, an observed p value of less than .05 would not really *mean* $p <$ .05 if the researcher conducted an analysis a dozen different ways and then settled on the one way that yielded a "significant" result.

So what is the solution to this general problem? Precisely because this problem can take many forms, there is no one solution. However, we briefly summarize a recent solution proposed by Simmons, Nelson, and Simonsohn (2011) that focuses on experimental and quasi-experimental research. We present their six suggestions for avoiding bias, by the way, not because we agree with all of them but because these authors were willing to address this often overlooked question. According to Simmons et al., researchers have to make so many decisions when conducting research and analyzing data that it is all too easy to show support for a hypothesis that has little or no validity. They point out, for example, that researchers have to decide exactly how to conduct experiments; choose which variable(s) to manipulate, measure, and report; and select a particular statistical analysis. At any step of this process, bias can creep in easily. For example, Simmons et al. model the extreme example of a researcher who runs one participant at a time in an experiment and stops to analyze his or her data after running each individual participant. As you might guess, if you run one analysis for every participant you run (even if it is the same analysis every time), it is easy to observe a "significant" result that has no real meaning. In light of issues such as this, Simmons et al. (2011, pp. 4–5) argue that authors who submit papers for publication should be required to do six things:

1. Authors must decide the rule for terminating data collection before data collection begins and report this rule in the article.

2. Authors must collect at least 20 observations per cell or else provide a compelling cost-of-data collection justification.

3. Authors must list all variables collected in a study.

4. Authors must report all experimental conditions, including failed manipulations.

5. If observations are eliminated, authors must also report what the statistical results are if those observations are included.

6. If an analysis includes a covariate, authors must report the statistical results of the analysis without the covariate.

We applaud Simmons et al.'s willingness to point out ways in which statistical analyses can be misleading. Furthermore, we have to appreciate their interest in minimizing the likelihood of false-positive biases in statistical testing. However, we are frankly appalled to see that they felt that some of these proscriptions were necessary in the first place. For example, consider the complete instructions they offered regarding their fourth point, which has to do with fully reporting all experimental manipulations.

> *4. Authors must report all experimental conditions, including failed manipulations.* This requirement prevents authors from selectively choosing only to report the condition comparisons that yield results that are consistent with their hypothesis. As with the previous requirement [in their Rule 3], we encourage authors to include the word *only* (e.g., "participants were randomly assigned to one of only three conditions").

The proscription against misleading readers by presenting a manipulation as if it were the only one in an experiment that included additional (failed) manipulations is simply a prohibition against lying. One would certainly have hoped that this proscription was completely unnecessary.

As another example, consider the prohibition against running and rerunning a statistical analysis every time you run a new participant. Does anyone (who doesn't also merely fabricate data) actually do this? Without waxing too cynical about Simmons et al.'s arguably high level of cynicism, let us express our appreciation for their thoughtful efforts to make explicit and concrete a set of rules for minimizing biases in statistical testing. We can all agree that this is a worthy goal. We are not sure, however, that imposing a rigid set of "one-size-fits-all" rules is the best solution to the kind of problems we have touched upon in this chapter. Consider, for example, Simmons et al.'s suggestion that researchers always run at least 20 participants per unique condition in experiments and quasi-experiments. Does this still apply when some unique experimental conditions simply involve trivial variations in stimuli that were included merely to test for generalizability (e.g., four randomly sampled soft drinks that all did or did not have a certain feature but were all expected to behave in the same way)? Does this mean that if you finish data collection in June with only 17 participants per cell in a conceptual replication of an effect you have documented many times in the past that you have to wait until October to get those remaining few participants per cell before analyzing your data—even when you know that you have 98% power to detect the very large effect your trying to document? Does the $n = 20$ rule apply to pilot testing as well as full-blown experiments? If I am pilot testing some new instructions, am I required to run 20 participants per cell before peeking at my manipulation checks to make sure the manipulations and instructions make sense to my sample? We're *not* arguing that bigger samples are not usually preferable over smaller samples, but a simple "rule of 20" may be too simple.

Four Simple Rules for Avoiding Bias in Data Analysis

Speaking of things that may be too simple, we'd like to suggest that researchers just follow four simple rules, each of which is briefer and less rigid than those suggested by Simmons et al. (2011). Here are those rules, which apply equally well to passive observational research as well as to lab experiments:

1. Don't capitalize on chance.
2. Be extremely transparent.
3. Don't lie or mislead.
4. Replicate.

We suspect that Simmons et al. might criticize our rules as lacking in rigor—as well as being shallow and overly simplistic. Regardless of whether one prefers their more detailed suggestions or our arguably simplistic suggestions, we heartily agree that it is all too easy to allow one's biases to creep into one's research in subtle ways that render traditional statistical analyses invalid. Regardless of what specific approach you adopt in your efforts to avoid bias, we hope we have communicated consistently in this text that statistics are powerful tools that require us to be rigorous and transparent as well as creative if we wish to track down the answers to important questions. Happy hunting.

References

Agresti, A., & Finlay, B. (1997). *Statistical methods for the social sciences.* Englewood Cliffs, NJ: Prentice Hall.

Aiken, L. S., & West, S. G. (1991). *Multiple regression: Testing and interpreting interactions.* Newbury Park, CA: Sage.

Allport, G. W. (1954). *The nature of prejudice.* Reading, MA: Addison-Wesley.

Anderson, C. A. (2001). Heat and violence. *Current Directions in Psychological Science, 10,* 33–38.

Baron, R. M., & Kenny, D. A. (1986). The moderator-mediator variable distinction in social psychological research: Conceptual, strategic, and statistical considerations. *Journal of Personality and Social Psychology, 51,* 1173–1182.

Baumrind, D. (1971). Current patterns of parental authority. *Developmental Psychology Monographs, 4*(1, Pt. 2), 1–103.

Baumrind, D. (1983). Rejoinder to Lewis's reinterpretation of parental control effects: Are authoritative families really harmonious? *Psychological Bulletin, 94,* 132–142.

Berkowitz, L. (1989). Frustration-aggression hypothesis: Examination and reformulation. *Psychological Bulletin, 106,* 59–73.

Billock, V. A., & Tsou, B. H. (2011). To honor Fechner and obey Stevens: Relationships between psychophysical and neural nonlinearities. *Psychological Bulletin, 137,* 1–18.

Blair, I. V., Judd, C. M., & Fallman, J. L. (2004). The automaticity of race and Afrocentric facial features in social judgments. *Journal of Personality and Social Psychology, 87,* 763–778.

Boninger, D. S., Gleicher, F., & Strathman, A. (1994). Counterfactual thinking: From what might have been to what may be. *Journal of Personality and Social Psychology, 67,* 297–307.

Bowlby, J. (1982). *Attachment and loss: Vol. 1. Attachment.* London: Hogarth.

Brown, J. D., & McGill, K. L. (1989). The cost of good fortune: When positive life events produce negative health consequences. *Journal of Personality and Social Psychology, 57,* 1103–1110.

Clancy, S. A., McNally, R. J., Schacter, D. L., Lenzenweger, M. F., & Pitman, R. K. (2002). Memory distortion in people reporting abduction by aliens. *Journal of Abnormal Psychology, 111,* 455–461.

Cohen, D., & Nisbett, R. E. (1994). Self-protection and the culture of honor: Explaining southern violence. *Personality and Social Psychology Bulletin, 20,* 551–567.

Cohen, D., Nisbett, R. E., Bowdle, B. F., & Schwarz, N. (1996). Insult, aggression, and the southern culture of honor: An "experimental ethnography." *Journal of Personality and Social Psychology, 70,* 945–960.

Cohen, J. (1994). The earth is round ($p < .05$). *American Psychologist, 49,* 997–1003.

Cohen, J., & Cohen, P. (1975). *Applied multiple regression/correlation analysis for the behavioral sciences.* Mahwah, NJ: Lawrence Erlbaum.

Cohen, S., Doyle, W. J., Skoner, D. P., Rabin, B. S. & Gwaltney, J. M. (1997). Social ties and susceptibility to the common cold. *Journal of the American Medical Association, 277,* 1940–1944.

Cronbach, L. J. (1951). Coefficient alpha and the internal structure of tests. *Psychometrika, 16,* 297–334.

Cronbach, L. J. (1955). Processes affecting scores on "understanding of others" and "assumed similarity." *Psychological Bulletin, 52,* 177–193.

Darley, J. M., & Gross, P. H. (1983). A hypothesis-confirming bias in labeling effects. *Journal of Personality and Social Psychology, 44,* 20–33.

Davies, H. T. O., Crombie, I. K., & Tavakoli, M. (1998). When can odds ratios mislead? *British Medical Journal, 316,* 989–991.

Deaton, A. (2008). Income, health and well-being around the world: Evidence from the Gallup World Poll. *Journal of Economic Perspectives, 22,* 53–72.

DeCarlo, L. T. (1997). On the meaning and use of kurtosis. *Psychological Methods, 2,* 292–307.

DeHart, T., Pelham, B. W., & Tennen, H. (2006). What lies beneath: Early experiences with parents and implicit self-esteem. *Journal of Experimental Social Psychology, 42,* 1–17.

Devine, P. G. (1989). Stereotypes and prejudice: Their automatic and controlled components. *Journal of Personality and Social Psychology, 56,* 5–18.

Eagly, A. H. (1978). Sex differences in influenceability. *Psychological Bulletin, 85,* 86–116.

Eagly, A. H., & Carli, L. (1981). Sex of researchers and sex-typed communications as determinants of sex differences in influenceability: A meta-analysis of social influence studies. *Psychological Bulletin, 90,* 1–20.

Erceg-Hurn, D. M., & Mirosevich, V. M. (2008). Modern robust statistical methods: An easy way to maximize the accuracy of your research. *American Psychologist, 63,* 591–601.

Erikson, R., & Torssander, J. (2008). Social class and cause of death. *European Journal of Public Health, 18,* 473–478.

Fabrigar, L. R., & Festinger, L. (1954). A theory of social comparison processes. *Human Relations, 7,* 117–140.

Fabrigar, L. Wegener, D. T, (2011). *Exploratory factor analysis (understanding statistics).* Oxford, UK: Oxford University Press.

Festinger, L. (1954). A theory of social comparison processes. *Human Relations, 7,* 117–140.

Field, A. P. (2005). *Discovering statistics using SPSS* (2nd ed.). London: Sage.

Fischhoff, B., Slovic, P., & Lichtenstein S. (1977). Knowing with certainty: The appropriateness of extreme confidence. *Journal of Experimental Psychology: Human Perception and Performance, 3,* 552–564.

Fisher, R. A. (1935). *The design of experiments.* Edinburgh, UK: Oliver & Boyd.

Fisher, R. A. (1938). *Statistical methods for research workers* (7th ed.). London: Oliver & Boyd.

Gilbert, D. T. (1991). How mental systems believe. *American Psychologist, 46,* 107–119.

Gladwell, M. (2008). *Outliers: The story of success.* New York: Little, Brown.

Goldberg, L. R. (1990). An alternative "description of personality": The Big-Five factor structure. *Journal of Personality and Social Psychology, 59,* 1216–1229.

Greenwald, Anthony G., McGhee, Debbie E., Schwartz, Jordan L. K. (1998). Measuring Individual Differences in Implicit Cognition: The Implicit Association Test. *Journal of Personality and Social Psychology, 74,* 1464–1480.

Hamilton, D. L., & Sherman, J. W. (1994). Stereotypes. In R. S. Wyer Jr. & T. K. Srull (Eds.), *Handbook of social cognition* (2nd ed., pp. 1–68). Hillsdale, NJ: Lawrence Erlbaum.

Harwell, M. R., Rubinstein, E. N., Hayes, W. S., & Corley, C. O. (1992). Summarizing Monte Carlo results in methodological research: The one- and two-factor fixed effects ANOVA cases. *Journal of Educational and Behavioral Statistics, 17,* 315–339.

Hastorf, A. H., & Cantril, H. (1954). They saw a game: A case study. *Journal of Abnormal and Social Psychology, 49,* 129–134.

Hedges, L. V. (1987). How hard is hard science, how soft is soft science? The empirical cumulativeness of research. *American Psychologist, 42,* 443–455.

Hetts, J. J., Sakuma, M., & Pelham, B.W. (1999). Two roads to positive regard: Implicit and explicit self-evaluation and culture. *Journal of Experimental Social Psychology, 35,* 512–559.

Hofstede, G., & Hofstede, G. J. (2005). *Cultures and organizations: Software of the mind* (Revised & expanded 2nd ed.). New York: McGraw-Hill.

James, W. (1890). *The principles of psychology.* New York: Holt.

Jones, J. M. (2009, May 18). GOP losses span nearly all demographic groups. Retrieved from http://www.gallup.com/poll/118528/GOP-Losses-Span-Nearly-Demographic-Groups.aspx

Judd, C. M., & Kenny, D. A. (1981). Process analysis: Estimating mediation in treatment evaluations. *Evaluation Review, 5,* 602–619.

Kahneman, D., & Deaton, A. (2010). High income improves evaluation of life but not emotional well-being. *Proceedings of the National Academy of Sciences, 107,* 16489–16493.

Kelley, H. H. (1971). Attribution in social interaction. In E. E. Jones, D. E. Kanouse, H. H. Kelley, R. E. Nisbett, S. Valins, & B. Weiner (Eds.), *Attribution: Perceiving the causes of behavior* (pp. 1–26). Morristown, NJ: General Learning Press.

Kline, R. B. (2005). *Principles and practice of structural equation modeling* (2nd ed.). New York: Guilford.

Kruger, J., Gordon, C. L., & Kuban, J. (2006). Intentions in teasing: When "just kidding" just isn't good enough. *Journal of Personality and Social Psychology, 90,* 412–425.

Lance, C. E., Butts, M. M., & Michels, L. C. (2006). The sources of four commonly reported cutoff criteria: What did they really say? *Organizational Research Methods, 9,* 202–220.

Levitt, S., & Dubner, S. J. (2005). *Freakonomics: A rogue economist explores the hidden side of everything.* New York: William Morrow/HarperCollins.

Lindman, H. R. (1974). *Analysis of variance in complex experimental designs.* San Francisco: Freeman.

Lowery, B. S., Hardin, C. H., & Sinclair, S. (2001). Social influence effects on automatic racial prejudice. *Journal of Personality and Social Psychology, 81,* 842–855.

MacKinnon, D. P. (2008). *Introduction to statistical mediation analysis.* Philadelphia: Psychology Press.

Maddox, K. B., & Gray, S. (2002). Cognitive representations of African Americans: Reexploring the role of skin tone. *Personality and Social Psychological Bulletin, 28,* 250–259.

McCrae, R. R., Kurtz, J. E., Yamagata, S., & Terracciano, A. (2011). Internal consistency, retest reliability, and their implications for personality scale validity. *Personality and Social Psychology Review, 15,* 28–50.

Morse, S., & Gergen, K. J. (1970). Social comparison, self-consistency, and the concept of self. *Journal of Personality and Social Psychology, 16,* 148–156.

Murray, S. L., Holmes, J. G., & Collins, N. L. (2006). Optimizing assurance: The risk regulation system in relationships. *Psychological Bulletin, 132,* 641–666.

Nuttin, J. M. (1985). Narcissism beyond gestalt and awareness: The name letter effect. *European Journal of Social Psychology, 15,* 353–361.

Nuttin, J. M. (1987). Affective consequences of mere ownership: The name letter effect in twelve European languages. *European Journal of Social Psychology, 17,* 381–402.

Payne, B. K., Cheng, C. M., Govorun, O., & Stewart, B. D. (2005). An inkblot for attitudes: Affect misattribution as implicit measurement. *Journal of Personality and Social Psychology, 89,* 277–293.

Pelham, B. W. (1991). On confidence and consequence: The certainty and importance of self-knowledge. *Journal of Personality and Social Psychology, 60,* 518–530.

Pelham, B. W. (1993). The idiographic nature of human personality: Examples of the idiographic self-concept. *Journal of Personality and Social Psychology, 64,* 665–677.

Pelham, B. W. (1995a). Self-investment and self-esteem: Evidence for a Jamesian model of self-worth. *Journal of Personality and Social Psychology, 69,* 1141–1150.

Pelham, B. W. (1995b). Further evidence for a Jamesian model of self-worth: Reply to Marsh (1995). *Journal of Personality and Social Psychology, 69,* 1161–1165.

Pelham, B. W., & Blanton, H. (2013). *Conducting research in psychology: Measuring the weight of smoke* (4th ed.). Pacific Grove, CA: Cengage.

Pelham, B. W., Carvallo, M., & Jones, J. T. (2005). Implicit egoism. *Current Directions in Psychological Science, 14,* 106–110.

Pelham, B. W., Koole, S. L., Hetts, J. J., Hardin, C. D., Seah, E., & DeHart, T. (2005). Gender moderates the relation between implicit and explicit self-esteem. *Journal of Experimental Social Psychology, 81,* 84–89.

Pelham, B. W., Mirenberg, M. C., & Jones, J. T. (2002). Why Susie sells seashells by the seashore: Implicit egotism and major life decisions. *Journal of Personality and Social Psychology, 82,* 469–487.

Pelham, B. W., & Neter, E. (1995). The effect of motivation on judgment depends on the difficulty of the judgment. *Journal of Personality and Social Psychology, 68,* 581–594.

Pelham, B. W., & Swann, W. B., Jr. (1989). From self-conceptions to self-worth: On the sources and structure of global self-esteem. *Journal of Personality and Social Psychology, 57,* 672–680.

Perdue, C. W., Dovidio, J. F., Gurtman, M. B., & Tyler, R. B. (1990). Us and them: Social categorization and the process of intergroup bias. *Journal of Personality and Social Psychology, 59,* 475–486.

Peterson, C., Park, N., & Seligman, M. E. P. (2005). Orientations to happiness and life satisfaction: The full life versus the empty life. *Journal of Happiness Studies, 6,* 25–41.

Pettigrew, T. F., & Tropp, L. R. (2008). How does intergroup contact reduce prejudice? Meta-analytic tests of three mediators. *European Journal of Social Psychology, 38,* 922–934.

Pinel, E. (1999). Stigma consciousness: The psychological legacy of social stereotypes. *Journal of Personality and Social Psychology, 76,* 114–128.

Preacher, K. J., & Hayes, A. F. (2004). SPSS and SAS procedures for estimating indirect effects in simple mediation models. *Behavior Research Methods, Instruments, & Computers, 36,* 717–731.

Rosenberg, M. (1965). *Society and the adolescent self-image.* Princeton, NJ: Princeton University Press.

Rosenthal, R., & Fode, K. L. (1963). The effect of experimenter bias on the performance of the albino rat. *Behavioral Science, 8,* 183–189.

Rosenthal, R., & Jacobson, L. (1966). Teachers' expectancies: Determinants of pupils' IQ gains. *Psychological Reports, 19,* 115–118.

Rosenthal, R., & Rosnow, R. L. (1991). *Essentials of behavioral research: Methods and data analysis* (2nd ed.). New York: McGraw-Hill.

Rosenthal, R., & Rubin, D. B. (1982). A simple general purpose display of magnitude of experimental effect. *Journal of Educational Psychology, 74,* 166–169.

Ross, L., Lepper, M. R., & Hubbard, M. (1975). Perseverance in self-perception and social perception: Biased attributional processes in the debriefing paradigm. *Journal of Personality and Social Psychology, 32,* 880–892.

Scheier, M. F., & Carver, C.S. (1985). Optimism, coping, and health: Assessment and implications of generalized outcome expectancies. *Health Psychology, 4,* 219–247.

Scheier, M. F., Carver, C. S., & Bridges, M. W. (1994). Distinguishing optimism from neuroticism (and trait anxiety, self-mastery, and self-esteem): A re-evaluation of the Life Orientation Test. *Journal of Personality and Social Psychology, 67,* 1063–1078.

Shimizu, M., & Pelham, B. W. (2004). The unconscious cost of good fortune: Implicit and explicit self-esteem, positive life events, and health. *Health Psychology, 23,* 101–105.

Shimizu, M., & Pelham, B. W. (2011). Liking for positive words and icons moderates the association between implicit and explicit self-esteem. *Journal of Experimental Social Psychology, 47,* 994–999.

Shimizu, M., Pelham, B. W., & Sperry, J. (2012). *The effect of subliminal priming on sleep duration.* Unpublished manuscript.

Simmons, J. P., Nelson, L. D., & Simonsohn, U. (2011). False-positive psychology: Undisclosed flexibility in data collection and analysis allows presenting anything as significant. *Psychological Science, 22,* 1359–1366.

Snyder, M., & Swann, W. B. (1978). Hypothesis-testing processes in social interaction. *Journal of Personality and Social Psychology, 36,* 1202–1212.

Snyder, M., Tanke, E. D., & Berscheid, E. (1977). Social perception and interpersonal behavior: On the self-fulfilling nature of social stereotypes. *Journal of Personality and Social Psychology, 35,* 656–666.

Sobel, M. E. (1982). Asymptotic confidence intervals for indirect effects in structural equation models. In S. Leinhardt (Ed.), *Sociological Methodology 1982* (pp. 290–312). Washington DC: American Sociological Association.

Stanton, J. M. (2001). Galton, Pearson, and the peas: A brief history of linear regression for statistics instructors. *Journal of Statistics Education, 9.* http://www.amstat.org/publications/jse/v9n3/stanton.html

Sternberg, R. J. (1988). *The triarchic mind: A new theory of human intelligence.* New York: Penguin.

Stevens, S. S. (1961, January 13). To honor Fechner and repeal his law. *Science, 133,* 80–86.

Swann, W. B., Jr. (1987). Identity negotiation: Where two roads meet. *Journal of Personality and Social Psychology, 53,* 1038–1051.

Swann, W. B., Jr. (1992). Seeking "truth," finding despair: Some unhappy consequences of a negative self-concept. *Current Directions in Psychological Science, 1,* 15–18.

Tabachnick, B. G., & Fidell, L. S. (2007). *Using multivariate statistics* (5th ed.). Boston: Allyn & Bacon.

Taylor, S. E., & Brown, J. D. (1988). Illusion and well-being: A social psychological perspective on mental health. *Psychological Bulletin, 103,* 193–210.

Thaler, R. (1985). Mental accounting and consumer choice. *Marketing Science, 4,* 199–214.

Triplett, N. (1898). The dynamogenic factors in pacemaking and competition. *American Journal of Psychology, 9,* 507–533.

Tversky, A., & Kahneman, D. (1972). Judgment under uncertainty: Heuristics and biases. *Science, 185,* 1124–1131.

Uchino, B. N. (2008). Social support and health: A review of physiological processes potentially underlying links to disease outcomes. *Journal of Behavioral Medicine, 29,* 377–387.

Vandello, J. A., & Cohen, D. (1999). Patterns of individualism and collectivism across the United States. *Journal of Personality and Social Psychology, 77,* 279–292.

Visher, P., & Nawrocki, M. (2002). *Jonah: A Veggie Tales Movie.* Flintstone, GA: Inside Joke Press.

von Hippel, P. T. (2005). Mean, median, and skew: Correcting a textbook rule. *Journal of Statistics Education, 13*(2). www.amstat.org/publications/jse/v13n2/vonhippel.html

Wason, P. C. (1971). Problem solving and reasoning. *British Medical Bulletin, 27,* 206–210.

Weinstein, N. D. (1984). Why it won't happen to me: Perceptions of risk factors and illness susceptibility. *Health Psychology, 3,* 431–457.

Weinstein, N. D. (1987). Unrealistic optimism about susceptibility to health problems: Conclusions from a community-wide sample. *Journal of Behavioral Medicine, 10,* 481–500.

Winer, B. J. (1971). *Statistical principles in experimental design* (2nd ed.). New York: McGraw-Hill.

Zimmerman, D. W. (1987). Comparative power of Student t test and Mann-Whitney U test for unequal sample sizes and variances. *Journal of Experimental Education, 55,* 171–174.

Zumbo, B. D., & Zimmerman, D. W. (1993). Is the selection of statistical methods governed by level of measurement? *Canadian Psychology, 34,* 390–400.

Author Index

Agresti, A., 69, 70
Aiken, L. S., 264, 271, 279, 288, 300, 301n1
Allport, G. W., 287, 398
Anderson, C. A., 177

Baron, R. M., 332, 334, 335, 349, 352
Baumrind, D., 360
Berkowitz, L., 334, 350
Berscheid, E., 397
Billock, V. A., 164
Blair, I.V., 90
Blanton, H., 127, 279, 356, 387
Boninger, D. S., 160n3
Bowdle, B. F., 207
Bowlby, J., 360
Bridges, M. W., 348
Brown, J. D., 221, 300
Butts, M. M., 136

Cantril, H., 163
Carli, H., 32
Carvallo, M., 123n1
Carver, C. S., 348, 350
Cheng, C. M., 226
Clancy, S. A., 210
Cohen, D., 150, 158, 159, 207
Cohen, J., 28, 316
Cohen, P., 316
Cohen, S., 332
Collins, N. L., 339
Corley, C. O., 170
Crombie, I. K., 258
Cronbach, L. J., 138, 148, 158

Darley, J. M., 398
Darlington, D., 160n3
Davies, H. T. O., 258
Deaton, A., 365
DeCarlo, L. T., 77
DeHart, T., 326, 360, 362, 364

Devine, P. G., 229
Dovidio, J. F., 208
Doyle, W. J., 332
Dubner, S. J., 233

Eagly, A. H., 32
Erceg-Hurn, D. M., 79, 371, 383
Erikson, R., 365

Fabrigar, L., 136
Fallman, J. L., 90
Festinger, L., 322
Fidell, L. S., 34, 77, 132, 370, 372
Field, A. P., 160n3
Finlay, B., 69, 70
Fischhoff, B., 180
Fisher, R. A., 15, 192, 209
Fode, K. L., 356, 398

Galton, F., 81
Gergen, K. J., 322
Gilbert, D., 42n3
Gilbert, D. T., 398
Gladwell, M., 373
Gleicher, F., 160N3
Goldberg, L. R., 39, 388
Gordon, C. L., 349, 351-353
Govorun, O., 226
Gray, S., 90
Greenwald, A. G., 230
Gross, P. H., 398
Gurtman, M. B., 208
Gwaltney, J. M., 332

Hamilton, D. L., 398
Hardin, C. H., 229, 230, 287
Harwell, M. R., 170
Hastorf, A. H., 163
Hayes, A. F., 334, 353n2
Hayes, W. S., 170

413

Hedges, L. V., 32
Hetts, J. J., 208
Hofstede, G., 225
Hofstede, G. J., 225
Holmes, J. G., 339
Hubbard, M., 398

Jacobson, L., 171, 356, 398
James, W., 277
Jones, J. M., 287
Jones, J. T., 102, 123n1
Judd, C. M., 90, 329

Kahneman, D., 121, 365
Kelly, H. H., 317
Kenny, D. A., 329, 332, 334, 335, 349, 352
Kline, R. B., 68, 77, 78, 132, 369
Koole, S. L., 39
Kruger, J., 349, 351-353
Kuban, J., 349, 351-353
Kurtz, J. E., 149

Lamm, E., 8–9
Lance, C. E., 136
Lenzenweger, M. F., 210
Lepper, M. R., 398
Levitt, S., 233
Lichtenstein, S., 180
Lindman, H. R., 252
Lowery, B. S., 229, 230

MacKinnon, D. P., 329, 332, 334, 350, 353n1, 353n2
Maddox, K. B., 90
McCrae, R. R., 149
McGill, K. L., 300
McNally, R. J., 210
Michels, L. C., 136
Mirenberg, M. C., 102
Mirosevich, V. M., 79, 371, 383
Morse, S., 322
Murray, S. L., 339

Nawrocki, M., 335
Nelson, L. D., 403
Neter, E., 122, 123
Nisbett, R. E., 207
Nuttin, J. M., 102

Park, N., 150
Payne, B. K., 226
Pearson, K., 81–82, 166
Pearson, Karl, 1

Pedhazur, E. J., 350
Pelham, B. W., 39, 102, 122, 123, 123n1, 127, 208, 214, 276, 277, 279, 287, 293, 300, 326, 356, 360, 387
Perdue, C. W., 208
Peterson, C., 150
Pettigrew, T. F., 287
Pinel, E., 157, 158
Pitman, R. K., 210
Preacher, K. J., 334, 353n3

Rabin, B. S., 332
Rosenberg, M., 88
Rosenthal, R., 29, 31, 171, 356, 398
Rosnow, L., 29, 31
Ross, L., 398
Rubinstein, E. N., 170

Sakuma, M., 208
Schacter, D. L., 210
Scheier, M. F., 348, 350
Schwarz, N., 207
Seligman, M. E. P., 150
Sherman, J. W., 398
Shimizu, M., 293, 300
Simmons, J. P., 403
Simonsohn, U., 403
Sinclair, S., 229, 230
Skoner, D. P., 332
Slovic, P., 180
Snyder, M., 397
Sobel, M. E.,
Spearman, C., 127, 128, 142
Stanton, J. M., 81
Sternberg, R. J., 128
Stevens, S. S., 164
Stewart, B. D., 226
Strathman, A., 160n4
Swann, W. B., Jr., 276, 277, 397, 398

Tabachnick, B. G., 34, 77, 370, 131, 132, 372
Tanke, E. D., 397
Tavakoli, M., 258
Taylor, S. E., 221
Tennen, H., 326, 360
Terracciano, A., 149
Thaler, R., 164
Torssander, J., 365
Triplett, N., 213
Tropp, L. R., 287
Tsou, B. H., 164
Tversky, A., 121, 122
Tyler, R. B., 208

Author Index

Uchino, B. N., 332

Vandello, J. A., 150, 158, 159
Visher, P., 335
von Hippel, P. T., 70

Wason, P. C., 396
Wegener, D. T., 136
Weinstein, N. D., 221

West, S. G., 264, 271, 279, 300, 301n3
West, S. W., 288
Wickens, T., 76, 369
Winer, B. J., 252

Yamagata, S., 149

Zimmerman, D. W., 170, 252
Zumbo, B. D., 170, 252

Subject Index

Absolute value, 187
Abstract characters, 3
Actual data, 47
Adjusted *r*-square, in multiple regression, 242–248
Afrocentrism, hypothetical correlational study of, 90–91
Aggression:
   gender and, study, 205–207
   mediational model, frustration and, 334–337
Alpha level:
   defined, 14
   type I and II errors and, 24–25
Alternative hypothesis, 14–15
Analysis:
   mediation, 41
   path, 41–42
   See also Analysis of covariance; Analysis of variance
Analysis of covariance (ANCOVA), 37, 304–305
   conducting on data sets, 312–314
   covariate-adjusted means (income and gender study), 305–309
   gender differences in income, 305–309
   generating predicted scores, 309–312
   smoking survey, 312–314
Analysis of variance (ANOVA), 23–24, 27
   defined, 36
   factorial, 36–37
   mixed model, 41
   repeated measures, 40–41, 213
   three-way and beyond, 207–208
   two-way, 23
   See also Analysis of variance (ANOVA), factorial; Analysis of variance (ANOVA), one-way
Analysis of variance (ANOVA), factorial:
   factorial design, efficiency of, 209
   hypothetical example of when and how, 201–205
   independent variables, 198–200
   planned contrasts, memory study, 210–211
   three-way ANOVAs, 207–208
Analysis of variance (ANOVA), one-way:
   contrasts and, 195–198
   levels, trouble with, 191–193
   understanding one-way, alcohol and, 193–195
ANCOVA. *See* Analysis of covariance
ANOVA. *See* Analysis of variance
APA style, 181–182
Ascending sort, 62
Associations, sorting unique, 263
Attribution theory, suppressor variables and, 317–318
Austin Powers (movie), 68

Beck Depression Inventory, 94, 95 (figure), 274
Behavioral confirmation, 397
Between-subjects design:
   about, 214–215
   mixed models and, 218–220
Bias:
   common errors and, 395–399
   data analysis, decision making and, 401–404
   four rules for avoiding, 405
   in human thinking, 395–399
   statistical, 399–401
Bimodal distribution, 10, 11 (figure)
Binomal effect size display, 30
Blind cola taste test, *t* test, 177–180
Bogus trait study, 214–215
Bolt, Usain, 269
Bone (graphic novel character), 280
Bootstrapping, 353n

Brandeis, Louis, 92, 96–99
Bush, George W., 226

Candyland game, 205–206
Cantril Self-Anchoring Striving Scale, 222
Carey, Drew, 269
Casablanca, 140
Casual analysis, problem with, 17
Categorical criterion variables, 251–253
Categorical predictors:
    multiple regression and, 279–283
    with multiple levels, 287–293
Causal plausibility, 328–331
Causal starting points vs. third variables, 326–328
Causes, competing, 325–326
Centering:
    defined, 270
    in multiple regression analysis, 268–271
Central tendency, 6, 8
Cervantes, 34
Ceteris paribus, 231–232
Cheating, study of, 20–24, 43n
Chi-square statistic, 113–116
    one-sample Chi-square test in SSPS, 189–190
    reporting results of, 120–123
Chi-square test of association, 35
Cliffhanger (game), 79, 395, 401
Coefficient:
    correlation. *See* Correlation coefficient
    phi, 113–116
Coefficient alpha, 139
Coefficient of determination, 29–30
Commands, in SSPS software:
    compute, 183
    "If," 183–184
    LG10 command (SSPS), 187
    means, 186–187
    recode, 185–186
    SD of SDX, 187
    SQRT, 187
    value labels command (SSPS), 187–188
Competing causes, disentangling, 325–326
Compute command, SSPS, 183
Computer, using to analyze data. *See* Statistical software, using
Confidence intervals, in logistic regression, 259–260
Confirmation bias, errors of, 399–401
Congruent, 229–230
Conscientiousness, 39

Continuous predictors, multiple regression and, 279–283
Contrasts:
    defined, 196
    finding meaning in means using, 195–198
    planned, 210–211
    quadratic, 196
Corrected item-total correlation, 139
Correlation:
    corrected item total, 139
    hypothetical study, Afrocentrism, 90–91
    hypothetical study, unfairness of life, 82–89
    multiple regression, multiple predictors and, 239–242
    point-biserial, 303–304
    running in SPSS software, 82–89
    split-half, comparing Cronbach's alpha and, 142–143
    study, interpersonal attraction, 116–118
Correlation coefficient:
    binomial effect size display and, 30
    degrees of freedom for, 88
    inferential statistics and, 22
    leptokurtic variable and, 76
    nominal variables and, 101–102
    Pearson as inventor of, xv, 1, 81–82
    Pearson's *r* as, 29
    phi coefficient and, 101–102
    product-moment, 81
Covariance:
    analysis of, 37. *See also* Analysis of covariance (ANCOVA)
    homogeneity of, 37
    income and gender, hypothetical study, 305–309
Covariate-adjusted means:
    hypothetical study, income and gender, 305–309
    interpret, 312
Critical values, 22
Cronbach's Alpha, 37–38
    internal consistency and, 138
    limitations of, 149–151
    split-half reliability and, 142–143
Crossed variables, 198
Cross-product term, 264–265
Cuneiform writing, 2–3
Curvilinear correlation:
    Brandeis's hypothesis, 96–99
    See also Linear/curvilinear correlation

Subject Index

Darlington, Dick, 127
Data:
    analysis, avoiding bias in, 405.
        See also Bias
    caveats regarding real, 131–133
    defined, 5
    merging and management, 387–394
    missing, variable values and, 52–55
    multivariate cleaning, 259–260
    practice with real, 273–276
    skewness/kurtosis and, 74–77
Data cleaning:
    about, 355–357
    missing data, 357–367
    multicollinearity, 383–385 (example)
    multivariate outliers and, 371–380
    outliers and, 368–380
    univariate outliers and, 371
    using skills, 380–382
David Letterman Show (television), 9
Decision making, judgment and, 121
Dependency regulation:
    in relationships, 339–340
    model, 342
Descriptive statistics:
    Brandeis's hypothesis, 96–99
    central tendency and dispersion, 6–9
    cheating, study of, 20–24
    correcting for skewness/kurtosis, 78–79
    defined, 5–6
    estimating spending in U.S. population, 51–52
    ethnic diversity of U.S., describing, 55–64
    hypothetical correlational study, of Afrocentrism, 90–91
    hypothetical study, unfairness of life, 82–89
    inferential statistics, 13–15
    in public opinion polls, 64–67
    kurtosis and, 68, 70, 73–79
    missing data, variable values and, 52–55
    outliers, power of impossible, 93–96
    predicting scores on Y from scores on X, 100
    probability theory, 16–20
    shape of distributions and, 9–12, 66–77
    skewness and, 68, 70–74
    study of freedom of press, 91–93
    survey, shoe sizes, 46–51
Dichotomous categorical variable, 252
Dispersion:
    defined, 6
    measures of, 7–8

Distribution:
    shape of, 9–12, 67–79
    types of, 10–12
Don Quixote, 34
Dragnet (television show), 127
D^2 statistic, 375
Dummy coded variables, 288–289

Edison, Thomas, 265, 272
Effect size:
    estimates of, 28–31
    significance tests and, 25
    table, 31
Eigenvalues, 130, 135–138
Empirical plausibility, 331–332
Errors:
    experiment-wise, 192
    margin of, 65
    reliability and, 154, 155
    standard. *See* Errors, standard
    type I and II, 24–25
Errors, standard:
    adjusted r-square, r-square and, 242–248
    calculating, 65
    mediation and, 337–339
    of the mean, 161–162
    of the proportion, 66
    See also Standard error
Estimates, of effect size, 28–31
Ethnic diversity of U.S.:
    describing, 55–64
    ethnic composition of three bogus states, (table), 57
Excel files, reading in SSPS, 177–178
Expectancies, creating, 171–173
Expected frequencies, 21, 22
Experimenter bias, 398
Experiment-wise error, 192
Extraversion, 39

Factor analysis, 127–133
    defined, 39–40, 125
    efficiency, 208
    principle components analysis and, 39–40, 133–138, 160n3
    reliability analysis and, 138–140
    reporting results of, 157–160
Factorial ANOVA:
    defined, 36–37
    independent variables and, 198–200
Factorial designs, 198, 199
Factor loadings, 136–137
Feedback interaction term, 265

Field experiment, 171
Freakonomics (Levitt and Dubner), 4, 233
Freedom House survey, 92
Freedom of the press, in Europe, study of, 91–93
Frequency analysis, 111
Friends (television show), 102
F test, 193–195

Gallup Organization, 92, 221, 313, 365
Gallup Worldview website, 373–374
Galton, Francis (Sir), 81
General Social Survey, 312
Gore, Al, 355

Happiness, bending rules about, study, 163–165
Healthways, 313
Heat and aggression, single/two-sample *t* test, 176–177
Heston, Charlton, 233
Hillary, Edmund, 265, 271, 289
History, human, numbers and language and, 2–5
Homogeneity of covariance, 37
Hypothesis:
 bias and, 396–397
 Brandeis, through curved lens, 96–99
 inferential statistics and, 13–15
 statistical testing of, 28–33
Hypothetical path model, positive beliefs and health, 347–349
Hypothetical studies:
 aggression among kids, 205–206
 Afrocentrism, 90–91
 generating predicted scores, 309–312
 income and gender, 305–309
 income and occupation, 113–116
 self-esteem, 143–145
 unfairness of life, 82–89

"If" command, SSPS, 183–184
Income and gender, hypothetical study, 305–309
Income and occupation, hypothetical study, 113–116
Incongruent, 229–230
Independent sample *t* tests:
 defined, 36
 two samples, 171–173.
 See also Two-sample *t* tests
Independent variables:
 factorial ANOVAs and, 198–200
 multiple regression analysis and, 263–264

Individual differences, controlling for. *See* Within-subjects design, mixed model analysis and
Inferential statistics:
 cheating and, 20–21
 defined, 5
 methods, 13–15
Intelligence, teacher expectancy study and, 171–174
Interactions:
 defined, 199
 detecting, 198–205
 main effects and, 205–207
 table, 266
 terms, 265
 three-way analysis of variance, (ANOVA), 207–208
Interactions, in multiple regression analysis:
 categorical/continuous predictors and, 279–283
 categorical predictors, multiple levels, 287–293
 centering, simple slope tests and, 268–271
 data, practice with real, 273–276
 median splits, isolating/analyzing subgroups, 272–273
 moderator effects, *R*-square values, 276–279
 moderators and, 264–268
 report results, two-way interaction, 300–301
 simple slopes, estimating technique, 283–287
 testing, interpreting three-way interaction, 294–299
 variable type, determines analysis type, 263–264
Internal consistency, 127
Interpersonal attraction, correlational study of, 116–118
Interrater agreement, 127
Interviewer's dilemma, 365

James, LeBron, 5, 372
Jones, Scott, 371–372
Jordan, Michael, 6
Judgment, decision making and, 121

Kerry, John, 226
Kurtosis:
 correcting for, 78–79
 issues in data and, 74–77
 shape of distribution and, 68, 70–74

Subject Index

Labels, descriptive, in SPSS, 83
"Ladder of Life," 222
Language, human history and, 2–5
Law of diminishing returns, 164
Levels, trouble with, ANOVA and, 191–193
LG10 command, SSPS, 187
Likert scales, 369
Linear/curvilinear correlation, 81–82
Little Rascals, The (television show), 294
Logistic regression:
   confidence intervals in, 259
   defined, 39
   how to use, 253–256
   multiple regression analysis and, 251–253
   odds ratios and, 256–258
   removing predictor in analysis, 258–259

Mahalanobis D^2 statistic, 78, 375, 376
Main effects, 205–207
   hypothetical study of aggression in children, 205–206
   interactions vs., 200, 209, 227
   lab study of self-pay, 206–207
   multiple regression analysis and, 263, 265, 267–268, 270–271, 275, 278, 294
   simple main effects test, 203
   statistical separation of, 36–37
Margin of error, 65
Mean(s):
   command, SSPS, 186–187
   covariate-adjusted, 305–309
   defined, 34–35
   contrasts and, 195–198
   Galton and, 81
Measurement error, significance tests and, 25–26
Measures:
   descriptive, 6–7
   of dispersion, 7–8
   reliability of, 125. *See also* Reliability analysis
MediaLab, software, 356
Median:
   defined, 34–35
   Galton and, 81
   in shoe size survey, 56
Median splits, 269, 272–273
Mediation analysis:
   analysis of teasing, study, 351–353
   associations, self-esteem/relationship, 339–340
   as specific case of path analysis, 341–343
   defined, 41
   model, frustration and aggression, 335–337
   moderation and, 333–334
Memory study, planned contrasts, 210–211
Merging variable, 390
Meta-analysis:
   defined, 28, 31
   statistical testing of hypothesis and, 31–33
Mixed model ANOVA:
   defined, 41
   mixed models, 218–220
   political attitudes study, 226–228
   sample results of study using, 229–230
Mode:
   defined, 34–35
   in shoe study, 56
Moderator analysis:
   defined, 38–39
   mediation and, 333–334
Moderator effects, R-square values and, 276–279
Moderator variables, interactions and, in multiple regression, 264–268
Multicollinearity:
   data cleaning and, 383
   defined, 160n
   example of, 384–385
   suppressor variables and, 323–324
Multiple predictor variables, correlation, multiple regression and, 239–242
Multiple regression analysis, 231–232
   confidence intervals, in logistic regression, 259–260
   correlation, multiple predictor variables and, 239–242
   data, consider more, 235–236
   defined, 38
   interactions in. *See* Interactions, in multiple regression analysis
   logic of, 233–234
   logistic regression analysis and, 251–258
   multivariate data cleaning, 259–260
   predicted scores, generating in, 309–312
   predictor/criterion variables, 232–233
   real-world application, 248–251
   repeat logistic regression, predictor removal, 258–259
   r-square, adjusted r-square, and standard errors in, 242–248
   statistical interaction and, 38–39
   suppressor variables and, 315–316

testing for, interpreting three-way
  interactions in, 294–299
using SSPS to check answers, 236–239
Multivariate data cleaning, 259–260
Multivariate outliers:
  identifying, dealing with, 371–380
  skewness/kurtosis and, 68, 78

Name-letter effect:
  pilot study of, 102–107
  second pilot study of, 107–113
  study, single-sample, two-sample *t* test, 174–176
Names, using, in SPSS, 83
Naming fallacy, 132
National Opinion Research Center, 312
National Rifle Association, 233
Natural confounds, 305
Negatively skewed distribution, 68–69
Nominal variables, tests involving, 101–123
  chi-square statistic, phi coefficients, odds ratios, 113–116
  correlational study of interpersonal attraction, 116–118
  correlation coefficient, 101–102
  from marriage to mental illness, hypothetical study, 118–120
  name-letter preference studies, 102–113
  report results of chi-square analysis of, 120–123
Nonparametric statistics, nominal variables, tests involving, 101–123. See also Nominal variables
Normal distribution:
  defined, 10
  figure, 11, 12, 68
Normally distributed variables, t tests and, 161–163
Null hypothesis, 13–14
Numbers:
  language, human history and, 2–5
  Roman numerals, 3–4

Obama, Barack (President), 4–5, 57, 288
Observed frequencies, 21, 22
OCEAN (Openness to Experience, Conscientiousness, Extraversion, Agreeableness, Neuroticism, 388–389)
Odds ratios, 113–116, 123n, 256–258
Omnibus test, 193–195
O'Neal, Shaquille, 372

One-way analysis of variance (ANOVA), 36. See also Analysis of variance (ANOVA)
Operational definitions, 23
Optimism, repeated measures study, 220–225
Outcome, simplifying in single/two-sample t tests, 166–171
Outliers:
  data cleaning and, 368–370
  defined, 69
  power of impossible, 93–96
  univariate, identifying, dealing with, 371

Paired-sample *t* test:
  defined, 40
  use of, 215–217
Participants, 32
Path analysis:
  analysis of teasing, study, 351–353
  defined, 41–42
  hypothetical path model, 347–349
  logic of, 343–347
  mediation analysis as specific case of, 341–343
  reading, additional, 350
  web pages, useful, 350
Pearson, Karl, 251
  quotation, 1, 2
  tribute, 81–82
Pearson's *r*, 35, 101–102
Pelham, Bill, 1–2
Phi coefficient:
  defined, 35
  use of, 113–116
Phineas and Ferb television show, 399–401, 401–404
PISA, Programme for International Student Assessment, 374
Planned comparison, 36
Planned contrasts, memory study, 210–211
Plato, 1–2
Plausibility:
  causal, 328–331
  empirical, 331–332
Point-biserial correlation, 303–304
Positively skewed distribution, 68–69
Positive test bias, 397
Predicted path model, 343
Predictor variables, 232–233
  categorical, multilevel interactions and, 287–293

categorical/continuous, multiple
    regression and, 279–283
multiple regression, correlation and, 239–242
practice using, 320–323
Press, freedom of, study of European, 91–93
Principal components analysis:
    defined, 39–40, 128
    with real data, 133–135
Principle of parsimony, 126
Principle of regression, toward
    the mean, 16–17
Probability judgment, 121
Probability theory:
    computing, 42n
    table, coin toss, 19
    use of, 16–20
Product-moment correlation coefficient,
    81–82
Proportion:
    calculate, 183
    standard error of the, 66
Psychology, cross-cultural, 221
Psych (television show), 280
Public opinion polls, descriptive statistics
    and, 64–67

Quadratic contrast, 196
Quasi, 68

Random assignment, 213
Random digit dialing, 313
Range, 8–9
    defined, 6
    restriction of, 79
    significance testing and, 26–27
Raw means, 312
Real-world application, multiple
    regression analysis, 248–251
Recode command, SSPS, 185–186
Rectangular distribution, 10, 11
Redundancy, eliminating, 128
Regression, toward the mean
    principle of, 16–17
Regression analysis, 38, 39
Reliability, 125
Reliability analysis, 37–38, 125
    analysis, 138–140
    as tool for item development, 145–149
    concept of reliability, 126–127
    Cronbach's alpha and, 149–151
    eigenvalues, checking, 135–138
    factor analysis and, 127–131
    principal components analysis, 133–135

psychological scales, 152–156
real data, caveats, 131–133
reporting results, 157–160
self-esteem study, 143–145
"trism" scale, 140–142
Repeated measures ANOVA:
    defined, 40–41, 213
    optimism study, 220–225
Replication, 19
Reporting experimental manipulations, 404
Residuals, 305
Restriction of range, 79
Roman numerals, 3–4
Rosenberg Self-Esteem Scale, 94, 95
    (figure), 183–188, 339, 383
R-square:
    in multiple regression and, 242–248
    values, moderator effects and, 276–279

Samples:
    independent, two-sample t test, 171–173
    size of, significance testing and, 26
    See also Single-sample t tests; Two-sample
        t tests
Scale:
    adding items together to make, 140–142
    reliability of psychological, 152–156
Scores:
    generating predicted scores in multiple
        regression, 309–312
    predicting scores on Y from scores
        on X, 100
    range of set, 6
    shape of distribution and, 9–12
Scree plot, 130–131, 136
SD or SD.X command SSPS, 187
Self-fulfilling prophecies, 398
Self-pay, lab study of, 206–207
Shape of distribution. *See* Distribution,
    shape of
Shimizu, Mitsuru, 356
Shoe size survey, 46–51
Significance tests:
    alpha levels, type I and II errors
        and, 24–25
    effect size and, 25
    of hypothesis, 28–33
    measurement error and, 25–26
    restriction of range and, 26–27
    sample size and, 26
Simple slopes:
    centering, multiple regression and, 268–271
    estimating technique, 283–287

Simultaneous multiple regression, 270
Single measure, 126
Single-sample t tests, 36, 161–171
Skewness:
   correcting for, 78–79
   defined, 68–69
   issues in data and, 74–77
   normal distribution and, 68–74
   positively skewed distribution, 69 (figure)
Sobel test, 335, 338
Software, statistical, using. See Statistical software, using
Spearman, Charles, 127–128
Spending, estimating, in U.S. population, 51–55
Split-half reliability, Cronbach's alpha and, 142–143
SQRT command SSPS, 187
SSPS. See Statistical software, using
Standard deviation, 161–162
   defined, 6–7, 35
   in ethnic diversity of U.S., 56
   in normal distribution, 12 (figure)
Standard error:
   calculating, 65
   in multiple regression, 242–248
   mediation and, 337–339
   of the mean, 161–162
   of the proportion, 66
   See also Errors, standard
Starting points, causal, third variables vs., 326–328
Statistical interactions. See Interactions
Statistical software, using, 16, 45
Statistical testing, of hypothesis:
   about, 28
   estimates of effect size, 28–31
   meta-analysis, 31–33
   specific tests and their uses, 34–42
Statistics, 5
Stereotyping, 398
Structural equation modeling, 42
Studies. See Hypothetical studies
Subgroups, isolating/analyzing, in multiple regression, 272–273
Subtraction, 270
Summation sign, 21
Suppression, 38
Suppressor variables. See Variables, suppressor
Surprise index, 21
Survey, telephone, 313
Syntax statements, SSPS and, 183–188

Tails, of distribution, 69
Teacher expectancy study, 171–174
Test-retest reliability, 127
Three Stooges, The (movie, TV show), 294, 297, 298, 299, 301n
Three-way analysis of variance, ANOVAs and, 207–208
Tolerance, 324
Trism, 137–138
True score, 154–155
t test, 27
   means and, 303–304
   See also Single-sample t tests; Two-sample t tests
Two-sample t tests, 171
   blind cola taste test, 177–180
   heat and aggression archival study, 176–177
   independent samples, 171–173
   name-letter preference study, 174–176
   reporting results in APA style, 181–182
   syntax statements and logical operands, 183–188
   teacher expectancy study, 171–174
Two-way interaction. See Interactions
Type I error, 24–25
Type II error, 24–25

Univariate outliers, 371
Unweighted effects coding, 301n
U.S. Department of Agriculture, website, 374

Value labels command, SSPS, 187–188
Variability:
   defined, 6
   restriction of range, significance testing an, 26–27
Variable breaks, 109
Variable labels, using in SPSS, 83
Variables:
   dichotomous categorical, 252
   independent, factorial ANOVAs and, 198–200
   merging, 390
   moderator, 39
   predictor/criterion, 232–233
   values, missing data and, 52–55
Variables, suppressor:
   attribution theory and, 317–318
   example of suppression, practice, 318–320
   multicollinearity, 323–324
   multiple regression and, 315–316
   practice, analyze data sets, 320–323

Subject Index

Variance, 35
    analysis of. *See* Analysis of variance (ANOVA)
Variance inflation factor (VIF), 324
Vetter, Eddie, 265, 271, 272

Weighting process, 271
Wickens, Thomas, 76, 369
Within-subjects design, mixed model analysis and, 213
    between subjects/within-subjects, combine, 218–220
    bogus subjects/traits, 214–215
    defined, 213
    mixed model design, 226–230
    repeated measures, optimism study, 220–225
    three within-subjects versions, 215–217
World Poll survey (Gallup), 92, 221
Writing, cuneiform, 2–3

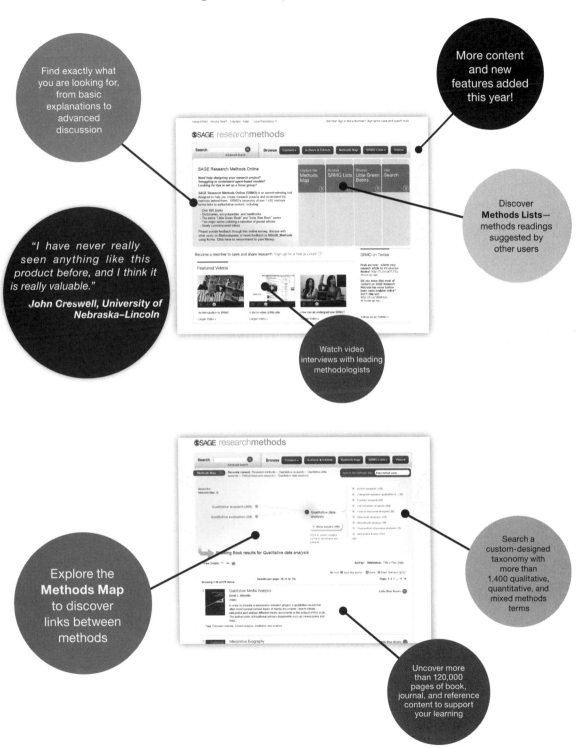